Structural Colors in the Realm of Nature

Structural Colors in the Realm of Nature

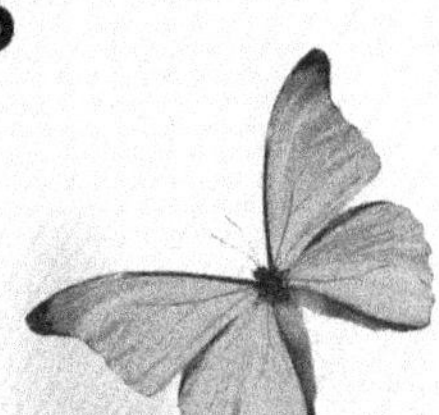

Shuichi Kinoshita
Osaka University, Japan

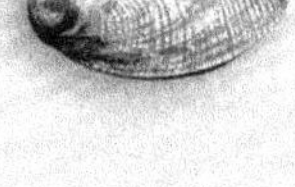

World Scientific
NEW JERSEY • LONDON • SINGAPORE • BEIJING • SHANGHAI • HONG KONG • TAIPEI • CHENNAI

Published by

World Scientific Publishing Co. Pte. Ltd.

5 Toh Tuck Link, Singapore 596224

USA office: 27 Warren Street, Suite 401-402, Hackensack, NJ 07601

UK office: 57 Shelton Street, Covent Garden, London WC2H 9HE

British Library Cataloguing-in-Publication Data
A catalogue record for this book is available from the British Library.

STRUCTURAL COLORS IN THE REALM OF NATURE

ISBN-13 978-981-270-783-3
ISBN-10 981-270-783-2

Typeset by Stallion Press
Email: enquiries@stallionpress.com

Printed in Singapore.

To my parents, Yoshio and Yasuko Kinoshita

To my wife, Sachiko

Preface

In recent years, structural colors have attracted much attention in a wide variety of research fields. This is because they are originated from complex interactions between light and sophisticated microstructures created in the natural world. In addition, their inherent regular structures are one of the most conspicuous examples of the nonequilibrium order formation. The structural colors are deeply connected with recent rapidly growing fields of photonics. Their mechanisms are, in principle, of a purely physical origin, which differ considerably from the ordinary coloration mechanisms such as in pigments, dyes, and metals, where the colors are anyway produced by virtue of the energy consumption of light. Since the first scientific descriptions on structural colors by Hooke in 1665, a tremendous number of research objects have been investigated using various scientific methodologies. I was somehow trying to summarize this entangled research field, which expanded over many research fields in biology, chemistry, physics, and engineering. It was when a publisher proposed me to write a book on this topic.

I began collecting papers and books on structural colors, which were soon found to be more than 500. Seeing these materials, I felt bewildered because too many research objects in a diverse world of nature were reported, in which most of them were far from clarifying their mechanisms. Most of the books or review papers appeared so far classified the structural colors through their mechanisms and took some species as the typical examples. I thought it difficult to follow this way because the mechanisms for most of the species were completely unknown or only partly known.

I decided to write this book on the basis of species and not of the mechanisms. In order to do that, however, it was necessary to know the kinds of species that have been investigated for these 350 years as the research objects. My student and I picked up all the scientific names appearing in the papers, and after revising them to current names, listed up according to the current classifications. It was really hard work for us (two of these results, butterflies and birds, are listed in Appendix A). After the classification was completed, I selected lepidopterans, beetles, birds, fishes, and plants as the main objects. Particularly, as the number of the researches concerning *Morpho* butterflies appeared extraordinarily large, I selected them as special objects to give a detailed description on this subject. In spite of that, some of the topics, such as moth-eye structure, had to get over the frame of classification. Moreover, in case of biological reflectors, which will occupy an important part of structural colors, I had to omit those related to vision, because they were too diversely distributed in animal worlds and also the vision itself was one of the largest research fields, which I felt exceeded the scope of this book. In a similar sense, we had to omit interesting descriptions concerning minerals and fossils, and also marine and atmospheric phenomena. If I get a chance again, I would like to write a continuation of this book in future.

According to the above policy, I began to write the manuscript and soon found it really difficult, because most of the research fields were actually unknown to me and I had to write without any real feeling. Then, in each field, I had to learn the classifications, fundamental concepts, and proper technical terms. The language problems sometimes became a large barrier. Further, species picked up in the papers were often completely unknown to me. Hence, I had to order a lot of books and also search over the Internet. Sometimes I went to zoological gardens, aquaria, and repositories of the museums to see real objects. However, I do not deny that many parts of this book were anyway borrowed from excellent papers and books published so far and that the misunderstandings of the contents will be anyway present. Thus, I strongly recommend that if one wishes to investigate the phenomena more deeply, he should read the original papers thoroughly. In spite of that, I am still satisfied with this book, because through the above policy, it becomes possible to write the whole story for each species.

As a physicist, I had a wish to write the physical basis for each phenomenon in order to give a clue to this diverse world. For this purpose, I prepared a section to explain the elementary optical processes that would at least partly contribute to the structural colors. In addition, I prepared more detailed descriptions as a new section, where I described the detailed derivations of the important formulas, which would help the readers to study the mechanisms. Completing this book, I have noticed that most of the researches have just commenced in a sense of physics and will offer quite interesting research objects. Thus, this book will be helpful even for physicists to search for new research objects on the sophisticated interactions between light and microstructures, and also on the peculiar self-organization processes undergoing in nature.

Before ending, I would like to express my gratitude to Dr Shinya Yoshioka, who has been extensively performing the experiments and has extended our research field to a variety of animal worlds including butterflies, beetles, and birds. I am also indebted to Naoko Arakawa, Dr Yoshihiko Fujimura, Dr Makoto Kambe, and Eri Nakamura for their direct and indirect supports for producing this book. This book is owed mostly to the researches done by my students, Kenji Kawagoe, Kazuhiko Ikeda, Tsutomu Hamada, Chieko Matsumiya, Naoko Okamoto, and Yasuhiro Fujii, in cooperation with Prof. Akira Saito (Osaka University) and Prof. Takahiko Hariyama (Hamamatsu University School of Medicine). Also, I would like to express my gratitude to Prof. Helen Ghiradella for her sincere support. I owe Dr Akihiro Yoshida (JT Biohistory Research Hall) and Prof. Kuniaki Nagayama (National Institute for Physiological Sciences) for leading me to this fascinating field. I also thank Mr Toshiyuki Suwa of Osaka University Life Sciences Library for kindly helping me in searching the literatures that were difficult to obtain.

Finally, I express my sincere gratitude to my wife, Sachiko, and to my parents for their continuous encouragements and supports. I also thank the publisher for giving me an opportunity to write this book.

2007.11.28
Shuichi Kinoshita

Contents

Chapter 1

Introduction

1.1 What Is Structural Color?

In nature, a tremendous number of orders and patterns are generated spontaneously, which enliven our surroundings. One of the most remarkable consequences of these processes reveals itself as the so-called *structural color*, which exhibits striking brilliancy owing to elaborate structures furnished with living creatures. They sometimes reflect surprisingly intense light in a wide angular range, while in another case prohibit any reflection of light. These functions are natural consequences of complicated interactions between light and microstructures through purely physical properties of light.

The scientific definition of structural color has not yet been settled and its characteristics are often referred to in contrast to pigmentary color. When a matter is illuminated with white light, we see a specific color if the reflected light of only a particular wavelength range is visible to our eyes. There are two ways to eliminate the other wavelengths of light (see Fig. 1.1): One is the case where the light is absorbed in a material, which is usually the case for ordinary coloration mechanisms in colored materials such as pigments, dyes, and metals. In these materials, the illuminating light interacts with electrons within the material and excites them to higher excited states by virtue of energy consumption of light. The color in this case is anyway caused by the exchange of energies between light and electron.

The other is the case where the light is reflected, scattered, and deflected not to reach the eyes under the presence of a specific structure. The coloration in this case is based on a purely physical operation of light that interacts with various types of spatial inhomogeneity. Thus, it does not essentially lose the energy of light. In this sense, even a simple prism should

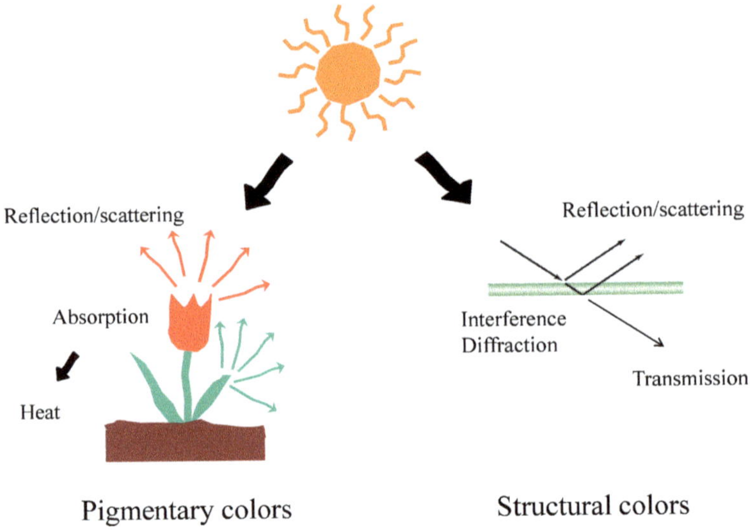

Fig. 1.1 Pigmentary colors and structural colors. (Illustrated by Dr. S. Yoshioka.)

be categorized into this type of structural color, because the light waves with different wavelengths are deflected differently by virtue of the dispersion of the refractive index. However, most of the structural colors appearing in nature somehow utilize special mechanisms to enhance the coloration. It is generally believed that the structural colors mostly come from the following five elementary optical processes and their combinations: (1) thin-film interference, (2) multilayer interference, (3) diffraction grating, (4) light scattering, and (5) photonic crystal.

Structural color and *iridescence* are two major keywords for these phenomena and seem to be used commonly in an equivalent sense. However, in relation to the above optical processes, the iridescence should be used somewhat in a restricted sense when the color apparently changes with viewing angles. For example, thin-film interference is generally iridescent, while light scattering is usually non-iridescent but of a structural origin.

Although many excellent books (Simon, 1971; Fox, 1976; Nassau, 2001; Parker, 2003; Kinoshita, 2005; Berthier, 2007) and review articles (Parker *et al.*, 1998b; Srinivasarao, 1999; Parker, 2000; Vukusic and Sambles, 2003; Parker, 2004; Kinoshita and Yoshioka, 2005b; Kinoshita *et al.*, 2008) have appeared so far, the field of structural colors is rapidly growing year by year so that new mechanisms and species appear almost every year. Thus, new information is always expected. Further, the above articles tend to classify the structural colors through their mechanisms and take some species as the typical examples. Thus, to overview the whole species-specific fields through an eye of "structural color" is quite difficult. In the present book, I dare to summarize the world of structural colors on the basis of species and not of the mechanisms. This clearly benefits us to treat them even when their mechanisms are not completely known or only partly known. It is also possible to describe the whole story for each species under species-specific background.

This book is organized as follows: In Sec. 1.2, we will describe the historical review of the structural colors that began with the observations by Hooke and Newton in the 17th century, which was then succeeded by Lord Rayleigh and Michelson, and flourishes nowadays in various scientific and industrial fields. In Chap. 2, a brief description of structure-based properties of light has been given, which will appear later as actual forms in various living creatures after a variety of modifications (Kinoshita and Yoshioka, 2005b, 2005c; Kinoshita *et al.*, 2008). In Chap. 4, we will show the structural colors in lepidopterans, with a particular emphasis laid on the famous *Morpho* butterflies. In Chaps. 5–8, we demonstrate the structural colors appeared as specific forms in beetles, birds, fishes, plants, and the other animals. In Chap. 9, we prepare the detailed derivations of the important formulas, which would be of help to the readers to study the physical mechanisms.

1.2 Historical Review

The study of structural colors has a long history. Probably, the oldest scientific description on the structural colors appeared in "*Micrographia*", written by Hooke (1665). In this book, he described the microscopic observation of the brilliant feathers of peacock and duck, and found that their colors were destroyed by a drop of water. He speculated that alternate layers of thin plate and air might strongly reflect the light. Newton (1704) described in "*Opticks*" that the colors of the iridescent peacock arose from the thinness of the transparent part of the feathers. In spite of these pioneering works on structural colors, further scientific development must have waited for the establishment of electromagnetic theory by Maxwell in 1873 and also for the experimental study on the electromagnetic wave by Hertz in 1884. The fundamental properties of light such as reflection, refraction, interference, and diffraction could be quantitatively treated thereafter, and the studies on structural colors proceeded quite rapidly.

However, there arose a significant conflict between two hypotheses: One was *surface-color*, which was proposed by Walter in 1895 (see Rayleigh (1919)) and was thought to originate from the reflection at a surface involving pigments. The other was *structure-color* that originated from purely physical operation of light. These two hypotheses drove the world of physics into two at that time. Walter explained the variation of colors with the incident angle of light as due to the change of polarization in the reflection at the absorption band edge. The idea of surface-color was then succeeded by Michelson (1911), who conducted the experiments of the reflectivity on seemingly metallic samples such as golden scarab beetle and *Morpho* butterfly, and described that they resembled the surface reflection from a very thin surface layer involving dye.

Lord Rayleigh, on the other hand, derived a formula to express the reflection properties from a regularly stratified medium using electromagnetic theory (Rayleigh, 1917), and considered it as the origin of colors of twin crystals, old decomposed glass and probably those of some beetles and butterflies. He surveyed the studies performed so far, and described that the brilliant colors, which varied greatly with the incident angle of light, were not due to the ordinary operation of dyes, but came from the structural colors (Rayleigh, 1919).

Many experimental works were performed, on an optical microscopic level, to clarify the relationship between brilliant colorations and the microstructures at the surface of iridescent, metallic, and whitish materials. Onslow (1923) observed more than 50 iridescent animals to settle the conflict between these hypotheses. Merritt (1925) measured the reflection spectra of tempered steel and *Morpho* butterfly, and interpreted in terms of thin-layer interference. In 1924–1927, Mason (1923a, 1923b, 1926, 1927a, 1927b) published a series of papers on various types of color-producing structures in animals investigated by a microscope and supported the interference theory. Thus, the interference of light enforced the power gradually, which finally kept away the interest of physicists.

The complete understanding of their structures was made after the invention of electron microscope. In 1939, the first attempt was made to clarify the mechanism of blue coloring in the bird feathers. Frank and Ruska applied the first marketable electron microscope, which developed in this year, and found spongy structure consisting of keratin and air at the inner wall of the medullary cell of blue feather of a bird pitta (Frank and Ruska, 1939; Frank, 1939). In 1942, Anderson and Richards (1942) and Gentil (1942) investigated the scales of the famous *Morpho* butterflies.

These observations revealed a surprisingly complicated structure on a tiny scale of the butterfly wing, which accelerated the structural study on a nanometer scale. Many biologists attempted to elucidate the structures causing iridescence and accumulated enormous data. In 1960, a sophisticated microstructure was reported on the feather of a humming bird (Greenewalt *et al.*, 1960a). After then, beautiful microstructures were discovered one after another in many species of birds such as peacock, humming bird, pheasant, and dove. In 1967, the structural coloration due to helicoidal structure analogous to the cholesteric liquid crystal was found in scarabaeid beetles (Neville and Caveney, 1969). Beautiful multilayered structure was found in a kind of a jewel beetle in 1972 (Durrer and Villiger, 1972). Highly reflecting structures were found in the 1960s within the integuments and eyes of fish and cephalopods, which are now known as animal reflectors (Land, 1972). On the other hand, regular modification of the surface was discovered to cause the anti-reflection effect, which is well known as moth-eye structure (Bernhard *et al.*, 1968). The motile nature of the iridescent cell in fish to the ambient illumination was clarified in 1982, which was due to a change in the distance between adjacent regularly arranged platelets (Lythgoe and Shand, 1982).

On the other hand, the developmental studies concerning the formation of such microstructures are quite limited even up to now. Good examples for these studies are those in butterfly scale and multilayered structure in beetle. Ghiradella reported a comprehensive study on the formation process of butterfly microstructures within a scale-forming cell using an electron microscope (Ghiradella *et al.*, 1972; Ghiradella, 1989, 1998). Schultz and Rankin (1985b) reported the development of multilayer structure before and after ecdysis of beetle.

In spite of the progress of the structural studies in biology, the physical interpretation of structural colors had not essentially proceeded, since Lord Rayleigh proposed the multilayer interference. Only recently, the structural colors have been a subject of extensive studies because their applications have been rapidly growing in many industrial fields related to vision such as painting, automobile, cosmetics, display, and textile. It has been soon noticed that simple multilayer interference no longer reproduces the actual appearance of the natural color. Furthermore, recent researches have revealed that even a very simple structure in nature has surprising multiple functions, which by far exceed our expectations. In the following chapters, we will show the beautiful microstructures produced in the natural world and will give a physical basis for their optical operations.

Chapter 2

Fundamentals of Structural Coloration

2.1 Fundamental Properties of Light

2.1.1 *Light as an Electromagnetic Wave*

It was during the 19th century that the scientists reached a conclusion that light was a kind of an electromagnetic wave. This was brought about by two great scientists who founded the basis of electromagnetic theory. The most important work was done by Maxwell who predicted the existence of electromagnetic wave through the famous Maxwell equations, while Hertz performed an experiment to confirm that electromagnetic wave was actually emitted from electric oscillations.

Maxwell equations for electric and magnetic fields in a medium are generally given as

$$\nabla \times \mathbf{E} = -\frac{\partial \mathbf{B}}{\partial t}, \tag{2.1}$$

$$\nabla \times \mathbf{H} = \mathbf{j} + \frac{\partial \mathbf{D}}{\partial t}, \tag{2.2}$$

$$\nabla \cdot \mathbf{D} = \rho, \tag{2.3}$$

$$\nabla \cdot \mathbf{B} = 0, \tag{2.4}$$

with $\mathbf{D} = \epsilon\mathbf{E} = \epsilon_0\mathbf{E} + \mathbf{P}$ and $\mathbf{B} = \mu\mathbf{H}$, where $\mathbf{E}$, $\mathbf{D}$, $\mathbf{H}$, $\mathbf{B}$, and $\mathbf{P}$ are electric, electric displacement, magnetic, magnetic flux density, and polarization vectors, respectively. $\mathbf{j}$ and ρ are current density and charge density, both of which will be set to zero, when a dielectric insulator not bearing true charge is employed as a medium. ϵ, ϵ_0, and μ are the permittivity of the medium and vacuum, and the magnetic permeability of the medium, respectively. For a uniform and isotropic medium, the above equations are further reduced to

$$\nabla \times \mathbf{E} = -\frac{\partial \mathbf{B}}{\partial t}, \tag{2.5}$$

$$\nabla \times \mathbf{B} = \epsilon\mu \frac{\partial \mathbf{E}}{\partial t}, \tag{2.6}$$

$$\nabla \cdot \mathbf{E} = 0, \tag{2.7}$$

$$\nabla \cdot \mathbf{B} = 0. \tag{2.8}$$

Eliminating **B** or **E** from the above equations, and using the vector relation

$$\nabla \times (\nabla \times \mathbf{E}) = \nabla(\nabla \cdot \mathbf{E}) - \nabla^2 \mathbf{E}, \tag{2.9}$$

and the relation $\nabla \cdot \mathbf{E} = 0$ or $\nabla \cdot \mathbf{B} = 0$, we obtain a set of wave equations:

$$\nabla^2 \mathbf{E} = \epsilon\mu \frac{\partial^2 \mathbf{E}}{\partial t^2}, \tag{2.10}$$

$$\nabla^2 \mathbf{B} = \epsilon\mu \frac{\partial^2 \mathbf{B}}{\partial t^2}. \tag{2.11}$$

For simplicity, we only focus on a wave equation for electric field (Eq. (2.10)). A particular solution for the above equation is expressed as a plane wave propagating toward an arbitrary direction, and is described as

$$\mathbf{E} = \mathbf{E}_0 \cos(\mathbf{k} \cdot \mathbf{r} - \omega t + \phi), \tag{2.12}$$

which means that a plane wave propagates toward the direction of **k** with the angular frequency ω and the initial phase ϕ. Inserting Eq. (2.12) into Eq. (2.10), we obtain $k^2 = \epsilon\mu\omega^2$, which leads to $k = \sqrt{\epsilon\mu}\omega = n\omega/c$ with the refractive index $n = \sqrt{\epsilon\mu/\epsilon_0\mu_0}$. Here, μ_0 and c are the magnetic permeability and the velocity of light in vacuum, respectively.

Usually, Eq. (2.12) is written in a complex form as

$$\tilde{\mathbf{E}} = \mathbf{E}_0 \mathrm{e}^{\mathrm{i}(\mathbf{k} \cdot \mathbf{r} - \omega t + \phi)}, \tag{2.13}$$

where $\tilde{}$ indicates, for a while, that a variable is expressed in a complex form. The real and complex forms are related with each other through the following relation:

$$\mathbf{E} = \frac{1}{2}(\tilde{\mathbf{E}} + \tilde{\mathbf{E}}^*) = \mathrm{Re}\{\tilde{\mathbf{E}}\}. \tag{2.14}$$

The product of two oscillating quantities, for instance $\mathbf{A}(t)$ and $\mathbf{B}(t)$, with the same frequencies that are on the order of light frequency is usually evaluated by its cycle average defined as

$$\overline{\mathbf{A}(t) \cdot \mathbf{B}(t)} = \frac{1}{T} \int_0^T \mathrm{d}t \mathbf{A}(t) \cdot \mathbf{B}(t) = \frac{1}{2} \mathrm{Re}\{\tilde{\mathbf{A}}(t) \cdot \tilde{\mathbf{B}}^*(t)\}, \tag{2.15}$$

where $T = 2\pi/\omega$ is a period of oscillation. The intensity of light I is then defined using a Poynting vector $\mathbf{S} = \mathbf{E} \times \mathbf{H}$ as

$$\begin{aligned} I \equiv \overline{|\mathbf{E} \times \mathbf{H}|} &= \frac{1}{2} E_0 H_0 \\ &= \frac{1}{2}\left(\frac{\epsilon}{\mu}\right)^{\frac{1}{2}} E_0^2 = \frac{1}{2}\epsilon c E_0^2, \end{aligned} \tag{2.16}$$

where E_0 and H_0 are the amplitudes of the electric and magnetic fields, respectively, and we have used the relation $E_0/H_0 = (\mu/\epsilon)^{1/2}$.

2.1.2 *Fresnel's Law*

If a plane wave is incident, with an incidence angle of θ_1, on an interface between two isotropic media, designated as 1 and 2, with different refractive indices of n_1 and n_2, then the incident wave will be reflected and refracted at the interface. Here, the tangential components of electric and magnetic fields are connected with each other at the interface such that $E_{1x} = E_{2x}$, $E_{1z} = E_{2z}$, $H_{1x} = H_{2x}$, and $H_{1z} = H_{2z}$, as shown in Fig. 2.1. The reflectivity and transmittance at the interface are dependent on the direction of the electric or magnetic field vector. We usually call the direction of electric field as the polarization direction. When the directions of incidence and reflection lie within the xy plane (a plane of incidence), the polarization directions perpendicular and parallel to the plane of incidence are called s- and p-polarization, respectively. Although the light waves having these

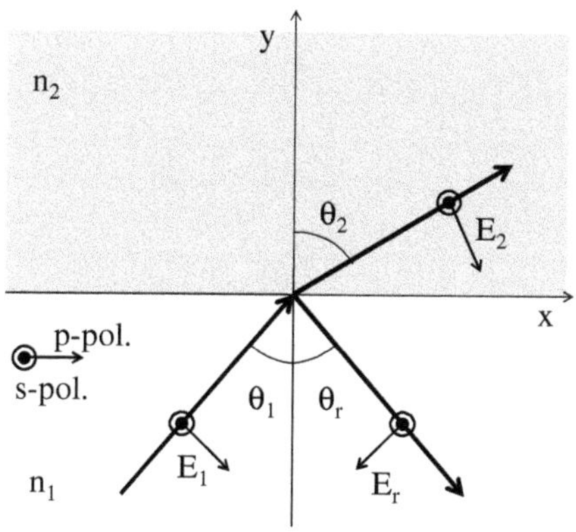

Fig. 2.1 Geometry for incidence, reflection and refraction of s- and p-polarized light at an interface between isotropic materials with the refractive indices of n_1 and n_2.

two polarizations generally mix together in a statistical way and generate unpolarized light, here we confine ourselves within the case where light is either in s- or p-polarization state such as in the case of laser light.

As described above, from the boundary condition at the interface for the plane wave expressed as Eq. (2.12) or (2.13), it is easy to show that three wave vectors of $\mathbf{k}_1$, $\mathbf{k}_2$, and $\mathbf{k}_r$ lie in the plane of incidence with $\omega_1 = \omega_2 = \omega_r$. Further, the law of reflection $\theta_1 = \theta_r$ and Snell's law $n_1 \sin\theta_1 = n_2 \sin\theta_2$ are proved to hold generally. The amplitudes of these three waves are related according to the following relations of Fresnel's law.

If we put the amplitudes of the electric fields for the incident, reflected, and refracted light as A_1, A_r, and A_2 such that $A_1 \equiv E_{1z}$, $A_r \equiv E_{rz}$, and $A_2 \equiv E_{2z}$ for s-polarization, and $A_1 \equiv \sqrt{E_{1x}^2 + E_{1y}^2}$, $A_r \equiv \sqrt{E_{rx}^2 + E_{ry}^2}$, and $A_2 \equiv \sqrt{E_{2x}^2 + E_{2y}^2}$ for p-polarization, then the amplitude reflectivity and transmittance are given as

$$r_s = \frac{A_r}{A_1} = \frac{n_1 \cos\theta_1 - n_2 \cos\theta_2}{n_1 \cos\theta_1 + n_2 \cos\theta_2}, \tag{2.17}$$

$$t_s = \frac{A_2}{A_1} = \frac{2n_1 \cos\theta_1}{n_1 \cos\theta_1 + n_2 \cos\theta_2}, \tag{2.18}$$

for s-polarization, and

$$r_p = \frac{A_r}{A_1} = \frac{n_2 \cos\theta_1 - n_1 \cos\theta_2}{n_2 \cos\theta_1 + n_1 \cos\theta_2}, \tag{2.19}$$

$$t_p = \frac{A_2}{A_1} = \frac{2n_1 \cos\theta_1}{n_2 \cos\theta_1 + n_1 \cos\theta_2}, \tag{2.20}$$

for p-polarization. Under normal incidence, these relations are reduced to

$$r_s = \frac{A_r}{A_1} = \frac{n_1 - n_2}{n_1 + n_2}, \tag{2.21}$$

$$t_s = \frac{A_2}{A_1} = \frac{2n_1}{n_1 + n_2}, \tag{2.22}$$

using the geometry defined for s-polarization. The resultant power reflectivity and transmittance are given as

$$R = |r_{s,p}|^2 \quad \text{and} \quad T = \frac{n_2}{n_1}|t_{s,p}|^2, \tag{2.23}$$

and the energy conservation is confirmed by the relation $R + T = 1$.

These relations, Eqs. (2.17) and (2.19) or (2.21), give the direct proof that the light incident from the medium of lower refractive index to that of higher one, changes the sign of the amplitude of the reflected light, meaning that the phase change occurs by 180°, while it does not in the inverse case. Further, if we transform Eq. (2.19) into

$$r_p = \frac{A_r}{A_1} = \frac{\tan(\theta_1 - \theta_2)}{\tan(\theta_1 + \theta_2)}, \tag{2.24}$$

which shows that complete depletion of reflection occurs when $\theta_1 + \theta_2 = \pi/2$ is satisfied. The incident angle in this case is expressed as

$$\tan\theta_1 = \frac{n_2}{n_1}, \tag{2.25}$$

and is called Brewster's angle. The extremely different reflectance between *s*- and *p*-polarizations around the Brewster's angle is often used for generating a polarized state from an unpolarized light in nature and also in laser technology. Figure 2.2 shows the typical power reflectance against the incident angle when light is incident from a material having the lower refractive index to the higher, and that in the inverse case. It is clear that only light with *p*-polarization shows Brewster's angle, the presence of which reduces the reflectivity over a whole angular range except for normal incidence and for large incidence angles. The completely different behaviors in the reflectivity at large incidence angles are found in case of $n_1 > n_2$. This is because total reflection occurs at incidence angles larger than $\theta_{1c} \equiv \sin^{-1}(n_2/n_1)$, where the relation $\sin\theta_2 = (n_1/n_2)\sin\theta_1 > 1$ formally holds.

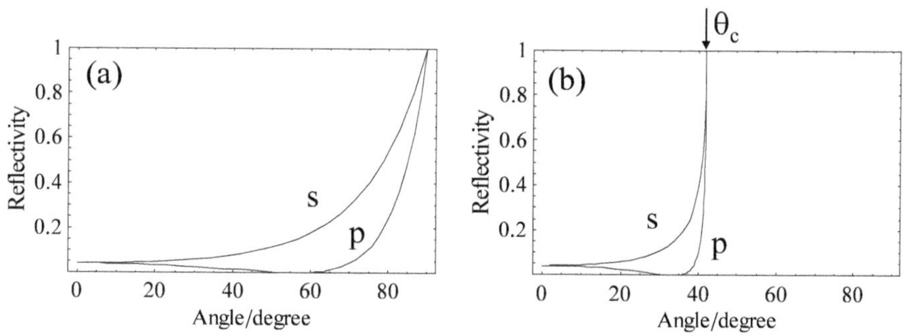

Fig. 2.2 Reflectivity for *s*- and *p*-polarized incidence with changing incident angle at an interface between two isotropic media with (a) $n_1 = 1.0$ and $n_2 = 1.5$, and (b) $n_1 = 1.5$ and $n_2 = 1.0$. θ_c is a critical angle for total reflection.

2.1.3 *Polarizations*

In general, the difference of the reflectivity between s- and p-polarizations affects only the amplitude of the reflected light, if the refractive indices are real quantities, and hence when the unpolarized light is incident on the interface, then the amplitudes for s- and p-polarizations change after the reflection. In this case, the polarization state after reflection is expressed by the sum of unpolarized light and linearly polarized light. On the other hand, when the refractive index has an imaginary part, the phase of the reflected light will be changed.

For example, we show an extreme case where light is incident from a medium having a real refractive index to an ideal metal having a pure imaginary index. In this case, the amplitude reflectivity is expressed simply as

$$r_{s,p} = \exp[\mathrm{i}\phi_{s,p}], \tag{2.26}$$

where

$$\sin\phi_s = -2n_1|n_2|\cos\theta_1\cos\theta_2/(n_1^2\cos^2\theta_1 + |n_2|^2\cos\theta_2), \tag{2.27}$$

and

$$\sin\phi_p = 2n_1|n_2|\cos\theta_1\cos\theta_2/(n_1^2\cos^2\theta_2 + |n_2|^2\cos\theta_1). \tag{2.28}$$

Thus, the amplitude does not change, while the phase does, and hence the polarization state may be different from a linear polarization state.

In order to express such polarization states generally, we consider a z-propagating plane wave in an isotropic medium, whose electric fields are assumed to be expressed as

$$\tilde{E}_x = E_{x0}\exp[\mathrm{i}(k_z z - \omega t + \phi_x)], \tag{2.29}$$

$$\tilde{E}_y = E_{y0}\exp[\mathrm{i}(k_z z - \omega t + \phi_y)]. \tag{2.30}$$

When the phase difference between y- and x-component satisfies the relation $\phi \equiv \phi_y - \phi_x = m\pi$ with m as an integer, the ratio $\tilde{E}_y/\tilde{E}_x$ is expressed by a real number as $\tilde{E}_y/\tilde{E}_x = \pm E_{y0}/E_{x0}$ irrespectively of time and position. The polarization state expressed by this relation is called linear polarization. Thus, when the linearly polarized light is incident on the interface and only the amplitudes of s- and p-polarization are changed differently, the reflected light also shows linear polarization, although its polarization direction will be changed. When $\phi = \pm\pi/2$ with $E_{y0}/E_{x0} = 1$, the amplitudes satisfy the relation $\tilde{E}_y/\tilde{E}_x = \pm\mathrm{i}$, which are called circular polarizations with plus sign for left-circular and minus sign for right-circular polarization. These two states, linear and circular polarizations, are special cases for the polarization states, and it is common that the ratio takes a value different from these extreme cases, which is generally called elliptic polarization.

On the contrary, in nature, a pure polarization state described above is rather rare, and in any way it is the mixture of polarized and unpolarized states. The unpolarized state should be understood in terms that the polarization direction is randomly and uniformly changing, the rate of which is much faster than the detector response. For example, direct light from the sun is almost completely unpolarized, while blue sky, the consequence of the scattering of unpolarized light, is partially polarized, i.e. the mixture of polarized and unpolarized light. To deal with polarized, unpolarized, and their mixed states consistently, Stokes parameters are often employed, which are defined as

$$S = \begin{pmatrix} I \\ Q \\ U \\ V \end{pmatrix} = \begin{pmatrix} \langle|\tilde{E}_x|^2\rangle + \langle|\tilde{E}_y|^2\rangle \\ \langle|\tilde{E}_x|^2\rangle - \langle|\tilde{E}_y|^2\rangle \\ 2\mathrm{Re}\langle\tilde{E}_x^*\tilde{E}_y\rangle \\ -2\mathrm{Im}\langle\tilde{E}_x^*\tilde{E}_y\rangle \end{pmatrix}. \tag{2.31}$$

Consider the case where the electric fields of z-propagating plane light are expressed as

$$\tilde{E}_x = E\cos\theta \cdot \exp[\mathrm{i}(k_z z - \omega t + \phi_x)], \tag{2.32}$$

$$\tilde{E}_y = E\sin\theta \cdot \exp[\mathrm{i}(k_z z - \omega t + \phi_y)]. \tag{2.33}$$

Then, each component of Stokes parameters is calculated as $I = E^2$, $Q = E^2\cos 2\theta$, $U = E^2\sin 2\theta\cos\phi$, and $V = E^2\sin 2\theta\sin\phi$, with $\phi = \phi_y - \phi_x$. Thus, for the linear polarization, V should be zero. The typical Stokes parameters are summarized in Table 2.1, which are calculated from the above results. In the mixture of polarized and unpolarized light, the electric fields are expressed as

$$\begin{aligned} \tilde{E}_x &= E_p\cos\theta \cdot \exp[\mathrm{i}(k_z z - \omega t + \phi_{x0})] \\ &\quad + E_{ux}\exp[\mathrm{i}(k_z z - \omega t + \phi_{x1}(t))], \end{aligned} \tag{2.34}$$

$$\begin{aligned} \tilde{E}_y &= E_p\sin\theta \cdot \exp[\mathrm{i}(k_z z - \omega t + \phi_{y0})] \\ &\quad + E_{uy}\exp[\mathrm{i}(k_z z - \omega t + \phi_{y1}(t))], \end{aligned} \tag{2.35}$$

where ϕ_{x0} and ϕ_{y0} are the definite phase shift for x- and y-components, while $\phi_{x1}(t)$ and $\phi_{y1}(t)$ are assumed to change randomly with time, and

Table 2.1 Stokes parameters

Linear[a]	Circular[b]	Unpolarized
$\begin{pmatrix} 1 \\ \cos 2\theta \\ \sin 2\theta \\ 0 \end{pmatrix}$	$\begin{pmatrix} 1 \\ 0 \\ 0 \\ \pm 1 \end{pmatrix}$	$\begin{pmatrix} 1 \\ 0 \\ 0 \\ 0 \end{pmatrix}$

[a] θ: inclination angle of electric vector, which is measured anticlockwise from the $+x$ direction.
[b] $\pm$: $+$ sign for right- and $-$ sign for left-circular polarization.

also E_{ux} and E_{uy} to be slowly varying random functions of time. The Stokes parameters become as follows: $I = E_p^2 + \langle E_{ux}^2 \rangle + \langle E_{uy}^2 \rangle$, $Q = E_p^2 \cos 2\theta + \langle E_{ux}^2 \rangle - \langle E_{uy}^2 \rangle$, $U = E_p^2 \sin 2\theta \cos \phi$, and $V = E_p^2 \sin 2\theta \sin \phi$. Since the relation $\langle E_{ux}^2 \rangle = \langle E_{uy}^2 \rangle \equiv \langle E_u^2 \rangle$ generally holds for unpolarized light, the Stokes parameters are simply expressed as

$$S = E_p^2 \begin{pmatrix} 1 \\ \cos 2\theta \\ \sin 2\theta \cos \phi \\ \sin 2\theta \sin \phi \end{pmatrix} + \langle E_u^2 \rangle \begin{pmatrix} 1 \\ 0 \\ 0 \\ 0 \end{pmatrix}, \tag{2.36}$$

where the first term in the right-hand side corresponds to the polarized component, while the second term to the unpolarized component. Thus, the degree of polarization is generally defined as

$$P = \frac{E_p^2}{E_p^2 + \langle E_u^2 \rangle} = \frac{\sqrt{Q^2 + U^2 + V^2}}{I}. \tag{2.37}$$

In Fig. 2.3, we apply this relation to the case where unpolarized light is obliquely incident and reflected at an interface between two isotropic media with different refractive indices. The degree of polarization of the reflected light increases with increasing incidence angle, takes a maximum value of unity at a Brewster angle, and decreases again. Thus, it is rather surprising that completely unpolarized light produces completely polarized light only through reflection. The polarization is known to play an important role in eye perception in animals. Recently, it is found that the anisotropic reflection for s- and p-polarizations has been found to affect significantly structure-based color mixing in some moths (Yoshioka and Kinoshita, 2007), and circularly polarized reflection is known to occur in some kinds of beetles (Neville and Caveney, 1969).

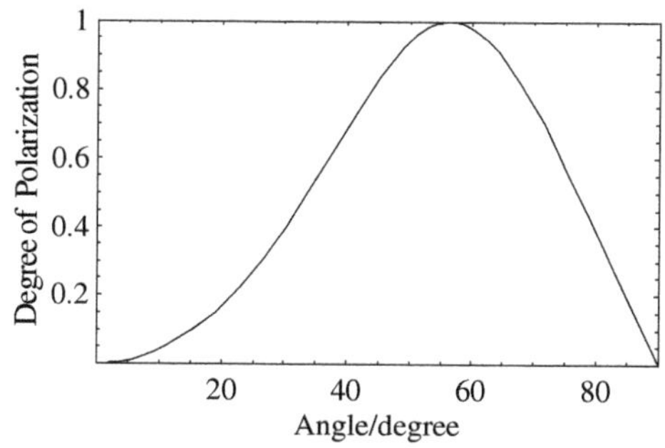

Fig. 2.3 Incidence-angle dependence of the degree of polarization, calculated using Stokes parameters. Unpolarized light is assumed to be obliquely incident and reflected at an interface between isotropic materials with the refractive indices of $n_1 = 1.0$ and $n_2 = 1.5$.

2.2 Thin-Film Interference

Thin-film interference is one of the simplest structural colorations, and yet it has a wide variety. Consider a case where a plane wave of light is incident on a thin film of thickness d and the refractive index n_b with the angle of refraction θ_b (Fig. 2.4). The light reflected at the two surfaces interferes with each other. In general, the interference condition differs whether the thin film is attached to a material having a higher refractive index or not. The former is the case for anti-reflective coating on glasses and binoculars. A

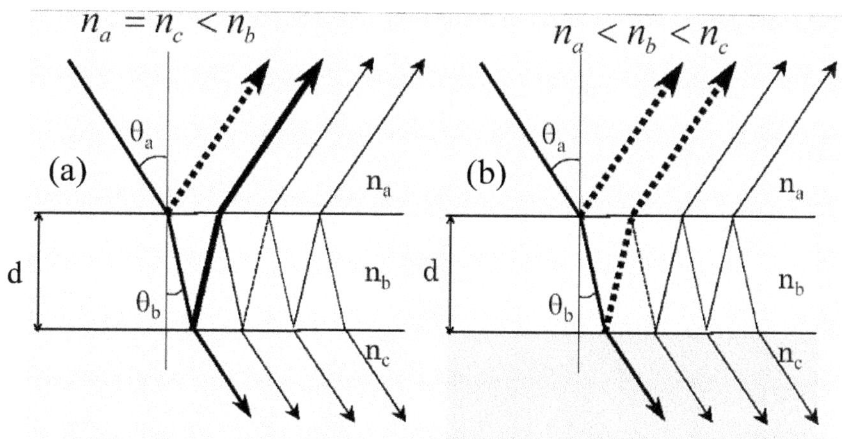

Fig. 2.4 Schematic diagram of thin-film interference. A thin film is attached to the material with (a) lower and (b) higher refractive indices.

typical example of the latter case is soap bubble (see Plate 1 in Appendix B). The reason for the difference is that when the light is incident from a material with a smaller refractive index to that with a higher one, the reflection at a surface changes its phase by 180°, while it does not in the inverse case, as described in the previous section (Sec. 2.1.2). For example, when light is incident on a soap bubble, the light passes through air ($n = 1.0$) $\rightarrow$ water ($n \sim 1.33$) $\rightarrow$ air ($n = 1.0$). Thus, the reflection at the first interface changes its phase, while that at the second one does not. In the case of the anti-reflective coating on glasses, the light passes through air ($n = 1.0$) $\rightarrow$ coating material ($1.0 < n < 1.5$) $\rightarrow$ glass ($n \sim 1.5$), in which the phase changes occur at both surfaces. Since both phase changes amount exactly to 180°, they do not eventually affect the interference condition.

In a soap-bubble case, the condition for constructive interference becomes

$$2n_b d \cos\theta_b = \left(m - \frac{1}{2}\right)\lambda, \tag{2.38}$$

where λ is a wavelength giving the maximum reflectivity and m is an integer expressing the order of interference. On the other hand, absolutely the same condition is applied to the anti-reflective coating where the destructive interference occurs. The constructive interference in the latter is obtained simply as

$$2n_b d \cos\theta_b = m\lambda. \tag{2.39}$$

In the above treatment, we consider only one-time reflection at each surface. However, when the material has high reflectance at the interface, multiple reflections more than two times considerably affect the interference. An exact calculation considering such multiple reflections gives the amplitude reflectivity and transmittance as

$$\begin{aligned} r &= r_{ab} + t_{ab}r_{bc}t_{ba}\mathrm{e}^{\mathrm{i}\phi} + t_{ab}r_{bc}r_{ba}r_{bc}t_{ba}\mathrm{e}^{2\mathrm{i}\phi} + \cdots \\ &= r_{ab} + t_{ab}r_{bc}t_{ba}\mathrm{e}^{\mathrm{i}\phi}\kappa, \end{aligned} \tag{2.40}$$

and

$$\begin{aligned} t &= t_{ab}t_{bc}\mathrm{e}^{\mathrm{i}\phi/2} + t_{ab}r_{bc}r_{ba}t_{bc}\mathrm{e}^{3\mathrm{i}\phi/2} + \cdots \\ &= t_{ab}t_{bc}\mathrm{e}^{\mathrm{i}\phi/2}\kappa, \end{aligned} \tag{2.41}$$

where

$$\kappa = 1/\left(1 - r_{bc}r_{ba}\mathrm{e}^{\mathrm{i}\phi}\right), \tag{2.42}$$

and

$$\phi = 4\pi n_b d \cos\theta_b/\lambda. \tag{2.43}$$

r_{ab} and t_{ab} are the amplitude reflectivity and transmittance at an interface for the light propagating from a to b, respectively, and are obtained from Fresnel's law (Eqs. (2.17)–(2.20)). Then, the (power) reflectivity and transmittance are given as $R = |r|^2$ and $T = (n_c/n_a)|t|^2$. It is easily verified that the conditions of Eqs. (2.38) and (2.39) correspond to $\kappa = 1$ with $t_{ab}t_{ba} = 1$ and $|r_{ab}| = |r_{bc}|$.

The examples calculated for the cases of soap bubble and anti-reflective coating are shown in Fig. 2.5, where we set the wavelength giving maximum or minimum reflectivity at $\lambda = 500\,\mathrm{nm}$ for $m = 1$. It is clear that the reflectivity is rather low in each case and smoothly changes with the wavelength. Thus, simple thin-film interference gives only weak dependence on the wavelength with low reflectance. The incidence-angle dependence is shown in Fig. 2.6(a). Since the reflectivity at an interface rapidly increases, according to the Fresnel's law, as the incident angle is increased, the reflection due to thin-film interference will be more vividly perceptible at larger angles. However, since the reflection spectrum is broad and is only weakly dependent on the incidence angle, the color of the reflected light will be only weakly perceptible.

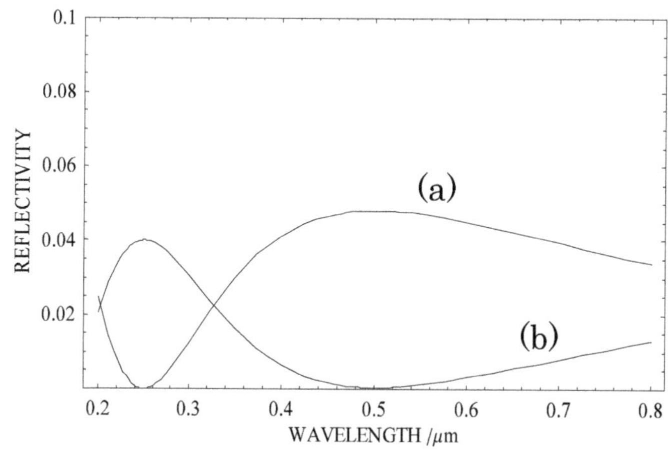

Fig. 2.5 Reflection property of thin-film interference satisfying (a) a soap-bubble (Eq. (2.38)) and (b) anti-reflection (Eq. (2.39)) relation. The refractive index and thickness of the film are set to be 1.25 and 0.1 μm, respectively. The film is assumed to be contacted with (a) air ($n = 1.0$) and (b) a material ($n = 1.5$), respectively.

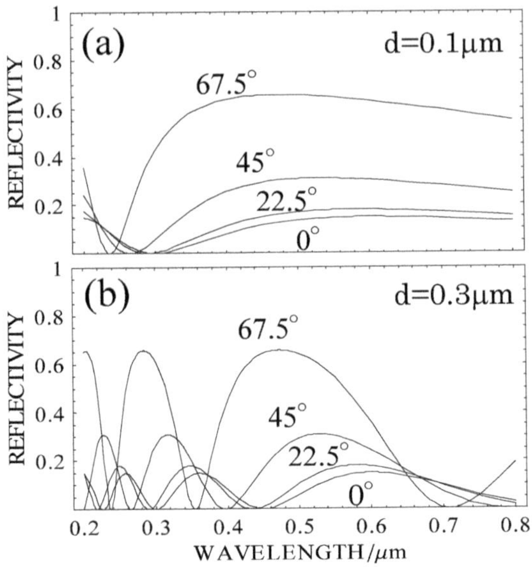

Fig. 2.6 Thin-film interference due to a film ($n = 1.5$) in air for various angles of incidence. The thicknesses of the films are set at (a) 0.1 and (b) 0.3 μm, respectively.

To enhance the reflectivity of the thin-film interference, it is necessary to increase the reflectivity at each interface. To realize this, a material having a higher refractive index such as TiO_2 ($n = 2.76$ for rutile type and 2.52 for anatase type) is often employed. Interference pearl luster pigment is one of such examples. This pigment involves a thin silica flake with both surfaces coated with TiO_2. The thickness of the coating is so adjusted that the thin-film interference occurs efficiently. When the reflectivity at the surfaces increases much more, the thin-film interference can be regarded as a kind of Fabry–Perot (FP) interferometer, because multiple reflections at both surfaces occur effectively. Using the relation given above (Eq. (2.40)), we calculate the reflectivity with changing refractive indices of the material, as shown in Fig. 2.7. At relatively low indices, low reflectivity with only weak dependence on the wavelength is observed. With increasing refractive indices, the overall reflectivity is increased and tends to be saturated. Furthermore, as in an FP interferometer, periodic dips appear at the wavelengths where the relation $2n_b d\cos\theta_b = m\lambda$ is satisfied. Thus to make the thin-film interference work efficiently as a coloring method, it is necessary to select appropriate reflectivity at the surfaces.

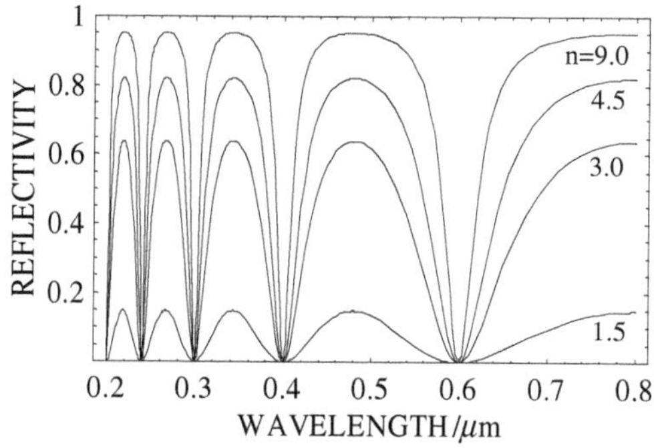

Fig. 2.7 Reflection property of thin-film interference in air for materials with various refractive indices. The thickness of the material is so chosen that the optical path length becomes 0.6 μm.

When a thin film is attached to a metal surface, similar coloration is attained if the proper choice of metal is performed. Typical examples corresponding to Ti and Al metals are shown in Fig. 2.8. When the reflectivity of metal is high enough as in the case of Al_2O_3–Al film, the total reflectivity is almost saturated and no particular coloration is expected. However, under appropriate reflectivity as in TiO_2–Ti system, the specific coloration and iridescence are apparent. Similar effect is attainable when both surfaces of a transparent film are coated by a metal layer with appropriate thickness. These types of pigment are called "optically variable pigments" and are widely used in industries.

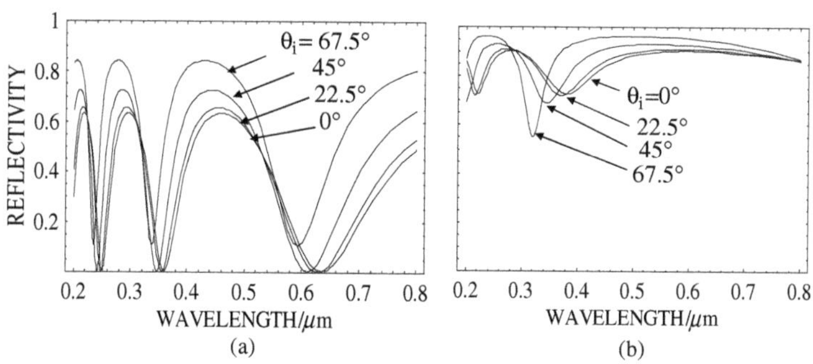

Fig. 2.8 Reflection spectra of a metal coated by an oxide thin film corresponding to (a) Ti and (b) Al, calculated for various angles of incidence. The thickness and refractive index of oxide film are set to be (a) 0.15 μm and 2.76, and (b) 0.15 μm and 1.761, respectively, while the complex refractive indices of the metal are assumed to be (a) 2.16 + 2.93i and (b) 0.68 + 4.80i.

When the thickness of the film becomes larger, regular multiple peaks appear in the reflection spectrum (Fig. 2.6(b)) owing to the higher order interference of $m > 1$. These peaks in the visible region cause a special effect called *non-spectral color* to human vision in contrast to *spectral color*, where a single wavelength of light is perceived by the eye. The non-spectral color is deeply connected with the spectral sensitivities of vision and is explained in terms that more than two color receptors among three are sensed simultaneously. Violet is a spectral color, but purple and magenta are non-spectral colors. White is also a non-spectral color, where all the three color-receptors are perceived. The multiple reflection peaks in thin-film interference is possible to be the origin of the non-spectral color, if its peak separation fits that of the response curves of the color receptors. It should be noted that the angular dependence is more pronounced for the higher-order interference than that for the first order owing to the narrow spectral widths (Fig. 2.6(b)), which will make its iridescence effect more vivid. A typical example has been recently found in the neck feather of rock dove, which specifically shows green/purple two-color iridescence (Yoshioka *et al.*, 2007).

2.3 Multilayer Interference

Multilayer interference is qualitatively understood in terms that a pair of thin layers piles periodically. Consider two layers designated as A and B with thickness d_A and d_B, and refractive indices n_A and n_B, respectively, as shown in Fig. 2.9. We assume $n_A > n_B$. If we consider a certain pair of AB layers, the phases of the reflected light at the upper and lower B–A interfaces change by 180°. Thus, the relation similar to the anti-reflective coating in thin-film interference of Eq. (2.39) is applicable:

$$2(n_A d_A \cos\theta_A + n_B d_B \cos\theta_B) = m\lambda \tag{2.44}$$

for the constructive interference with the angles of refraction in A and B layers as θ_A and θ_B. On the other hand, if we consider only A layer within AB layer, the phase of the reflected light does not change at an A–B interface. Thus, if in addition to Eq. (2.44), a soap-bubble relation of Eq. (2.38),

$$2n_A d_A \cos\theta_A = \left(m' - \frac{1}{2}\right)\lambda, \tag{2.45}$$

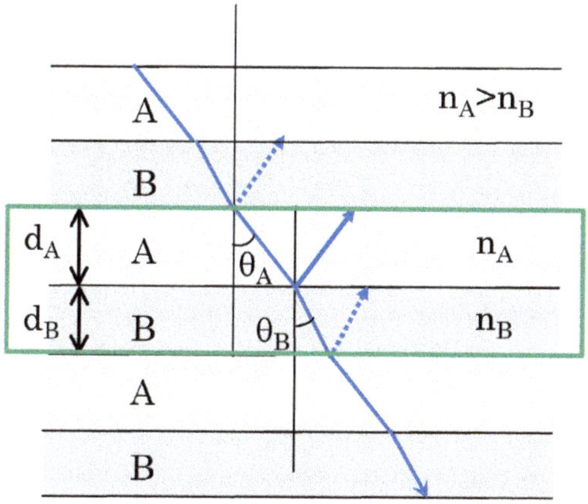

Fig. 2.9 Schematic illustration of multilayer interference.

is further satisfied, the reflected light from the A–B interface adds to that from the B–A interface, so that the multilayer will give the maximum reflectivity. Here, $m' \leq m$ should be satisfied because of the restriction of the thickness. In particular, Eqs. (2.44) and (2.45) for $m = 1$ and $m' = 1$, respectively, correspond to the lowest-order case, where the optical path lengths, defined as the length multiplied by the refractive index, in the A and B layers are equal to each other. Land (1972) called this case *ideal multilayer*. On the other hand, if the thickness of A layer does not satisfy Eq. (2.45), while the sum of A and B layers satisfies Eq. (2.44), then the reflection at the A–B interface works destructively so that the peak reflectivity will decrease. This case is called *non-ideal multilayer*.

The above interpretation, however, is too simple to understand the mechanism of the multilayer interference. In fact, it is only applicable when the difference in the refractive indices of the two layers is small enough. Otherwise, the multiple reflections modify the interference condition to a large extent. Detailed discussion on this point will be described later in conjunction with one-dimensional photonic crystal (Secs. 2.5 and 9.1.4). Thus, the quantitative evaluation of the wavelength-dependent reflectivity is rather complex in a general case, but is important to understand the principle of structural colors in nature, because most of the structural colorations are originated from the multilayer interference or its analogue.

The method to calculate reflectivity for multilayered structure has been frequently described since Lord Rayleigh (1917) presented the paper. Nowadays, the reflectivity and transmittance are often calculated through several conventional methods including transfer matrix (Hecht, 2002), iterative, and Huxley's method (Huxley, 1968). Some of these methods are suitable for the evaluation of actual structural colorations in nature, because they are applicable to the multilayer with arbitrary refractive indices and thickness without any periodicity.

The procedures to calculate reflectivity according to the above three methods will be described in Sec 9.1. Here we show only the results of the calculations. First, we show, in Fig. 2.10, the reflectivity from an ideal multilayer with varying number of layers for normal incidence. This calculation corresponds to the case where the difference of the refractive indices of the two layers is very small ($\Delta n = 0.05$), which is a typical case of the multilayer composed of polymer materials. Since the optical path lengths of both layers are set at $0.125\,\mu$m, the peak wavelength of the reflectivity should be 500 nm. With increasing number of layers, the peak reflectivity increases rapidly, while the bandwidth decreases gradually. The bandwidth in this case can be estimated from the difference of the wavelengths at the minimum reflectivity in the nearest oscillations. These wavelengths are estimated from the condition that the reflected light from $(N+1)$ interfaces interferes destructively and is expressed as $2\hat{d} = \{m \pm 1/(N+1)\}\lambda$, where $\hat{d}$ is the optical path length of one layer and m is an integer, which is obtained

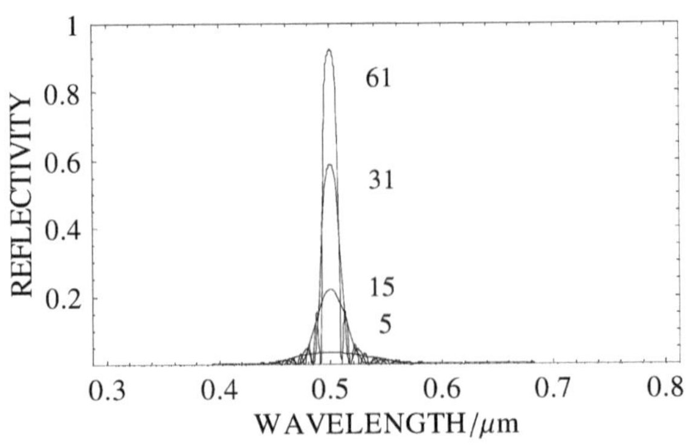

Fig. 2.10 Reflection spectra for ideal multilayers consisting of alternate layers with the refractive indices of 1.6 and 1.55 embedded in a material having the index of 1.55. The thicknesses of the layers are 78.1 and 80.6 nm, respectively. The numbers of the layers are indicated in the figure.

under the assumption of only one-time reflection at each interface. Using this relation, the bandwidth is estimated as

$$\begin{aligned}\Delta\lambda &\approx (\lambda_+ - \lambda_-)/2 = 4(N+1)\hat{d}/[2\{m^2(N+1)^2 - 1\}] \\ &\approx 2\hat{d}/(m^2 N),\end{aligned} \tag{2.46}$$

for large N. Thus, the bandwidth is inversely proportional to the number of layers. Therefore, in order to obtain high reflectivity for the multilayer having small refractive-index difference, it is necessary to pile up many layers, which inevitably decreases the bandwidth.

When the difference of the refractive indices of the two layers increases, the maximum reflectivity reaches unity for several layers. In Fig. 2.11, we show the reflection spectrum of the multilayer for various numbers of layers with the difference of the refractive indices of 0.6. It is found that only 13 layers are enough to make the maximum reflectivity reach almost unity. It is clear that the bandwidth in this case is considerably broader compared with that of the small difference of the refractive indices. Since the bandwidth for the large difference in refractive indices is rather difficult to show intuitively using the approximation of one-time reflection, we will describe it quantitatively in Sec. 9.1.4, by regarding a multilayer as a kind of one-dimensional photonic crystal.

For an ideal multilayer, the maximum reflectivity under normal incidence occurs where the relation $2n_A d_A = 2n_B d_B = (m' - 1/2)\lambda$ is satisfied (hereafter, we call the wavelength satisfying this relation for $m' = 1$ "fundamental wavelength"). The amplitude reflectivity at this wavelength is

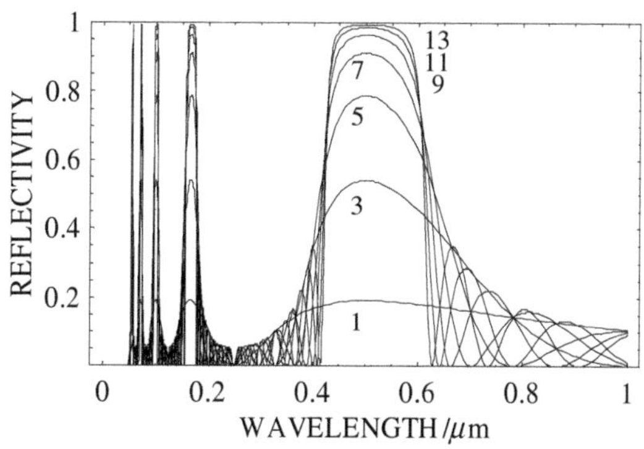

Fig. 2.11 Reflection spectra for ideal multilayers consisting of alternate layers with the refractive indices of 1.6 and 1.0 under normal incidence. The thicknesses of these layers are 78.1 and 125 nm, respectively. The numbers of the layers are indicated in the figure.

obtained using a transfer matrix method (Eqs. (9.6) and (9.7)) as $r = (1-\gamma)/(1+\gamma)$, where γ satisfies the following relation: $\gamma = (n_t/n_i)\cdot(n_A/n_B)^N$ for N of an even number and $\gamma = n_A^2/(n_i n_t)\cdot(n_A/n_B)^{N-1}$ for N of an odd number. Thus, if the multilayer is with large difference in n_A and n_B, the reflectivity easily reaches unity even for small N. On the contrary, in the wavelength at which $m' - 1/2$ becomes an integer, the reflection from A–B and B–A interfaces destructively interferes. The reflectivity at this wavelength becomes $r = (n_i - n_t)/(n_i + n_t)$ and becomes null when the refractive indices of the incident and transmitted regions are the same. We can see this effect in Fig. 2.11, where the reflectivity at $1/2, 1/4, 1/6, \ldots$ of the fundamental wavelength disappears, while those of $1/3, 1/5, 1/7, \ldots$ form a peak.

Next, we will show the result for a nonideal multilayer. In Fig. 2.12, we show the calculated result corresponding to Fig. 2.11 with 11 layers. The ratio of the optical path length of A layer with higher refractive index increases or decreases, while the sum of the optical path lengths of A and B layers is kept constant. When the deviation from the ideal multilayer becomes remarkable, the bandwidth reduces prominently in addition to the decrease in the peak reflectivity. At the same time, the peak position shifts toward longer and shorter wavelengths as the width of the A layer decreases and increases. It is also noticed that the reflectivity at the wavelength of 1/2 of the fundamental wavelength restores quickly. It is clear that the simple relation of Eq. (2.44) on the analogy of thin-film interference no longer holds, since the peak wavelength shifts definitely. An extremely nonideal multilayer in nature is actually found in the iridophore of tropical fish,

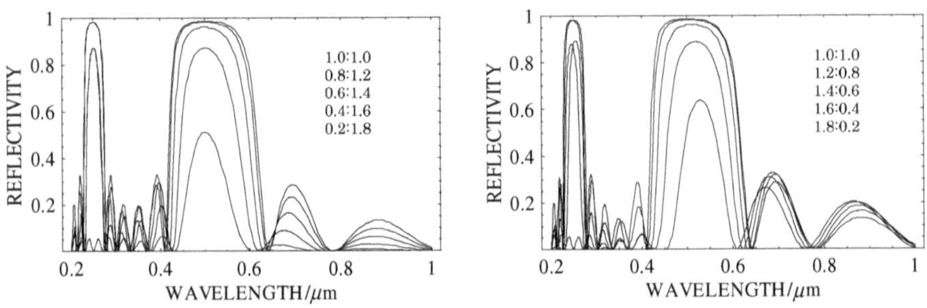

Fig. 2.12 Reflection spectra of nonideal multilayers of 11 alternate layers having the refractive indices of 1.6 and 1.0, respectively. The optical path length of the sum of these two layers is set constant at 0.25 μm. The ratio of the optical path lengths for the higher refractive index to the lower one is indicated in the figure.

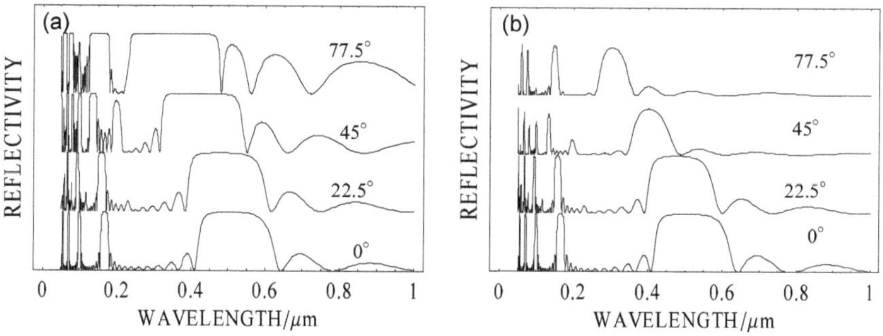

Fig. 2.13 Incidence-angle dependence of the reflection spectra from an ideal multilayer of 11 alternate layers having the refractive indices of 1.6 and 1.0, respectively, under the illumination of (a) s- and (b) p-polarized light. The thicknesses of these layers are set at 78.1 and 125 nm, respectively.

neon tetra, where very thin guanine crystals form nonideal layers of 1:20 (Sec. 6.3).

In Fig. 2.13, we show the angular dependence of the reflection from a multilayer for large difference of the refractive indices. The calculations for s- and p-polarizations show ordinary peak shifts with changing angles of incidence. In addition, the change in the bandwidth seems to be quite different for these two polarizations, namely, the reflection bandwidth under s-polarization remarkably increases, while that under p-polarization rather decreases. This is due to the difference of the variation of reflectivity at the interfaces and also due to the deviation from the ideal multilayer with changing angle of incidence.

In Figs. 2.14(a) and 2.14(b), we show examples for the reflectivity and transmittance for the multilayer, where one of the two layers shows the absorptive behavior. With increasing imaginary part of the refractive index, the reflectivity naturally decreases with narrowing bandwidth and diffuses the side-lobe structures without much change in the peak wavelength. However, the most remarkable is that the transmittance of the background region decreases quickly. This behavior is more easily seen in the total light energy calculated by the sum of the reflected and transmitted light intensities, as shown in Fig. 2.14(c). With increasing imaginary part, the fraction of the light energy naturally decreases from unity. The remarkable fact is that only light energy that belongs to the reflective band tends to remain. Thus, the introduction of absorptive material contributes mainly to the background reduction in multilayer reflection. Since the transmitted light more or less contributes to the backscattering of light due to underlying

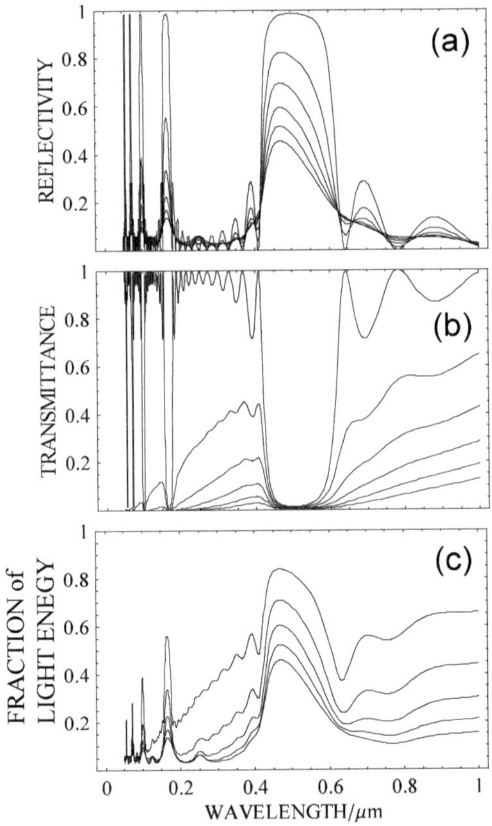

Fig. 2.14 Effect of absorption on the multilayer interference under normal incidence, where 11 alternate layers of the widths of 78.1 nm and of 125 nm are employed. The refractive index of the former is varied as 1.6, 1.6 + 0.05i, 1.6 + 0.1i, 1.6 + 0.15i, 1.6 + 0.2i, and 1.6 + 0.25i (from top to bottom in both figures), while that of the latter is kept constant at 1.0. (a) Reflectivity and (b) transmittance are shown. (c) Energy of light calculated from the sum of the transmittance and reflectivity. Notice that it is exactly 1.0 for the case without absorption.

structures, it is quite important for iridescent animals to visually emphasize the reflection color by reducing the background.

2.4 Diffraction of Light and Diffraction Grating

Diffraction is another wave-like character of light and is often explained using Huygens' principle. The propagation of wave is considered as the superposition of secondary waves produced by a propagating primary wave. In a three-dimensional case, the secondary wave is expressed by a spherical wave, and the envelope of all the secondary waves forms a new wave front. Using this principle, the diffraction is explained as follows: If an obstacle hinders the wave from producing a secondary wave, then the incomplete superpositions of remaining waves make the light propagation to be dispersive. Thus, even if a plane wave travels along a definite propagation direction, the obstacle makes the wave to be deflected toward various directions.

The problem of light diffraction was strictly treated by Kirchhoff in 1882–1883 (Kirchhoff, 1882, 1883) (see also Born and Wolf (1975)). The essential part of his theory is expressed as

$$\tilde{u}(P) = \int_{\Sigma} \frac{A e^{ikr_S}}{r_S} \cdot \frac{B e^{ikr_P}}{r_P} d\sigma, \tag{2.47}$$

where $\tilde{u}(P)$ is the complex amplitude of the wave at a point P, where a detector is placed, and r_S and r_P are the distances from a light source and a detector to a point on the plane of the orifice Σ, respectively, as shown in Fig. 2.15. B is called "declination factor" and is expressed as

$$B = \frac{1}{2i\lambda} \{\cos(\mathbf{n}, \mathbf{r}_P) - \cos(\mathbf{n}, \mathbf{r}_S)\}, \tag{2.48}$$

with $\mathbf{n}$ a unit vector normal to the plane of the orifice at that point, and $(\mathbf{n}, \mathbf{r})$ the angle between vectors $\mathbf{n}$ and $\mathbf{r}$. A is related to the light source. When the positions of the source and detector are far enough from the orifice, r_S and r_P in the denominator do not depend on the selection of

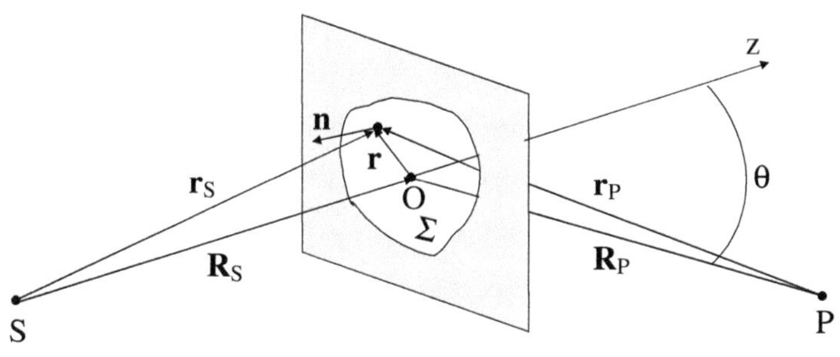

Fig. 2.15 Geometry for the diffraction of light.

a point on the plane of the orifice. Then, the following approximation is derived:

$$\tilde{u}(P) \approx AB\frac{e^{ik(R_S+R_P)}}{R_S R_P}\int_{\Sigma} e^{i(\mathbf{k}_S-\mathbf{k}_P)\cdot\mathbf{r}}d\sigma, \tag{2.49}$$

where R_S and R_P are the distances shown in Fig. 2.15 with $\mathbf{k}_S = k\mathbf{R}_S/R_S$ and $\mathbf{k}_P = -k\mathbf{R}_P/R_P$. For simplicity, we put B as a constant. The diffraction expressed by this approximation is called Fraunhofer diffraction. $\mathbf{K} \equiv \mathbf{k}_P - \mathbf{k}_S$ is often called a scattering vector.

For instance, if the orifice takes the form of a slender slit of the width a, and a plane wave is incident normally to a plane involving a slit, then

$$\tilde{u}(P) \propto \int_{-\infty}^{\infty} dx A(x) e^{iKx}, \tag{2.50}$$

where $K = k\sin\theta$ with θ the angle of diffraction and $A(x)$ a slit function. Putting

$$A(x) = \begin{cases} 1, & \text{for} - a/2 < x < a/2, \\ 0, & \text{otherwise,} \end{cases} \tag{2.51}$$

we obtain

$$\tilde{u}(P) \propto \int_{-a/2}^{a/2} dx\ e^{iKx} = \frac{2}{K}\sin\frac{Ka}{2}. \tag{2.52}$$

If an orifice takes a circular form with the radius a, the diffracted light is expressed as

$$\tilde{u}(P) \propto \frac{2\pi a}{K} J_1(Ka) \tag{2.53}$$

with $J_1(Ka)$ the first-order Bessel function. The experiment performed on a single slit illuminated by a He–Ne laser with changing slit widths is shown in Plate 2 (Appendix B). The diffraction patterns calculated for the above two cases are shown in Fig. 2.16. The pattern in Fig. 2.16(a) is characterized by a strong central line (0th order), and the surrounding higher-order lines. The dark lines appear at $K = \pm 2m\pi/a$, while the $\pm$1st-order lines appear near $K = \pm 3\pi/a$. The width of the 0th-order line is estimated from a half of the separation of the two dark lines and is expressed as $\Delta K \approx 2\pi/a$, which corresponds to the angular broadening of $\Delta\theta \approx \lambda/a$ with the wavelength of light λ. For a circular hole, the pattern also shows concentric circles as shown in Fig. 2.16(b). The exact radii of the dark circles are obtained

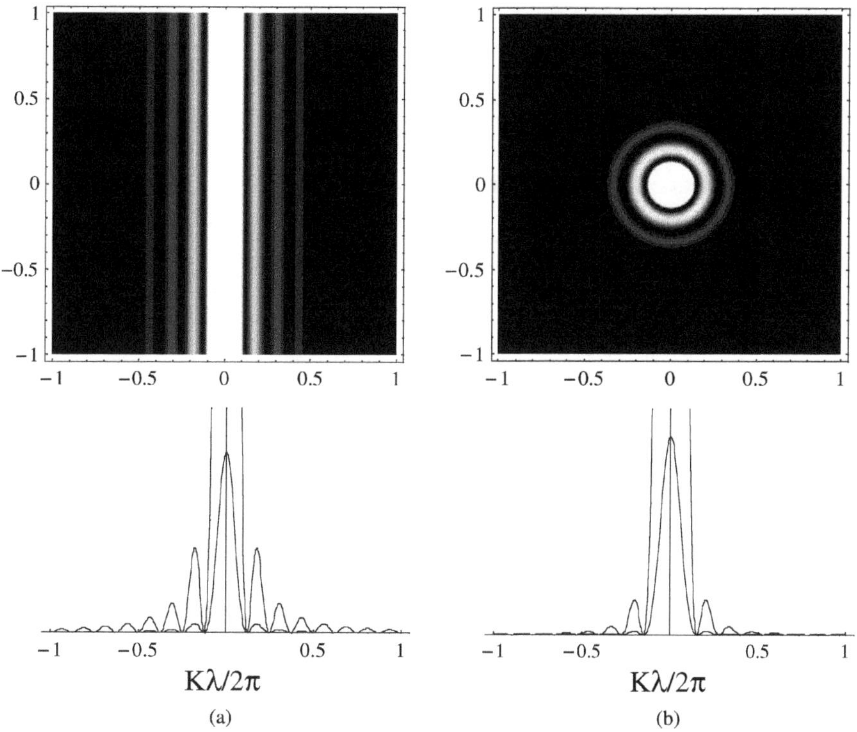

Fig. 2.16 Light diffraction images and their intensity profiles due to a single slit (a) and a circular hole (b). The width of the slit and the radius of the hole are set at 4.0 and 2.0 μm with the wavelength of light of 0.5 μm.

from the well-known values of the Bessel function, and are $K = 0.610 \times 2\pi/a$, $1.116 \times 2\pi/a$, and so on. The diameter of the 0th-diffraction spot is then roughly determined as $\Delta K \approx 0.610 \times 2\pi/a$, which corresponds to the angular broadening of $\Delta\theta \approx 0.610\lambda/a$. This value is often employed to estimate the resolution of optical system using a lens.

When N such structures are periodically aligned in one dimension, the diffracted light from those structures interferes with each other and a specific diffraction pattern appears. In general, the functional form of such a structure can be expressed as

$$\int d\mathbf{r}' A(\mathbf{r}-\mathbf{r}') \sum_{j=0}^{N-1} \delta(\mathbf{r}' - jd\mathbf{v}), \tag{2.54}$$

where $A(\mathbf{r})$ is a function expressing the form of an orifice, $\mathbf{v}$ a unit vector expressing the direction of the alignment, and d periodicity. Then the calculation of the electric field yields the intensity at a point P as

$$\tilde{u}(P) \propto \int_{-\infty}^{\infty} \mathrm{d}x \int_{-\infty}^{\infty} \mathrm{d}x' A(x - x') \sum_{j=0}^{N-1} \delta(x' - jd) \mathrm{e}^{\mathrm{i}K_v x}, \tag{2.55}$$

where $K_v = (\mathbf{k}_S - \mathbf{k}_P) \cdot \mathbf{v}$. If the shape of the orifice is expressed as an infinitely long slit, the intensity of diffracted light becomes

$$I(P) \propto \frac{1}{2}|\tilde{u}(P)|^2 = \frac{2\sin^2(K_v a/2)}{K_v^2} \frac{\sin^2(K_v N d/2)}{\sin^2(K_v d/2)}, \tag{2.56}$$

where the first part of the term in the right-hand side indicates the diffraction due to a single slit, while the second part originates from its repetition. Thus, the light diffraction from the periodic structures is expressed simply by the product of the diffraction from a single structure and the term related to the periodicity.

In Fig. 2.17, we show a typical example of light diffraction from single and multiple slits illuminated by a plane wave of light. In a single slit, light is diffracted most strongly toward a forward direction with a few sidebands at both sides. Between them, a low-intensity portion is observed, the position of which is determined by the relation of $ka \sin\theta = 2m\pi$. With increasing number of slits, another structure is superimposed on the diffraction pattern. The presence of sharp lines is one of the characteristics of the diffraction grating and is called diffraction spots that are characterized

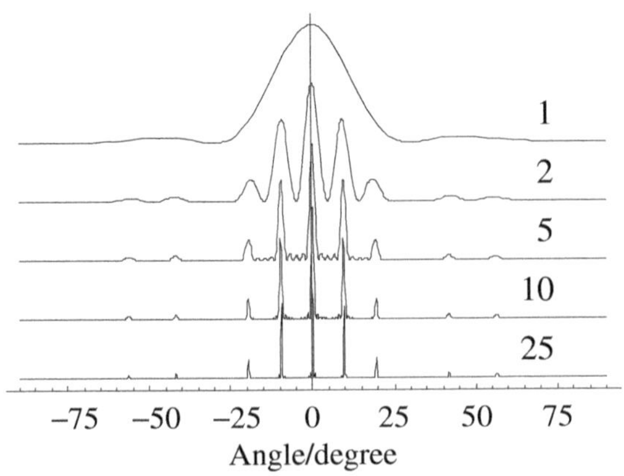

Fig. 2.17 Angular dependence of diffracted light intensity for various numbers of slits. The width of the slit is set at $1\,\mu$m with the interval of $3\,\mu$m for the wavelength of light of $0.5\,\mu$m.

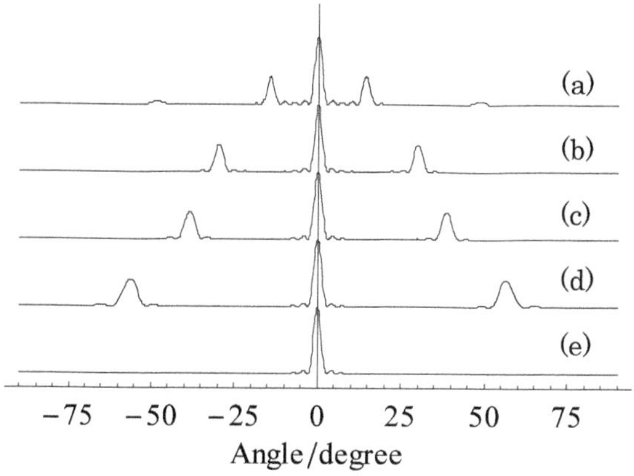

Fig. 2.18 Diffraction properties due to periodic multislits with reducing slit separation. A plane wave with the wavelength of $0.5\,\mu$m is assumed to illuminate the slits within a limited range of $10\,\mu$m in width. The ratio of the separation of the neighboring slits to the slit width is set to be constant at 2. The separations of two neighboring slits are set at (a) 2.0, (b) 1.0, (c) 0.8, (d) 0.6, and (e) $0.4\,\mu$m, respectively.

by the relation $kd\sin\theta = 2m\pi$. In Fig. 2.18, we show the diffraction from a diffraction grating with reducing periodicity. Since the diffraction spots are determined by the relation $kd\sin\theta = 2m\pi$ and hence $\sin\theta$ is inversely proportional to the periodicity, the angle of diffraction increases accordingly. And when $d < \lambda$, the diffraction spots other than 0th-order disappear and the diffraction grating behaves like a flat mirror, which is often called *zeroth-order grating* (Fig. 2.18(e)).

If small particles are regularly distributed on a plane, a two-dimensional (2D) grating will be formed. In such a case, the amplitude of the diffracted light in far field for $N \times N'$ 2D grating is expressed as

$$\tilde{u} \propto \mathrm{e}^{\mathrm{i}(\mathbf{k}_S+\mathbf{K})\cdot\mathbf{r}} \sum_{m=0}^{N-1}\sum_{n=0}^{N'-1}\sum_{j} \mathrm{e}^{\mathrm{i}\mathbf{K}\cdot(\mathbf{R}_{mn}+\mathbf{r}_j)}, \tag{2.57}$$

where $\mathbf{R}_{mn} = m\mathbf{a} + n\mathbf{b}$ with unit lattice vectors $\mathbf{a}$ and $\mathbf{b}$, and $\mathbf{r}_j$ is a vector expressing the position of a particle j in a unit cell. Here, for simplicity, we have omitted the term due to the diffraction from each particle. The intensity of the diffracted light is then proportional to

$$|\tilde{u}|^2 = \frac{\sin^2(N\mathbf{K}\cdot\mathbf{a}/2)}{\sin^2(\mathbf{K}\cdot\mathbf{a}/2)} \frac{\sin^2(N'\mathbf{K}\cdot\mathbf{b}/2)}{\sin^2(\mathbf{K}\cdot\mathbf{b}/2)} \left|\sum_j \mathrm{e}^{\mathrm{i}\mathbf{K}\cdot\mathbf{r}_j}\right|^2. \tag{2.58}$$

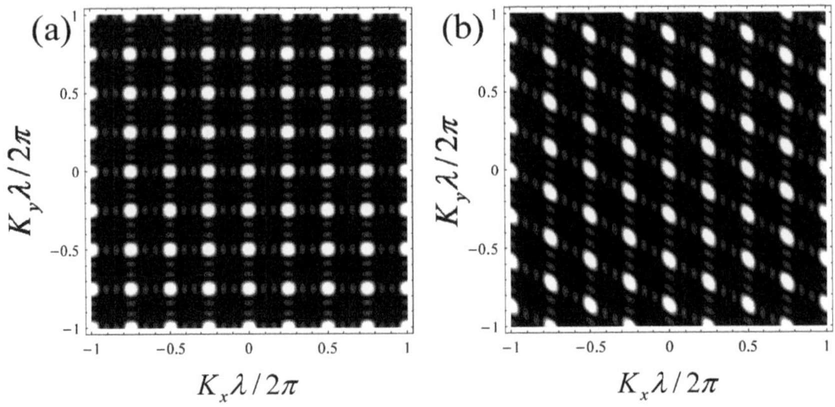

Fig. 2.19 Diffraction patterns from 4×4 2D grating with (a) square and (b) triangle lattices. The nearest particle–particle distances are set to be $2\,\mu$m in both cases and the wavelength of light to be $0.5\,\mu$m.

Typical diffraction patterns obtained for a square and triangle lattice are shown in Fig. 2.19. The size of each diffraction spot is related with the total number of lattices and the distribution of spots directly depends on the lattice structure.

The most familiar diffraction grating is CD or DVD, where we can easily observe the diffraction spots by light illumination (see Plate 2 in Appendix B). Actually, CD or DVD is a typical case of 2D grating, whose periodic structure is arranged in a concentric circle. As explained in the above discussion, a periodic structure is essential, irrespective of its individual structure, to observe the diffraction spots. In this sense, the diffraction grating can become the most familiar structural color. However, in animal and plant kingdoms, it is rather rare to find the grating-type structural color. One of the possible biological reasons is that the diffraction grating does not show specific color in itself, but gives only a mixture of various colors.

2.5 Photonic Crystals

If small identical particles are regularly arranged like a crystal as shown in Fig. 2.20(c), light scattered from each particle interferes and radiates secondary emission to regular directions (see Plate 3 in Appendix B). The theoretical treatment of this type of scattering was first developed in the

(a)

(b)

(c)

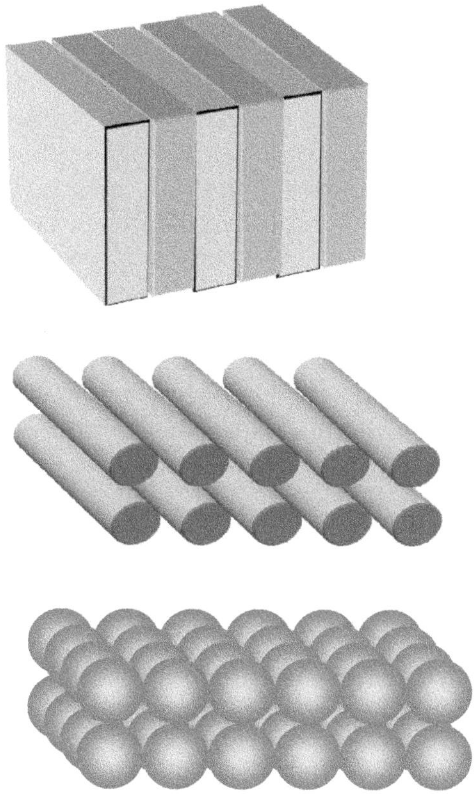

Fig. 2.20 Examples of (a) one-, (b) two-, and (c) three-dimensional photonic crystals.

field of X-ray diffraction (James, 1965), in which the theory considering only one-time scattering from each particle is called "kinematical theory," whereas that considering multiple scattering is called "dynamical theory." The latter is further categorized into two according to the treatment of atomic dipoles. Ewald considered the ensemble of discrete dipoles and derived the approximate expression for the scattering using electromagnetic theory, whereas Laue considered the continuous function of dielectric constant with the assumption of small periodic modulations. Anyway, rigorous solution of the scattering problem was not obtained, and two-beam approximation, which involved only the incident and scattered waves, was often employed.

These approaches have been extended to a system consisting of much larger particles with various crystal structures, and a new research field called photonics or photonic crystal has emerged. The interaction of electromagnetic wave with periodic microstructures has been extensively studied in conjunction with the upcoming new photon technology, which will prevail over prosperous electronics based on semiconductor technology. However, theoretical treatments have not been changed basically from the dynamical theory of X-ray diffraction except that accurate results are obtained through computer calculations. Detailed descriptions of photonic crystal have been amply reported nowadays (Sakoda, 2001). In the present section, we will describe one- and two-dimensional cases, which will offer a deep insight into the optical properties of photonic crystal found in nature.

We first derive a general relation for the propagation of light under periodically space-modulated permittivity. (Permittivity is sometimes called dielectric constant.) Using Maxwell's equation of Eqs. (2.1)–(2.4), the electric field vector $\mathbf{E}(\mathbf{r})$ satisfies the following relation:

$$\nabla \times (\nabla \times \mathbf{E}(\mathbf{r})) = \mu_0 \epsilon(\mathbf{r}) \omega^2 \mathbf{E}(\mathbf{r}). \tag{2.59}$$

When the permittivity $\epsilon(\mathbf{r})$ is periodically modulated, its reciprocal can be expanded in terms of reciprocal lattice vectors $\mathbf{G}$s as

$$\frac{1}{\epsilon(\mathbf{r})} = \sum_{\mathbf{G}} b_{\mathbf{G}} e^{i\mathbf{G}\cdot\mathbf{r}}, \tag{2.60}$$

where $b_{\mathbf{G}}$s are the expansion coefficients. According to Bloch's theorem, the electric field under periodic modulation on susceptibility is also expanded in a similar manner as

$$\mathbf{E}(\mathbf{r}) = \sum_{\mathbf{G}} \mathbf{E}_{\mathbf{G}} e^{i(\mathbf{k}+\mathbf{G})\cdot\mathbf{r}}. \tag{2.61}$$

Inserting Eqs. (2.60) and (2.61) into Eq. (2.59) yields

$$\mu_0 \omega^2 \mathbf{E}_{\mathbf{G}} = -\sum_{\mathbf{G}'} b_{\mathbf{G}-\mathbf{G}'} (\mathbf{k}+\mathbf{G}') \times ((\mathbf{k}+\mathbf{G}') \times \mathbf{E}_{\mathbf{G}'}). \tag{2.62}$$

A set of linear equations indexed by $\mathbf{G}$ gives the electric field in the modulated material.

First, we show a simple case of one-dimensional band structure calculated by means of plane-wave expansion method in Fig. 2.21, where we show four dispersion curves for various layer thicknesses of high permittivity keeping the period a constant. This figure corresponds to a multilayer of an infinite number of alternative layers with the refractive indices of

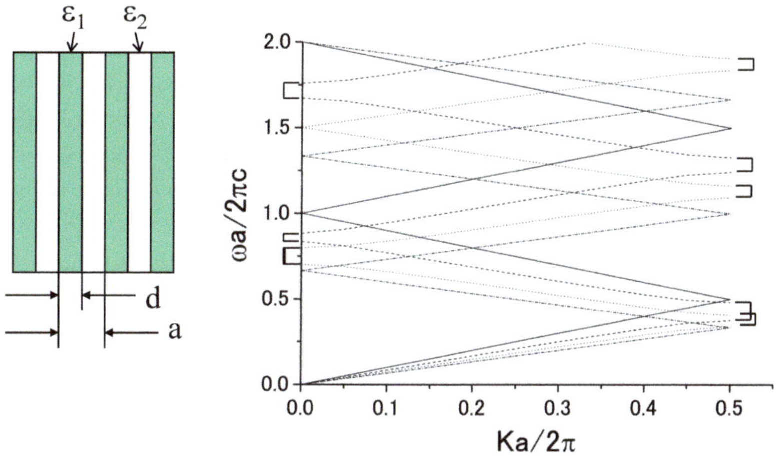

Fig. 2.21 One-dimensional photonic crystal and its photonic band structures for $d/a =$ 0, 1/3, 2/3 and 1 (from top to bottom). ϵ_1 and ϵ_2 are set at 2.25 and 1.0, respectively (Fujimura, 2008).

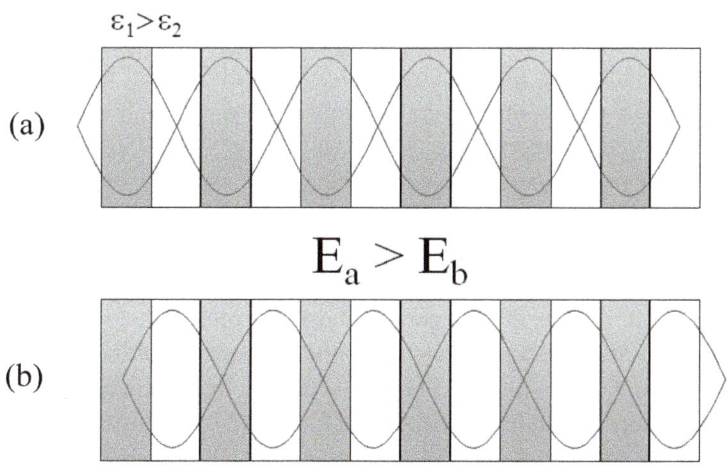

Fig. 2.22 Explanation for the origin of band gap in one-dimensional photonic crystal (see text), where E_a and E_b are energies of electromagnetic waves for cases (a) and (b), respectively.

1.5 and 1.0. At a glance, we easily see that the curves corresponding to $d/a = 0$ and 1 have no gap at $Ka/(2\pi) = 0.5$ that corresponds to the first Brillouin zone boundary, where $K \equiv |\mathbf{k}|$ in this case. This is because the multilayer is homogeneously occupied by a material of either lower or higher refractive index. On the other hand, the other two have a clear gap at $Ka/(2\pi) = 0.5\,m$, where m is an integer. Note that for convenience, the dispersion curves are intentionally bent at $Ka/(2\pi) = 0$ and 0.5 for large m. It is soon understood that the gradient of a straight line starting from an origin is inversely proportional to the average refractive index in composite systems, which are exactly equal to 1.0 and 1.5 for $d/a = 0$ and 1, respectively. The concept of the average refractive index is useful because it offers an easy way to estimate the reflection band centers by extrapolating the straight line to the zone boundaries (see Sec. 9.1.4), and also offers an approximate method of analyzing the optical properties of complex structures. The approximate methods to obtain the average refractive index and its limitation will be discussed in Sec. 9.4.

To further understand the origin of the band gap in an intuitive way, it is convenient to consider standing wave of light within a periodically modulated dielectric medium as shown in Fig. 2.22 (Fujimura, 2008). At the first zone boundary ($Ka/(2\pi) = 0.5$), the wavelength of light is just twice the periodicity. Thus, it is possible to make the wave of light in the medium match the loop (antinode) either at the layer of higher or lower permittivity. Since the electromagnetic energy is determined by the product of permittivity and the square of the electric field, the difference of the permittivity makes an energy difference between the two states, which corresponds to the energy gap. Thus, the energy gap will be the largest when the layer thicknesses are comparable with each other, while it becomes minimum when either layer has extremely small thickness. This roughly explains the difference of the reflection bandwidth in ideal and nonideal multilayers.

In Fig. 2.23, the result of a two-dimensional photonic crystal consisting of a square lattice of cylinder with an infinite length is shown. The dispersion curves between Γ and X indicate those of light propagating along a square lattice plane, while those between Γ and M do at 45° to this. The dispersion curves between X and M correspond to the angular change between them. It is clear that with the smallest radius of 0, no gap is seen at the zone boundary. With increasing radius of the cylinder, a clear gap is seen, in addition to the decrease of the gradient of the dispersion curve near the origin. On the other hand, even for the largest radius of $d/a = 1$, a clear gap is still seen. This is because the material is not completely homogeneous since cylinders having a circular section should be packed into a square lattice.

In general, it is difficult to express the characteristic optical quantities from these figures, and the transmission/reflection spectra are often

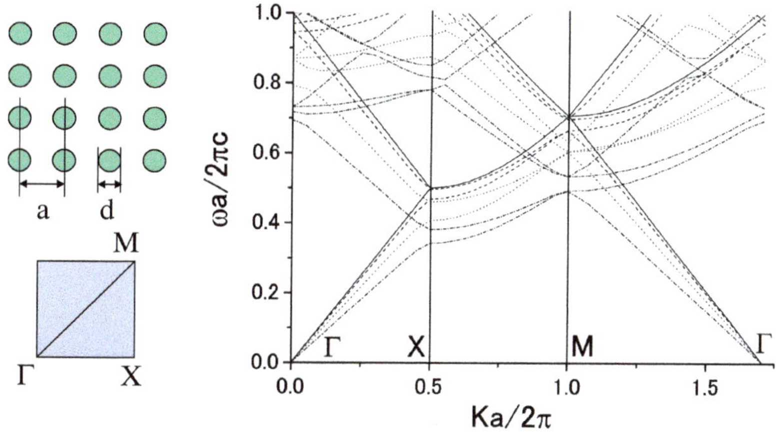

Fig. 2.23 Two-dimensional photonic crystal and its photonic band structures for $d/a =$ 0, 1/3, 2/3 and 1 (from top to bottom) in case of p-polarization (TE). The permittivity of a cylinder is set at 2.25, while that of a medium is set at 1.0 (Fujimura, 2008).

calculated numerically in a separate method. We show later an approximate method to estimate the width of the band gap in a one-dimensional crystal taking only two relevant modes into account (Sec. 9.1.4) and also show the average refractive index deduced by the straight curve in the low wave number region (Sec. 9.4). In relation to structural colors in nature, the photonic band calculations were performed for an iridescent spine of the sea mouse *Aphrodita* sp. (Parker *et al.*, 2001) and for the peacock feathers (Zi *et al.*, 2003). However, in general, it seems difficult to understand their true optical properties solely from the band calculations, because the incompleteness of the structures such as defect, disorder, and finiteness, may play an important role in natural creatures.

2.6 Light Scattering

Light scattering is another important source of structural colors and has been fully described since the first paper presented by Lord Rayleigh (1871a, 1871b). The light scattering by a particle much smaller than the wavelength of light is called Rayleigh scattering, and is easily derived by using the electromagnetic theory as a dipole radiation.

Consider a plane wave of light with the wave vector and angular frequency $\mathbf{k}$ and ω, respectively, is incident on a small particle located at a point $\mathbf{r}_0$ (Fig. 2.24). Then, an oscillating dipole is induced on the particle,

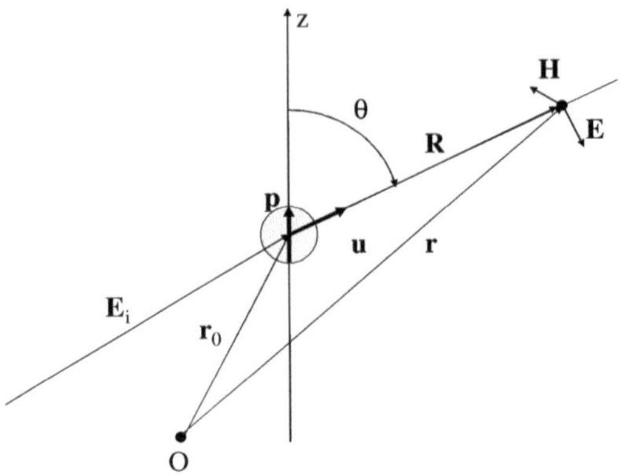

Fig. 2.24 Geometry for Rayleigh scattering.

which is expressed as

$$\mathbf{p}(\mathbf{r}_0, t) = \alpha \mathbf{E}_0 e^{i(\mathbf{k}\cdot\mathbf{r}_0 - \omega t)} \equiv \mathbf{p}_0 e^{i(\mathbf{k}\cdot\mathbf{r}_0 - \omega t)}, \tag{2.63}$$

where α is the polarizability tensor of the particle, and $\mathbf{E}_0$ is the amplitude of the electric field of the incident light. The induced dipole is oscillating in synchronization with the incident field, and radiates the secondary emission of light. The electric field at a point $\mathbf{r}$ far from the oscillating dipole is described as

$$\mathbf{E}(\mathbf{r}, t) = \frac{\mathbf{u} \times (\mathbf{u} \times \ddot{\mathbf{p}}(\mathbf{r}_0, t'))}{4\pi\epsilon c^2 |\mathbf{r} - \mathbf{r}_0|}, \tag{2.64}$$

where $\mathbf{u}$ and t' are unit vectors in the direction of $\mathbf{r} - \mathbf{r}_0$ and a retarded time expressed as $t' = t - |\mathbf{r} - \mathbf{r}_0|/c$ with the velocity of light c. ϵ is the permittivity of the medium immersing the particle. If we take the direction of the induced dipole as z-axis and employ a polar coordinate, the calculation of Eq. (2.64) results in

$$E_\theta = -\frac{|\mathbf{p}_0|\omega^2}{4\pi\epsilon c^2 R} e^{i(kR - \omega t)} \sin\theta, \tag{2.65}$$

$$E_\phi = 0, \tag{2.66}$$

with $|\mathbf{r} - \mathbf{r}_0| \equiv R$ and θ the polar angle. The light intensity in the direction of $\mathbf{u}$ is then calculated as

$$I \equiv \overline{|\mathbf{E} \times \mathbf{H}|} = (1/2)\epsilon c |E_\theta|^2 = \frac{\omega^4}{32\pi^2 \epsilon c^3 R^2} |\mathbf{p}_0|^2 \sin^2\theta, \tag{2.67}$$

where $\bar{A}$ means a cycle average of A.

Thus, the features of Rayleigh scattering are summarized as follows: (1) The scattering intensity is proportional to the fourth power of the light frequency and then of the inverse of wavelength; (2) The radiation is the most intense in the plane normal to the direction of the induced dipole, and is dependent as $\sin^2\theta$ on the polar angle; (3) The scattering intensity is inversely proportional to the second power of the distance between the induced dipole and the point of observation; (4) Scattered light is completely polarized if the polarizability is isotropic. As a result, the feature (1), i.e. blue light is more efficiently scattered than red light, holds irrespective of the polarizability, and thus the material containing many scatterers bears blue color, even if it is illuminated with white light.

When the size of the particle becomes comparable or larger than the wavelength of light, the above approximation of a point dipole no longer holds and another theory called Mie scattering should be applied (Mie, 1908). In this theory, a plane wave of light illuminates a dielectric sphere, by which light is scattered after reflection, refraction, and diffraction. Although this problem is complex, the evaluation of the boundary condition of the electromagnetic wave at the surface of the sphere gives an exact result for the electric field outside the sphere in a polar coordinate as follows (Fig. 2.25) (Born and Wolf, 1975):

$$
\begin{aligned}
E_r^{(s)} &= \frac{1}{k^I}\frac{\cos\phi}{r^2}\sum_{l=1}^{\infty} l(l+1)({}^eB_l)\zeta_l^{(1)}(k^{(I)}r)P_l^{(1)}(\cos\theta),\\
E_\theta^{(s)} &= -\frac{1}{k^{(I)}}\frac{\cos\phi}{r}\sum_{l=1}^{\infty}\Big\{({}^eB_l)\zeta_l^{(1)'}(k^{(I)}r)P_l^{(1)'}(\cos\theta)\sin\theta,\\
&\quad -\mathrm{i}({}^mB_l)\zeta_l^{(1)}(k^{(I)}r)P_l^{(1)}(\cos\theta)\frac{1}{\sin\theta}\Big\},\\
E_\varphi^{(s)} &= -\frac{1}{k(I)}\frac{\sin\phi}{r}\sum_{l=1}^{\infty}\Big\{({}^eB_l)\zeta_l^{(1)'}(k^{(I)}r)P_l^{(1)}(\cos\theta)\sin\theta,\\
&\quad -\mathrm{i}({}^mB_l)\zeta_l^{(1)}(k^{(I)}r)P_l^{(1)'}(\cos\theta)\sin\theta\Big\},
\end{aligned}
\tag{2.68}
$$

with

$$
\begin{aligned}
{}^eB_l &\equiv \mathrm{i}^{l+1}\frac{2l+1}{l(l+1)}\frac{\hat{n}\psi_l'(q)\psi_l(\hat{n}q)-\psi_l(q)\psi_l'(\hat{n}q)}{\hat{n}\zeta_l^{(1)'}(q)\psi_l(\hat{n}q)-\zeta_l^{(1)}(q)\psi_l'(\hat{n}q)}\\
{}^mB_l &\equiv \mathrm{i}^{l+1}\frac{2l+1}{l(l+1)}\frac{\hat{n}\psi_l(q)\psi_l'(\hat{n}q)-\psi_l'(q)\psi_l(\hat{n}q)}{\hat{n}\zeta_l^{(1)}(q)\psi_l'(\hat{n}q)-\zeta_l^{(1)'}(q)\psi_l'(\hat{n}q)},
\end{aligned}
\tag{2.69}
$$

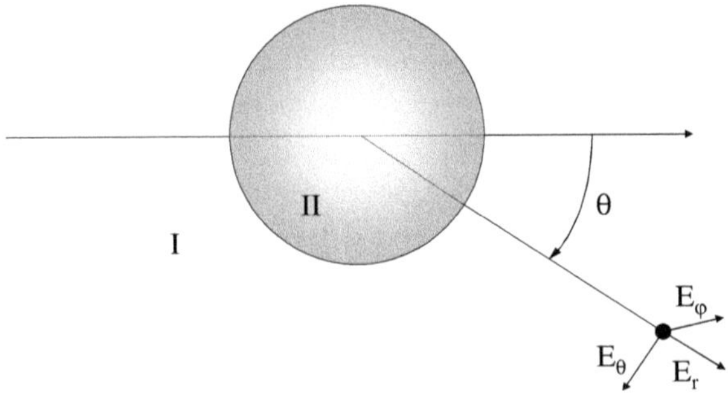

Fig. 2.25 Geometry for Mie scattering.

where $P_l^{(m)}(\cos\theta)$ is an associated Legendre polynomial and $\zeta_l^{(1)}(x) = \psi_l(x) - i\chi_l(x) = \sqrt{\pi x/2}H_{l+\frac{1}{2}}^{(1)}(x)$ with $H_{l+\frac{1}{2}}^{(1)}(x)$ a Hankel function. q and $\hat{n}$ are parameters expressed by $q = 2\pi a/\lambda^{(I)}$ and $\hat{n} = n^{(II)}/n^{(I)}$, respectively, where the medium surrounding the sphere is denoted as I, while inside the sphere as II, such that $k^{(I)} = n^{(I)}\omega/c_0$ with $n^{(I)}$ the refractive index of I.

van de Hulst showed the approximate expression for the efficiency factor Q_{ext} to the extinction of light by a sphere of the radius a in vacuum, in which Q_{ext} is defined by the ratio of the extinction cross section (corresponding to scattering cross section for a transparent material) to the geometrical cross section πa^2 (van de Hulst, 1957). The expression is very simple as in the following:

$$Q_{\text{ext}} = 2 - \frac{4}{\rho}\sin\rho + \frac{4}{\rho^2}(1 - \cos\rho), \tag{2.70}$$

with $\rho \equiv 2q|\hat{n} - 1|$. This expression is one of the most useful among complicated formulae appearing in the Mie theory.

Figure 2.26 shows the scattering-angle dependence of the Mie scattering for various radii of a dielectric sphere. It is clear that with increasing radius, the scattering becomes anisotropic and the forward scattering is dominant, while for a small radius, the scattering is essentially equivalent to Rayleigh scattering. In Fig. 2.27, the extinction cross section is calculated for various radii. When the radius is sufficiently small, the scattering intensity obeys ω^4 rule, which strictly corresponds to Rayleigh scattering, while with increasing radius, a peak appears in a visible region. Thus, Mie scattering contributes to the coloration, which is dependent on the size of a sphere.

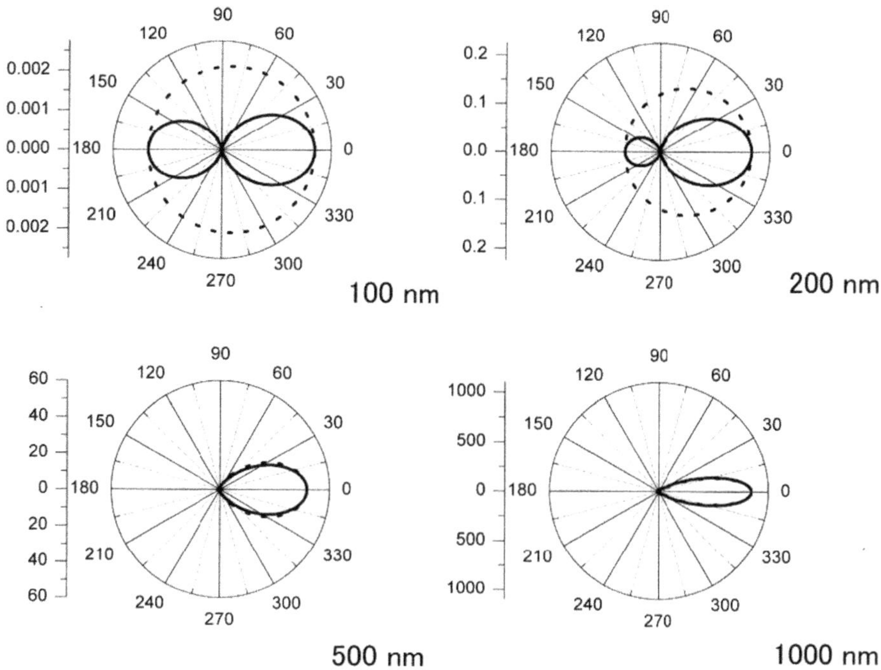

Fig. 2.26 Angular dependence of Mie scattering intensity for various radii of dielectric spheres. The refractive index of the sphere is set at 1.5, while that of the medium is at 1.33, with the wavelength of light 0.5 μm. Dashed and solid lines indicate the scattering intensities for the incident polarization perpendicular and parallel to a plane of this page.

In view of coloration mechanisms, scattering of light by particles is completely different from thin-layer interference and photonic crystals, because it has a root in the irregularity of the structure. The coloration due to light scattering is also known as non-iridescent color in contrast to iridescent colors due to the other mechanisms. Light scattering has been thought as a cause of blue coloring in a wide variety of random media. The origin of blue sky was explained first by Lord Rayleigh as due to the light scattering by atmospheric molecules (Rayleigh, 1871a,b) (see Plate 3 in Appendix B). The color of the suspension involving small colloidal particles is called Tyndall blue. The blue color in these cases is essentially based on the fact that the scattering cross section is dependent on the fourth power of the frequency of light and that the scattered light at right angle is polarized to a large extent. With increasing size of the particle, Mie scattering occurs, whose wavelength dependence differs considerably from Rayleigh scattering. Thus, different coloration should appear. Light scattering has

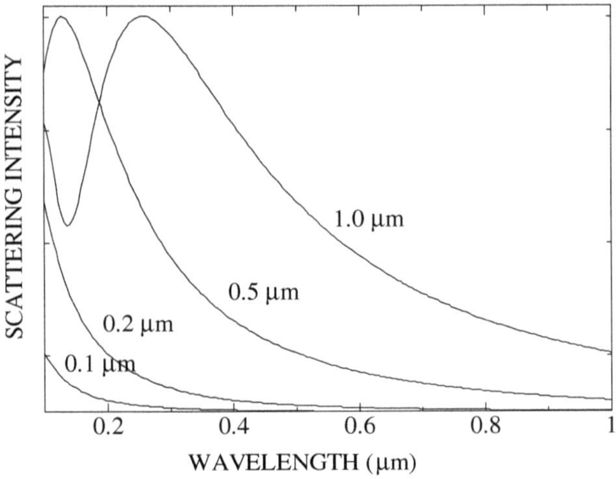

Fig. 2.27 Wavelength dependence of Mie scattering intensity for various radii of dielectric spheres with the refractive index of 1.5. The refractive index of the medium is set at 1.33.

been believed in the major causes of non-iridescent blue colorings in the animal world as well. However, recent extensive studies cast doubt upon these thoughts, because most of them show dense particle distribution and also because clear reflection peak appears in spite of small scatterers. The work concerning this subject will appear in Secs. 4.3 and 5.6.

Chapter 3

Butterflies and Moths

3.1 General Descriptions

3.1.1 *Phylogeny of the Lepidoptera*

Lepidoptera (butterflies and moths) is probably the most conspicuous and attractive group in insects, to which people's great attention has been successively paid in artistic and scientific activities since the dawn of civilization. The number of the lepidopteran species is now estimated as 150,000 and will approach finally 300,000–500,000. Thus, the classification of such a large taxon into subgroups is quite complicated and hence is changing year by year. Here, we will follow that employed in a book edited by Kristensen (1999, 2003). In this book, the order Lepidoptera is classified into 46 superfamilies (Fig. 3.1). A current name "butterfly" is now devoted to the following three superfamilies, Papilionoidea, Hesperioidea, and Hedyloidea, and it is strange enough that all the other species are called "moth."

The various colored lepidopteran species, whose wings are covered with mosaic tapestry of scales, are one of the most extensively studied insects from an optical aspect, such as pigmentary and structural colorations, visual mechanisms, antireflection effects on ommateum and wing, and so on. If we restrict ourselves within the structural colorations, about nine-tenth of the researches have been devoted solely to butterflies, although the number of butterflies, estimated as 17,500, is only one-tenth of the total lepidopteran species. We show the number of species and papers devoted to the researches on structural colors for each superfamily, together with a phylogeny of the superfamilies in Fig. 3.1. This figure clearly shows how butterflies are attractive for researchers and how moths have been avoided although various shining moths are known to be present.

As described above, butterfly species are anyway belonging to the three superfamilies (Ackery *et al.*, 1999). Hesperioidea includes only one

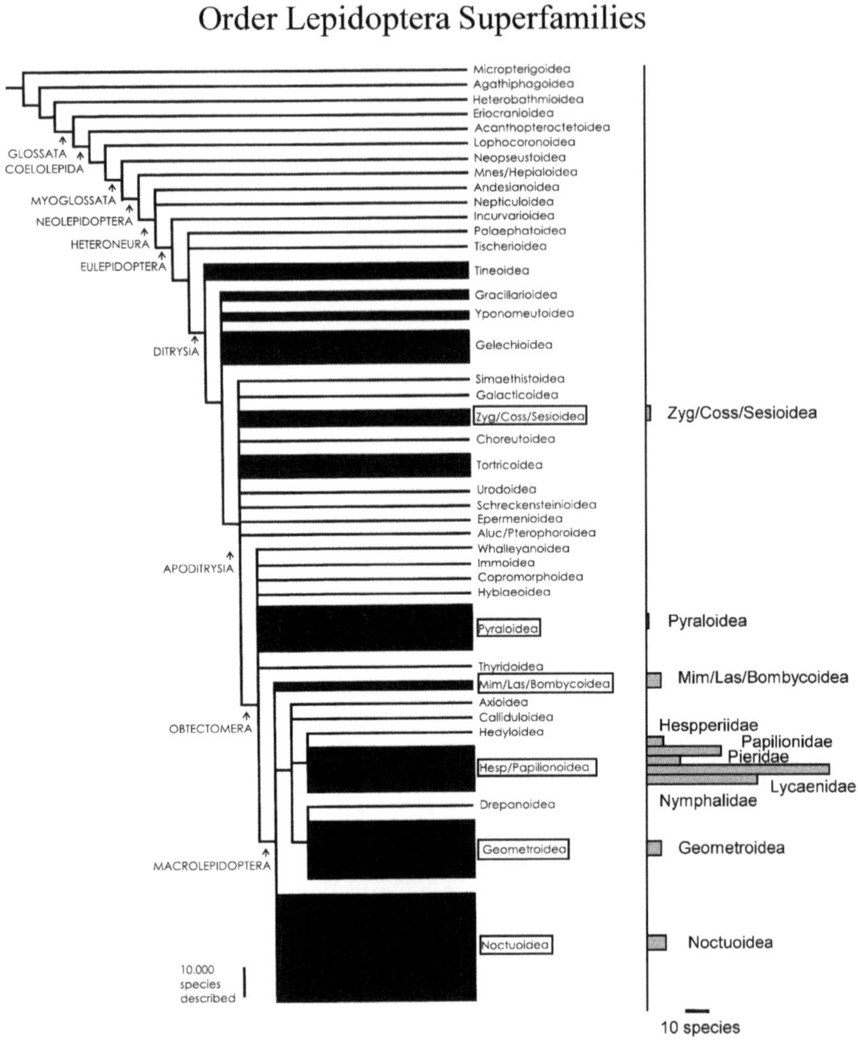

Fig. 3.1 Classification of Lepidoptera and their structural colorations. (Reproduced from Appendix I of Kristensen and Skalski (2003) with permission of Walter de Gruyter.) Histogram in the right-hand side shows the number of species whose microstructures were investigated under electron microscopy.

family Hesperiidae (skippers), under which five subfamilies are classified: Coeliadinae, Pyrrhopyginae, Pyrginae, Heteropterinae, and Trapezitinae. About 3500 species are belonging to this family. The Papilionoidea ("true" butterflies) is a large superfamily and 14,000 species are belonging to. This group comprises four families, Papilionidae (swallowtails), Pieridae (whites and sulphurs), Lycaenidae (blues, coppers, and metalmarks), and Nymphalidae (browns, fritillaries, admirals and monarchs). The Hedyloidea (moth-butterflies) is a sistergroup with the other two and used to be classified into moth. It comprises only a single subfamily Hedylidae with 40 species.

In Table 3.1, we list the numbers of species and papers, which anyway concerned with the structural colorations for each subfamily of the

Table 3.1 Classification of butterfly superfamilies.

Superfamily	Family	Subfamily	Species[a]	Species investigated[b]	Papers[c]
Hesperioidea	Hesperiidae	Coeliadinae	75	1(0)	1
		Pyrrhopyginae	150	2(2)	2
		Pyrginae	1000	5(4)	5
		Heteropterinae	150		
		Trapezitinae	60		
		Hesperiinae	2000	5(1)	4
Papilionoidea	Papilionidae	Baroniinae	1		
		Parnassiinae	70	2(1)	2
		Papilioninae	550	65(30)	50
	Pieridae	Pseudopontiinae	1		
		Dismorphiinae	100	1(0)	1
		Pierinae	700	17(6)	16
		Coliadinae	250	29(8)	24
		Riodininae	1250	11(6)	12
	Lycaenidae	Poritiinae	530	1(1)	1
		Miletinae	150	1(1)	1
		Curetinae	18	1(1)	1
		Lycaeninae	4000	110(67)	41
	Nymphalidae	Libytheinae	12		
		Heliconiinae	400	15(3)	15
		Nymphalinae	350	19(4)	17
		Limenitinae	1000	16(2)	10
		Charaxinae	400	3(0)	4
		Apaturinae	430	5(0)	9
		Morphinae	230	33(29)	83
		Satyrinae	2400	9(2)	7
		Calinaginae	8		
		Danainae	450	12(6)	5
Hedyloidea	Hedylidae		40		

[a]Total number of species (Ackery *et al.*, 1999).
[b]Number of species whose structural colors are investigated. The number of species, whose microstructures are investigated using electron microscopy, is shown in parentheses.
[c]Number of papers which refer to the structural colors in the butterfly species.

three butterfly superfamilies. We notice the researches are rather confined with specific groups such as Papilioninae, Coliadinae, Lycaeninae, and Morphinae. Particularly, the number of researches on the Morphinae are extraordinarily large, because this group includes the famous *Morpho* butterflies. On the contrary, the researches on the Riodininae, to which many structurally colored butterflies may belong, seem to be rare. Thus, this table will supply an excellent guide for the directions of the studies on structural colors from now on.

3.1.2 *Lepidopteran Scales*

It will be helpful to explain the most important characters in the lepidopteran structural colors. Since there are excellent works on lepidopteran scales by Ghiradella (1998) and also good books have been already published (Nijhout, 1991; Honda and Kato, 2005), readers will be better to refer to these works to understand more deeply. Here, we briefly review these works for the better understanding of lepidopteran structural colors.

Lepidopteran species are characterized by their wings covered with scales. The scales are of a thin plate-like form, whose typical dimensions are 0.2 mm in length, 0.1 mm in width, and 0.003 mm in thickness, which cover the wing like tiles on a roof or a dense tapestry. The butterfly wing is typically covered with two types of scales (sometimes three types are present): one is a cover scale (glass scale) and the other a ground scale (basal scale), which are shown in Fig. 3.13. The cover scales are variously specialized in shape and play a variety of roles on the wing, whereas the ground scale usually has a rectangular form and seems to be less specialized.

A scale is considered to be a flattened form of a bristle that has a long, cylindrical shape with a hollow inside. Both the scale and bristle have a complicated interior structure and are called macrochaete, which is in contrast to a simple hair called microchaete. The bristles are distributed in all orders of insects, whereas the scales are restricted in several orders: Coleoptera (beetles), Diptera (flies, horseflies), Psocoptera (booklices), Trichoptera (caddisflies), and Lepidoptera.

The scales of butterflies or moths are originated from a single-layered epithelial cell in a developing wing (Kühn, 1955). Some of the epithelial cells have scale primordia and become scale mother cells that are scattered irregularly on the wing. They will soon line up in rows with regular spacing. A scale mother cell divides twice (see Fig. 3.2): the first one is oriented perpendicular to the wing plane and the resulting daughter cell located at the

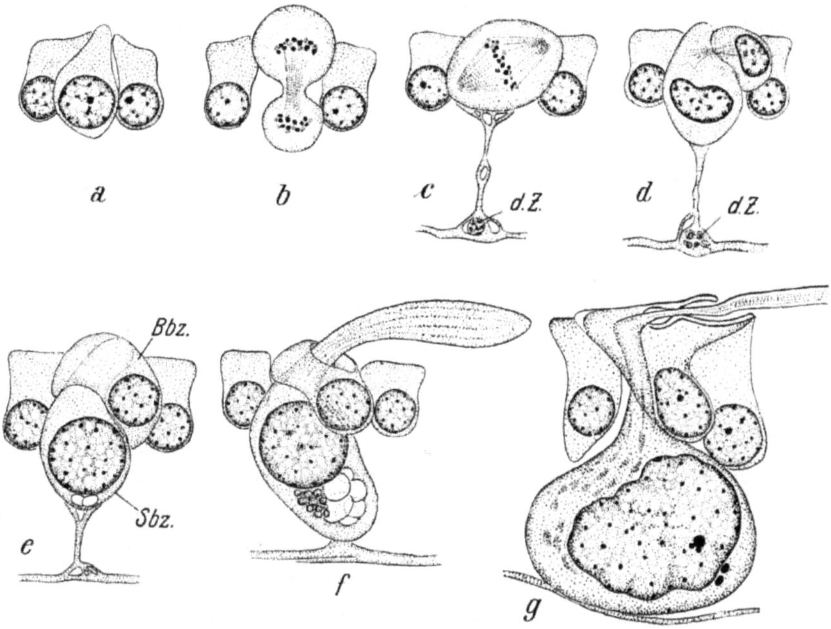

Fig. 3.2 Illustration of the developmental process of a scale in wing membrane of a flour moth, *Ephestia kuehniella*. Sbz: scale-forming cell, Bbz: socket-forming cell, and d.Z: degenerating cell (Kühn, 1955). (With kind permission of Springer Science and Business Media.)

inner side degenerates. On the other hand, the second one takes place at an angle of 45° to the surface, which results in a scale-forming cell at the proximal part and a socket-forming cell at the distal part. The scale-forming cell pushes a narrow cylindrical projection above the plane of the wing, while the socket-forming cell extends the projection around the neck of the scale projection. The scale projection grows and then flattens to form a scale (Overton, 1966). When it is fully grown, cuticle secretion begins along the scale.

A fully developed scale is of a thin, plate-like shape and is clearly asymmetric with respect to its surface: a surface faced to wing membrane is flat and featureless, while the obverse surface is decorated with elaborate structures. In Fig. 3.3, we show the morphology of a typical lepidopteran scale. In the frontal surface, parallel running ridges, which are decorated by an oblique lamellar structure, partly overlapping with each other. At the side of the ridge, regular pleated structures, called microrib, are running nearly perpendicular to the lamellar structure. Between the ridges, crossribs

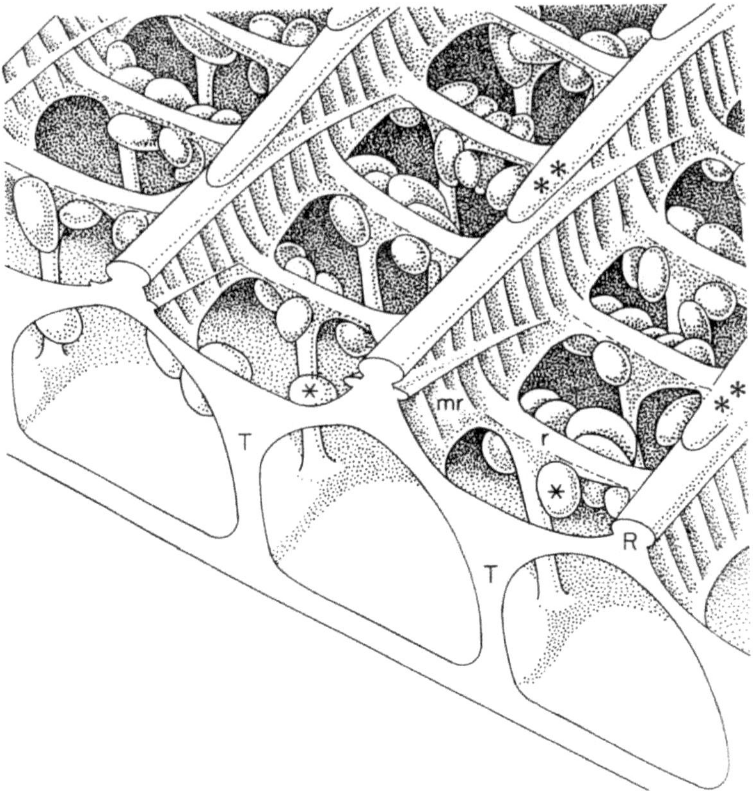

Fig. 3.3 Schematic view of lepidopteran scale. R: ridges, T: trabeculae, mr: microribs, r: cross ribs, *: ellipsoidal beads observed in Pieridae, **: ridge-lamellae (Ghiradella (1991) with permission).

connect the neighboring ridges together. There is a window between the crossribs, which opens an interior to an open space. Below a ridge, pillars, called trabeculae, support the form of a whole scale. The structural colors in lepidopteran scale is eventually expressed by the elaborations in various parts of the interior structure of the scale such as ridge-lamella, microrib, window, and trabeculae, which will be described in detail later.

3.2 *Morpho* Butterflies

3.2.1 *General Remarks*

The *Morpho* butterflies (see Fig. 3.4 and Plates 7 and 8 in Appendix B) are one of the most familiar iridescent creatures living in South and Central

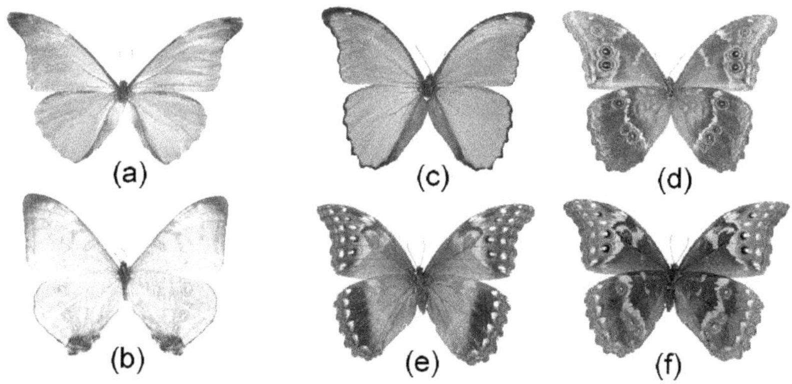

Fig. 3.4 Left: dorsal sides of male (a) *Morpho rhetenor* and (b) *M. sulkowskyi*. Right: sexual dimorphism in *M. didius*. Dorsal and ventral sides of (c, d) male and (e, f) female specimens are shown. See Plates 7 and 8 in Appendix B.

America, and have been extensively studied for over a century. Before describing a physical approach to its iridescence, here, we briefly mention the phylogenetic features of the *Morpho* species. Since the classification of butterflies has been changing year by year, the definite classification has not yet been settled for the *Morpho* butterflies. In a book edited by Kristensen, the *Morpho* species are classified into the family Nymphalidae, which consists of several subfamilies including Morphinae that is closely related to the subfamily Satyrinae (Ackery *et al.*, 1999). The Morphinae is classified further into three tribes, Morphini, Brassolini, and Amathusiini. The tribe Morphini includes two subtribes: one is Antirrheiti, to which the genera *Caerois* and *Antirrhea* are belonging, and the other is Morphiti involving only one genus *Morpho*. On the other hand, in Harvey's classification, Brassolinae is treated as an independent subfamily (Nijhout, 1991).

In the past, the classification of the *Morpho* butterflies was performed by Fruhstorfer as a form of genus *Morpho* (Fruhstorfer, 1913). He divided the genus *Morpho* into two groups, *Iphimedeia* and *Morpho*, and reported four species for the former and 27 species for the latter, with 101 additional subspecies. Le Moult and Réal published a catalog and reported 75 species under eight subgenera (Le Moult and Réal, 1962, 1963). After then, Blandin improved these classifications and added one more subgenus (Blandin, 1988, 1993). Thus, the number of subgenera became nine at this stage as shown in Fig. 3.5.

Recently, Penz and DeVries published a phylogenetic analysis for the *Morpho* butterflies by selecting 27 species under nine subgenera and investigated totally 120 characters including flight height, morphology, genitalia,

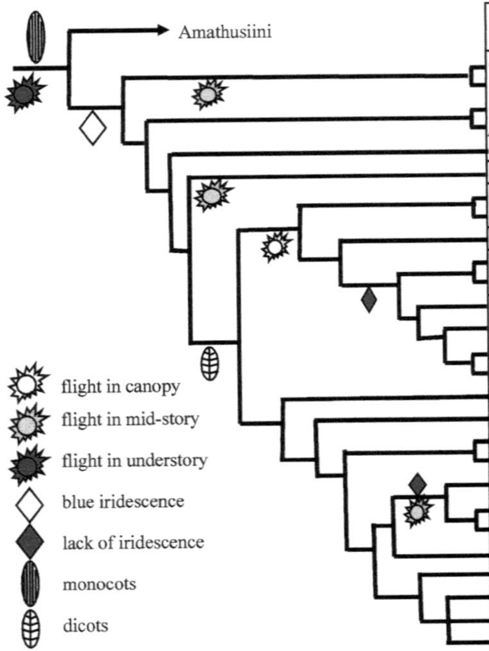

species	subgenus	iridescence	gloss	flight height	host plant	size of cover scale
adonis	*Cytheritis*	2	O	1	?	
eugenia		2	O	1	?	
aega		2	O	?	0	
portis		2	O	0	0	
sulkowskyi		1	O	0	0	
aurora	*Balachowskyna*	1	O	1	?	
cypris	*Cypritis*	2	O	2	1	
rhetenor		2	O	2	1	
anaxibia	*Iphixibia*	2	×	2	1	
hecuba	*Schwartzia*	0	×	2	1	
cisseis		0	×	2	1	
hercules	*Iphimedeia*	0	×	2	1	
amphitrion		0	×	2	?	
perseus		0	×	2	?	
theseus		0	×	2	?	
godarti	*Grasseia*	1	×	?	?	
menelaus		2	O	0	1	
didius		2	×	0	1	
amathonte		2	×	0	1	
laertes	*Pessonia*	0	×	?	1	
catenarius		0	×	1	1	
polyphemus		0	×	1	1	
deidamia	*Morpho*	2	×	0	1?	
granadensis		2	×	0	1	
peleides		2	×	0	1	
achilles		2	×	0	1	
achilleana		2	×	0	1	

1.0

Fig. 3.5 Phylogeny of the genus *Morpho* for 27 species under nine subgenera with flight height (understory (0), midstory (1), and canopy (2)) and larval host plant (monocotyledonous (0) and dicotyledonous (1)). (Reproduced from Fig. 12 and Appendix 2 of Penz and DeVries (2002) with permission.) The wing appearances on the iridescence (blue (2), whitish blue (1), and none (0)) and gloss (glossy (open circle) and mat (cross)) of the dorsal side are added in this table. The last column is the rough estimate of the areal size of cover scale obtained from the ratio of the area created by the distances of adjacent socket positions within the the row and between the rows. The ratio of 1.0 is shown below.

venation, wing color pattern, and larval host plant (Penz and DeVries, 2002). They constructed a phylogenetic tree from the analysis of these characters and compared the subgenera reported before. They found some of the subgenera were actually monophylic, while the others were rather polyphylic, and they proposed to abandon the subgeneric classification of the *Morpho*.

In Fig. 3.5, we show the *Morpho* phylogeny reproduced from their paper. Although the classification of the *Morpho* is really complicated and confusing both in specific and subgeneric levels, their investigations are really implicative to consider the evolutional and environmental meanings of the blue shining wings of the *Morpho* butterflies. From this table, we can read as follows: butterflies flying in understory with the larval host plant of monocotyledon were initially living, which had a common root with Brassolini. However, once the special character of blue iridescence was acquired, the group diverged off from the other and further continued divergences: those initially branching off possessed relatively small and somewhat pale blue wings. Among these, one group was still flying in understory (e.g. *M. aega*), while the other was flying out into mid-story (e.g. *M. adonis*), both possessing anyhow ancient character of blue iridescence. Then, a drastic change occurred, which made them to change the larval host plant from monocotyledon to dicotyledon, and at the same time, one group flied out vigorously into canopy with large wings. This change was so vigorous that one-half of the group even lost the blue iridescence and rather attained various colored wings (e.g. *M. perseus*). Another half strengthened the blue coloring by losing the cover scales and hence by exposing the strongly iridescent ground scales completely (e.g. *M. rhetenor*). The other group was staying in understory with the blue coloring rather suppressed by the presence of large cover scales. One group was changed into whitish by losing the iridescence and flied into mid-story (e.g. *M. catenarius*). The possibly most evolved group kept the blue coloring only as a broad stripe (e.g. *M. achilles*). Thus, the present genus *Morpho* has been established. Of course, this is only a fictitious story that is readable from Fig. 3.5. But if we attempt to investigate the mechanism of their blue coloring from an evolutional standpoint, such stories may help us to proceed as an excellent guide for the research both from physical and biological viewpoints.

Penz and DeVries also mentioned the biological meaning of the blue coloring in these butterflies (Penz and DeVries, 2002). The *Morpho* butterflies usually show the strong sexual dimorphism (Fig. 3.4), and the females are normally dull-colored and may retain the coloration of Brassolini ancestors.

Strongly blue iridescence is thus only a male character, which reminds us of female preference for brightly colored male. It is also known that the male *Morpho* butterfly has a habit to approach blue shining paper, and thus male-to-male interactions seem to be also important. Moreover, the relation with the predators is quite important in butterflies, and also the blue coloring left in female should be considered. Thus, the determination of the biological meaning of one character always leads us to a labyrinth of complicated roads.

3.2.2 *Basic Observations*

Before entering into the detail, we will show the most remarkable features of the *Morpho* butterflies through simple experiments: (1) The first one is the viewing-angle dependence of the wing color. When one sees the *Morpho* butterfly in front of the wing, the wing color is surely blue. But when the view direction is inclined obliquely while maintaining the direction perpendicular to the wing veins, one notices that the color does not change largely at first, but when the angle becomes large enough, the wing color changes into violet or dark blue rather suddenly, as shown in Fig. 3.6 and Plate 4 (Appendix B). On the contrary, if one sees the wing keeping the direction parallel to the wing veins, the blue color abruptly distinguishes and the wing turns black. The degree of color changing is dependent on the species.

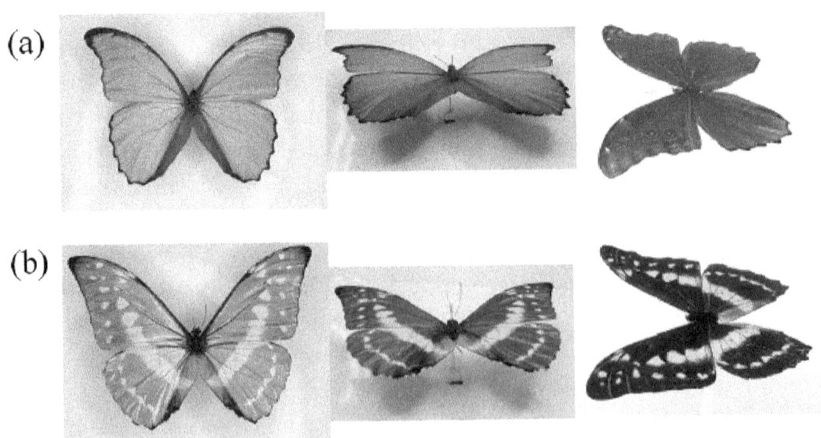

Fig. 3.6 Change of the wing color with changing viewing angle. (a) *M. didius* and (b) *M. cypris*: front (left), oblique front (center), and oblique side view (right). Photographed by Dr. S. Yoshioka. See Plate 4 in Appendix B.

For example, in *M. cypris*, the wing color changes very quickly into black, while in *M. didius*, its change is rather dull. Thus, one can easily understand the peculiar feature of iridescence in this butterfly: blue coloring in a wide angular range with an abrupt change into violet at large viewing angles, and the anisotropic reflection dependent on the viewing direction. All these are quite in contrast with ordinary iridescent animals and manufactured products.

(2) The second experiment is carried out by immersing the wing into a liquid. Let us use ethyl alcohol ($n = 1.359$) as a trial liquid. Surprisingly, its color changes into green with slightly dull shining and further the oblique view shows blue instead of violet (Fig. 3.7 and Plate 5 in Appendix B). The use of water is inappropriate, because it does not soak into the wing. The change of color is dependent on the refractive index of liquid employed. If we employ a liquid with much higher refractive index such as toluene of 1.497, then what happens. The wing changes into dark brown with no shining. These changes in color restore quickly when the liquid is evaporated. Thus, the change of the color in liquid is another typical feature.

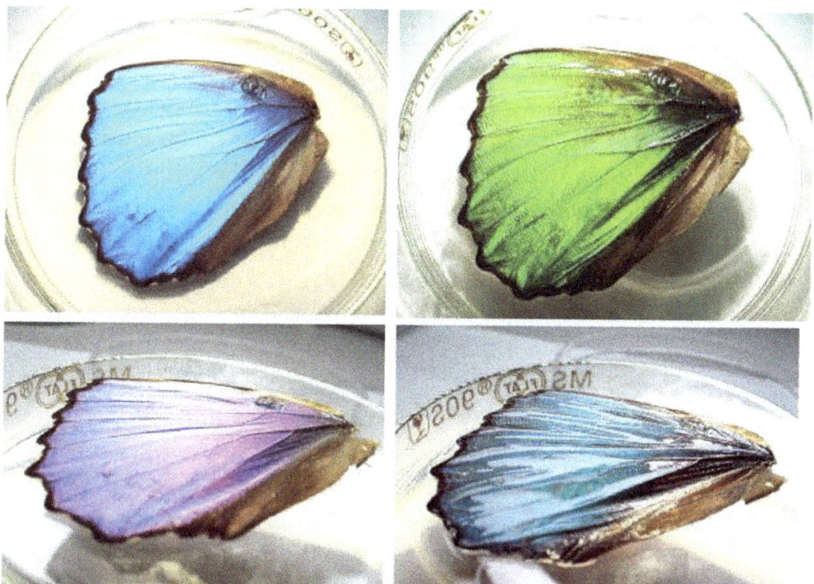

Fig. 3.7 Color change of the hindwing of *M. didius* when immersed into a liquid: in air (left) and in ethyl alcohol (right), and normal view (upper) and oblique view (lower). Photographed by N. Okamoto. See Plate 5 in Appendix B.

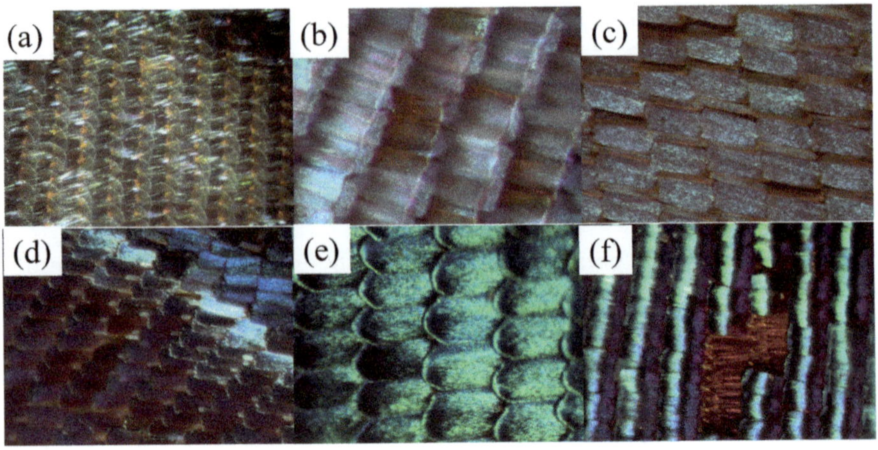

Fig. 3.8 Magnified views of the *Morpho* wings: (a) *M. adonis*, (b) *M. didius*, (c) *M. sulkowskyi*, (d) *M. aega*, (e) *M. rhetenor*, and (f) *M. anaxibia*. Photographed by N. Okamoto.

(3) The third experiment is to observe the wing using a stereomicroscope. Under a microscope, we can see regular rows of thin plates, as shown in Fig. 3.8 and Plate 5 (Appendix B). These plates are so-called scales, which show shiningly blue when the wing is illuminated from the upper side. If the light source can be detached, try to illuminate the wing from various directions. It shows that the blue shining is quite sensitive to the direction of illumination. It is shining only when the wing is illuminated from the direction perpendicular to the longer side of the scale. This experiment is another evidence for the anisotropic reflection described above. If the scale is not flat but is curved such as in *M. anaxibia* (Fig. 3.8(f)) or *M. achilles*, the dependence on the illumination direction is more peculiar. The shining area looks like a line on a scale and its position is changeable with the direction of illumination.

To magnify further, we have to use an optical microscope. It is not necessary to prepare an expensive epi-illumination fluorescence microscope, but an ordinary microscope with transmitted illumination is sufficient. In the latter case, the direct observation of the wing needs another removable light source. Holding the light source in hand, under rather high magnification, say 100× or 400×, we can clearly observe that the source of shining is a scale itself. Then, let us observe only a scale by removing it from the wing. One may notice that various forms of scales are present depending on the position of the wing and on the species. In blue iridescent region of the wing, two types of scales are usually present. One is called cover scale, below which the other type of scales called ground scale or basal scale are present. The ground scale usually takes a form of a rectangular shape, whose typical size is 0.2 mm in length and 0.1 mm in width. Both cover and ground scales are attached to the wing membrane alternatively through holes named socket.

On the other hand, the shapes of cover scales have a variety from species to species. In *M. adonis*, the cover scale is large and round (Fig. 3.8(a)): in *M. didius*, the size is nearly the same as a ground scale and its shape is rectangular (Fig. 3.8(b)). In *M. sulkowskyi*, the cover scale is slender (Fig. 3.8(c)), and in *M. aega*, its length is less than the half of a ground scale (Fig. 3.8(d)). In *M. rhetenor* and *M. cypris*, the cover scale is so small that one sometimes thinks the cover scales are lacking, but a close inspection shows that it is actually present at the root of the ground scale (Fig. 3.9(a)). If we employ the transmitted illumination, the cover scale usually looks transparent, while the ground scale looks dark brown. Thus, the presence of the pigment is easily confirmed. In *M. sulkowskyi*, both the cover and ground scales are nearly transparent, which makes the wing of this butterfly whitish. With further magnification, one can see that a scale has a lot of lines running along the longer side of the scale as shown in Fig. 3.9(b). This line is called ridge and becomes the most important part of the blue coloring, which will be more clearly seen when we employ a scanning electron microscope as described later (see Fig. 3.13).

(4) Finally, it is instructive to observe the reflection pattern from the *Morpho* wing under the illumination of white or blue light. As a light source,

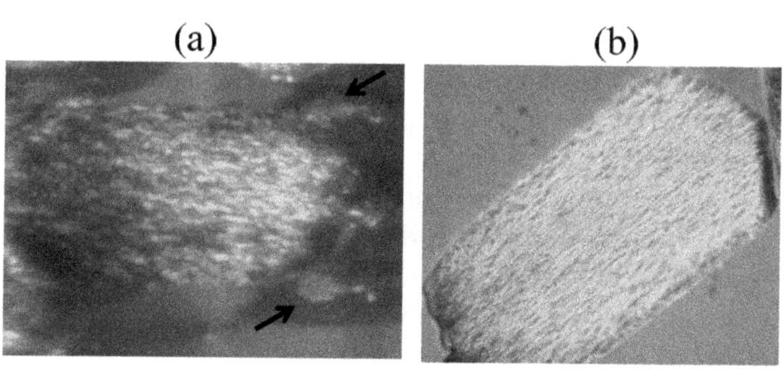

Fig. 3.9 Magnified views of the *M. rhetenor* scales: (a) a ground scale and two cover scales (arrow), and (b) a ground scale showing many ridges running along the longer side of scale. Photographed by N. Okamoto and Dr. S. Yoshioka.

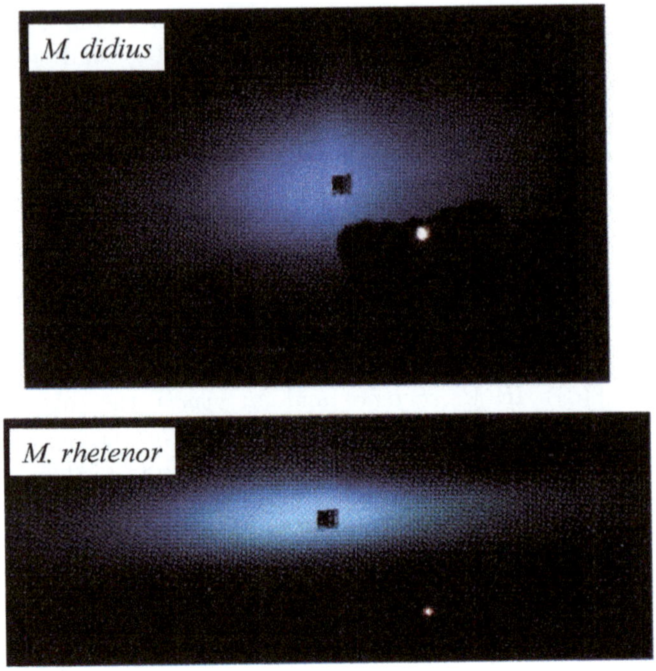

Fig. 3.10 Reflection pattern from the wings of *M. didius* and *M. rhetenor*. White light from a light source is passed through a square hole in a screen and then hits the wing where a light spot is seen (Yoshioka and Kinoshita, 2004).

a xenon lamp with high luminosity or a laser is preferable, but a flashlight using white light emitting diodes may also help us partly. The reflection pattern is projected onto a sheet of white paper. Then a long and narrow blue band extending perpendicular to the wing vein will be seen as shown in Fig. 3.10. The degree of the slenderness depends on the species: in *M. rhetenor*, it is much slender, while in *M. didius*, it is rather oval. This characteristic is quite different from a diffraction grating such as CD or DVD, in which diffraction spots are clearly seen as shown in Plate 2 (Appendix B). It is soon understood that the mechanism not to make any diffraction spot is quite important for this butterfly, because the butterfly should be strongly perceived from any direction.

Thus, these simple experiments described above have already revealed the most important part of the structural colors in these butterflies. We will show how and why the above-mentioned characteristics come out from such a tiny scale on the wing.

3.2.3 *History of the Morpho Studies*

The first scientific observation on *Morpho* butterflies appeared in a book written by Walter in 1895. Although it is now difficult for us to acquire his book, we can catch a glimpse of his description partly through Lord Rayleigh's paper in 1919 (Rayleigh, 1919). According to this, Walter performed the experiment using *M. menelaus* and observed *the scales, which glitter green-blue in air, become in ether* ($n = 1.36$) *a pure green, shining less strongly, again weak in chloroform* ($n = 1.45$) *a yellowish green and now decidedly weaker than in ether. In benzol* ($n = 1.52$) *and in bisulphide of carbon* (carbon disulfide) ($n = 1.64$) *the weak yellow-green lustre is perceptible only with direct sunshine in a dark room.* He speculated that owing to the accordance of the refractive indices of chitin involving dyes and the solvent, the reflection was prevented at the interface.

From the end of the 19th century to the beginning of the 20th century, the possible mechanisms accounting for the metallic color fell into the following categories: (1) diffraction of light due to a grooved structure, (2) thin-film interference, (3) multilayer interference, (4) selective reflection of light such as metals and colored crystals, and (5) surface colors. Walter insisted that the metallic color in nature was due to the reflection from the surface involving dyes and called "surface-colors." This idea was succeeded by Michelson (1911). He reported the measurement on the amplitude and phase change during the reflection of light at various surfaces and described that the reflection from the wing of *M. alga* (probably *M. aega*) resembled the surface reflection from a very thin surface layer less than 100 nm, which involved dyes. He also noticed the possibility of diffraction due to the presence of very thin hairs whose diameter was less than the wavelength of light (possibly it indicated the ridges on the scale) might somehow influence the blue coloring.

In 1917, Lord Rayleigh derived the general expression for the reflection from a regularly stratified medium using the electromagnetic theory, and deduced it as the origin of the colors of some beetles and butterflies (Rayleigh, 1917). He strongly suggested that the brilliant colors of humming-birds, butterflies, and beetles, which varied greatly with the incident angle of light, were not due to the ordinary operation of dyes, but came from the "structure-colors." Onslow reported the microscopic inspections on over 50 iridescent species to offer the experimental evidence to criticize the adaptability of the above two theories (Onslow, 1923). He investigated three *Morpho* butterflies and considered that the surface layer, which consisted of plates of chitin, acted as a grating and that the diffraction/interference was a major cause of the color. Süffert reported the angular dependence of reflection color for *M. aega* and *M. achilles* (Süffert, 1924). Merritt measured the reflection spectra of tempered steel and *M. menelaus*, and interpreted in terms of thin-film interference (Merrit, 1925).

On the other hand, Mason, a microscopist, conducted detailed investigations on various types of structural color in nature and published a series of papers. He described that in *M. menelaus* and *M. sulkowskyi*, there were two types of scale covering the wing and both had straight vanes (ridge) separated by about 1 μm with the height of 2–3 μm (Mason, 1927a). The vanes in outer scale (cover scale) are sparsely distributed with a separation of 1.5–2.0 μm and the iridescent blue was observed even for a vane separated from a scale. Further, the reflecting plane was found to incline by 10–20° from a base plane. The inner scale (ground scale) was equipped with much dense vanes and pigmented. He described that the scales in *Morpho* butterflies were characterized by their vaned scales, and the iridescence was

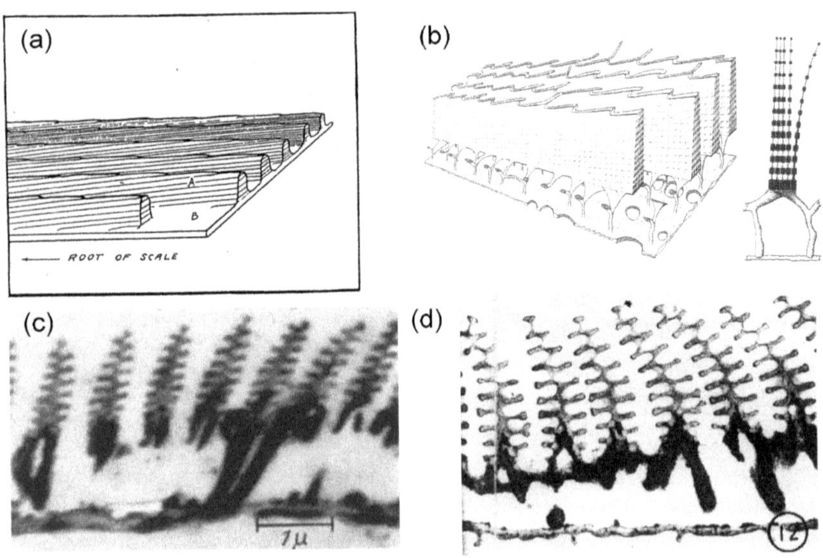

Fig. 3.11 Illustrations given by (a) Mason for a scale of *M. menelaus* (Reprinted with permission from (Mason, 1927a). Copyright (1927) American Chemical Society.) and (b) Anderson and Richards, Jr. for a scale of *M. cypris* (Anderson and Richards, 1942) (Reused with permission from Thomas F. Anderson and A. Glenn Richards, Jr., *J Appl Phys* **13**: 748, 1942. Copyright 1942, American Institute of Physics.). TEM images reported by (c) Lippert and Gentil for *M. aega* (Lippert and Gentil, 1959) (With kind permission from Springer Science + Business Media.) and (d) Ghiradella for *M. menelaus* (Ghiradella, 1978) (Reprinted with permission of John Wiley & Sons, Inc.).

localized in the individual vanes. His observation is accurate enough so that the illustration of a *Morpho* scale is able to pass even in the 21st century (see Fig. 3.11(a)).

The complete understanding of its structure was made after the invention of electron microscope. The work on the *Morpho* butterfly was done independently by Anderson and Richards (1942), and by Gentil (1942). Anderson and Richards performed the electron microscopic investigation on the wing of *M. cypris* and found hundreds of "vanes" possessing linear thickenings of 0.125 μm apart on the scale (see Fig. 3.11(b)). They deduced that the diffuse and multicolored reflection was due to a variation of their intervals and thicknesses. Moreover, they found no difference between the pigmented and nonpigmented scales, the latter of which appeared as a white stripe on the dorsal side of the wing. Gentil observed the scales of three *Morpho* butterflies, and found that a wavy lamellar structure in each vane was responsible of the iridescence. Later, Lippert and Gentil obtained much clearer microscopic images to show the cross-section of the iridescent scale in *M. aega*, *M. sulkowskyi*, and *M. menelaus*, and they found a definite lamellar structure in each vane (see Fig. 3.11(c)) (Lippert and Gentil, 1959). Ghiradella reported the microscopic images for a ground scale of several *Morpho* species (see Fig. 3.11(d)) (Ghiradella, 1984). She examined various lepidopteran scales and classified the iridescent scales morphologically into (1) thin-film lamella, (2) diffraction lattice, (3) Tyndall blue, (4) microrib-satin, (5) microrib thin-film, and (6) lamellar thin-film. She categorized *Morpho*-type scales into (6) (Ghiradella, 1984, 1998).

In contrast to structural and biological progress, optical and/or physical approach to the iridescence of the *Morpho* wing was somewhat off the pace. Wright measured the angular dependence of the reflected light from the wing (data not shown) and found that the reflection was not caused by simple multilayer interference due to lamellar structure nor simple diffraction due to the regular rows of vane (Wright, 1963). He speculated that a kind of three-dimensional diffraction grating might function the iridescence, deduced from the structure reported by Anderson and Richards (1942). Pillai also reported the angular and wavelength dependence (species not shown), and deduced that the multilayer interference and diffraction phenomenon were interconnected complicatedly (Pillai, 1968). After then, no physical approach had appeared until 1990s. Since the exact analysis for the optical response from the complicated structure had not been proposed, a quantitative model for *Morpho* coloring was solely restricted within a multilayer interference model based on the alternate layers of cuticle and air, which was then used to predict the wavelength of maximum reflection.

3.2.4 *Progress in the Morpho Studies After the 1990s*

After 1990s, the scientific research on the *Morpho* coloring has rapidly progressed, and various elaborate measurements and sophisticated calculations have been performed (Cowan, 1990; Kumazawa *et al.*, 1994; Tabata *et al.*, 1996a,b; Kumazawa *et al.*, 1996; Vukusic *et al.*, 1999; Vukusic, 2005; Mann *et al.*, 2001; Gralak *et al.*, 2001; Kinoshita *et al.*, 2002a,b; Yoshioka and Kinoshita, 2004; Kinoshita and Yoshioka, 2005a; Yoshioka and Kinoshita, 2005a, 2006a,b; Berthier *et al.*, 2003, 2006; Gonzato and Pont, 2003; Plattner, 2004; Wickham *et al.*, 2006; Banerjee *et al.*, 2007; Prum *et al.*, 2006) in addition to the detailed observations on their structure (Ghiradella, 1991, 1994, 1998, 2005; Bingham *et al.*, 1995) (see the reviews Srinivasarao (1999); Vukusic and Sambles (2000); Parker (2002); Tayeb *et al.* (2003); Vukusic (2003); Vukusic and Sambles (2003); Kinoshita and Yoshioka (2005b); Parker and Martini (2006); Berthier (2007)). This is partly because the structural color attracts great attention in various industrial researches particularly in painting, automobile, cosmetics, and textile industries, because its color is gentle to the eye and also to the environment while possessing strong stimulus to human perception.

With this tide, Tabata *et al.* investigated quantitatively the wings of *M. adonis* and *M. sulkowskyi* by optical measurements using an angle- and wavelength-resolved reflection spectrometer (Tabata *et al.*, 1996a). They employed the incident angle of $-45°$ and the detection angle of $+30°$ along the longer side of the sçale, while the tilt angle of the scale was varied from -90 to $90°$ in a plane perpendicular to the ridge. They obtained the wavelength giving the maximum reflection of 460 nm with broad angular spreading for *M. sulkowskyi*, while that of 482 nm with narrow angular spreading for *M. adonis*. They analyzed the high reflectivity of the wing using a multilayer interference model consisting of cuticle ($n = 1.5$) and air ($n = 1.0$) layers. They commented that the visual impression of the wings of *M. sulkowskyi* was due to the combined contributions of subjective gloss based on the structural color and the degree of the surface undulation due to the scale structure with a pitch of 107 μm and a height of 35 μm. They also measured the fluorescence properties of the *Morpho* wing and found that L-erythro biopterin was a major component of blue fluorescent pigment (fluorescence peak at 470 nm) in *M. sulkowskyi* (Kumazawa *et al.*, 1996). Later, they mimicked the multilayer in *Morpho* scale and incorporated 61 layers consisting of alternate layers of nylon 6 and polyester into a flattened fiber and demonstrated the structurally colored fiber (Iohara *et al.*, 2000), which will be described more in detail in Sec. 3.2.6.

The measurement of the transmission and reflection properties of a single scale was for the first time reported by Vukusic *et al.* (1999). The significance of the single-scale measurement lay in the fact that the measurement on the large wing areas suffered from various effects due to the scale alignment and from the extraneous effect due to noniridescent scales and wing substrate. They chose a ground scale of *M. rhetenor*, and a cover and ground scale of *M. didius* as samples. A single-scale fixed on a needle was illuminated by Ar^+ or He–Ne laser with a spot size of 30 μm. The reflected and transmitted light was collected by a lens with a vertical slit and a detector, which was able to scan over 360° while recording the integrated intensity. They found that the reflection pattern from the *M. rhetenor* ground sale was divided into two distinct lobes, while that from the cover and ground scale of *M. didius* showed broad lobing (Fig. 3.12). They attributed this difference to the distribution of the tilt angle of ridges. In the transmission side, they found the clear diffraction spots, which coincided with the spacing of the ridges.

Vukusic *et al.* investigated the absolute reflectivity and found that the reflectivity from a scale of 70% at 450 nm for *M. rhetenor*, and of 5–10% and 40% for a cover and ground scale of *M. didius*. In the transmission side, the transmission of 3–5% at all wavelengths was obtained for *M. rhetenor*, which was attributed to the scattering by the structure beneath the multilayering and to the absorption by melanin pigment. They analyzed the high reflectivity of the *Morpho* scale using multilayer interference model

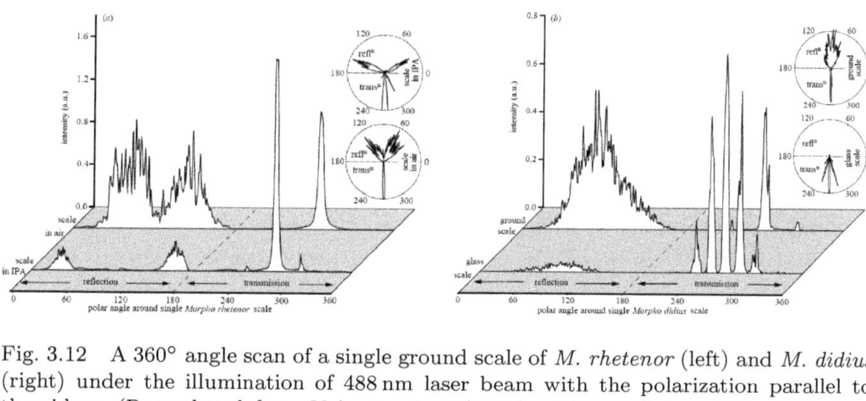

Fig. 3.12 A 360° angle scan of a single ground scale of *M. rhetenor* (left) and *M. didius* (right) under the illumination of 488 nm laser beam with the polarization parallel to the ridges. (Reproduced from Vukusic *et al.* (1999) with permission of Royal Society Publishing.)

and the angular broadening by taking account of the tilt-angle distribution of the multilayer. From the fitting with the experimental data, they obtained the refractive index of 1.56–0.056i at 450 nm for a ground scale of *M. rhetenor* with the Gaussian distribution of ±15° of surface tilts. Thus, they considered the essential cause of the blue coloring as the multilayer interference and of the angular broadening as the tilt-angle distribution.

On the other hand, our group paid special attention to irregular structures inherent to natural products and noticed that the interplay between regular structure within ridge and irregularity among ridges was extremely important to understand the strongly blue and yet remarkably diffusive nature of the *Morpho* wing (Kinoshita *et al.*, 2002a,b; Kinoshita and Yoshioka, 2005a,b). They proposed a simple model to explain the essential part of the *Morpho* coloring and its validity was confirmed by fabricating *Morpho* substrates, which only mimicked the principle of *Morpho* coloring (Yoshioka *et al.*, 2004; Saito *et al.*, 2004). It was shown that largely anisotropic character of the reflection from the wing was due to a slender shape of the shelf on the ridge, which was interestingly modified by placing wavelength-selective, anisotropic light-diffusing scales on the iridescent ground scales (Yoshioka and Kinoshita, 2004). Moreover, it was clarified that the role of melanin pigment in the structurally colored wing was to enhance the blue coloring in one hand, while in the other hand, it was to show the wing whitish using special structural modifications in addition to the removal of the pigment (Yoshioka and Kinoshita, 2005a, 2006b). The details of our work will be described in the following section.

Berthier *et al.* performed the optical measurement on the wing of *M. menelaus* and found the reflection plane inclined by 11° from the measurement of the angular dependence of the reflection spectrum (Berthier *et al.*, 2003). They also measured the polarization-dependent reflection spectrum and analyzed it by assuming that the juxtaposed ridges constituted the multilayer consisting of alternate layers of cuticle-rich and air-rich layers. They estimated the refractive indices of these layers using an effective medium model under the assumption of Maxwell–Garnett model taking account of the anisotropic ridge structure. Recently, they extended their work to totally 14 species of the genus *Morpho* and classified them according to the size of cover scale, the tilt angle of the shelf, the lamellar number, and the bidirectional reflectance distribution pattern obtained by scanning the direction of detection hemispherically over a wing (Berthier *et al.*, 2006). They found a photonic-crystal type periodic structure consisting of shelves within one and adjacent ridges and vertical trabeculae between shelves.

They also explained that the structural color in *M. adonis* came from the noncoherent interference effect among overlapping cover scales in addition to thin-film interference within a scale.

Owing to recent computational development, it has been possible to calculate directly electromagnetic field around microstructures and in far field as exactly as possible. The finite-difference time-domain (FDTD) algorithm is an effective method to calculate the scattering problem due to such a complicated structure by evaluating the field amplitudes on space-time grids. Plattner employed this method to extract the scattering feature using four types of model structures, a rectangular lattice with and without tapering, and an alternate lattice of rectangular elements with and without tapering (Plattner, 2004). He found that a strong extinction of the specular reflection occurs for an alternate lattice and also a tapered model proved the enhancement of the first-order diffraction spots by reducing the specular reflection. Since he calculated the field for a regular array of ridge, diffraction spots will appear inevitably, which differs considerably from the actual wing. He also presented the measurement of the reflection spectrum using supercontinuum light generated through a photonic crystal fiber, but the comparison between the theoretical and experimental results was not presented thoroughly.

Banerjee *et al.* employed a more sophisticated algorithm of nonstandard FDTD method to calculate a ridge structure mimicking *M. sulkowskyi* scale (Banerjee *et al.*, 2007). They employed a refractive index for cuticular material as $1.56 + 0.06i$ and calculated the reflection spectrum for unpolarized and polarized incidence, which was then compared with our experimental result. However, the regularly arranged ridge structure that they assumed naturally produced the diffraction spots, which altered largely the angular dependence of the reflection.

These computational methods are actually powerful to solve the complicated scattering problem, because they provide exact results within a framework of Maxwell's electromagnetic theory. It is clear that the closer the model approaches an actual structure, the more exact the results becomes. Thus in the limit, it is completely equivalent to what we call "measurement," i.e. the direct measurement of the scattered light from an actual wing, and hence there is no meaning to calculate. Therefore, it is extremely important to find out a proper model, which reduces the complicated structure to an essential part of the phenomenon, and to find out what is the most important to understand the features of the structural color as a whole.

3.2.5 *Physical Interpretation of the Morpho Coloring*

3.2.5.1 *Microscopic Observations*

In the present subsection, we will show the mechanism of the iridescence of the *Morpho* butterfly according to our detailed inspections of the macroscopic and microscopic structures on the wing. We will then propose a simple model based on the structural observations and the result of the optical measurement (Kinoshita *et al.*, 2002a,b; Kinoshita and Yoshioka, 2005a; Yoshioka and Kinoshita, 2004; Yoshioka and Kinoshita, 2006b).

At first, we will describe the microscopic observation on the *Morpho* wing. As described above, the scale structures of the *Morpho* butterflies have a wide variety and their physical mechanisms are not truly clarified except for a few species. Here, we will concentrate only on such well-known species, whose mechanisms of iridescence are known to some extent. Nevertheless, it will be sufficient to show their sophisticated structures and ingenious mechanisms. For this purpose, we choose three *Morpho* species, *M. didius*, *M. sulkowskyi*, and *M. rhetenor*, as typical examples.

As shown in Fig. 3.13, the dorsal wing surface of the male *M. didius* is covered with two kinds of scales: cover scales of a slender shape and slightly overlapping ground scales of a rectangular shape. Each of the cover and ground scales has a lot of minute ridges with a pitch of 1.4 and 0.6–0.7 μm, respectively. The cover scale is transparent, while the ground scale is pigmented so that under the transmitted illumination, the latter looks dark brown. The wing of the male *M. sulkowskyi* consists of alternatively arranged, transparent ground and cover scales, each having a normal

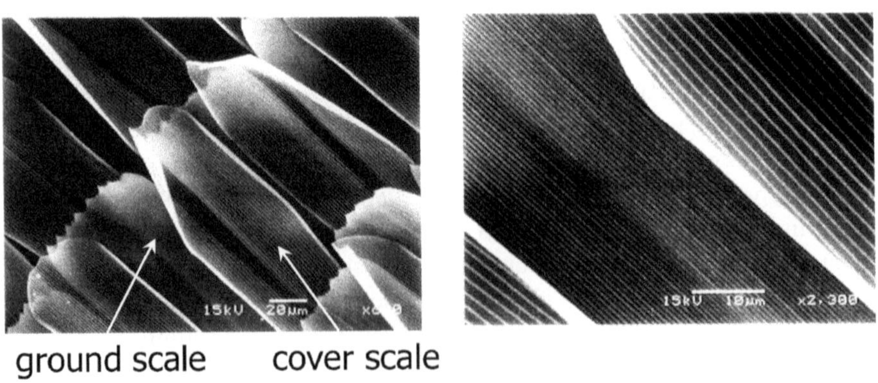

Fig. 3.13 Scanning electron microscopic images of the dorsal wing of *M. didius* (Kinoshita and Yoshioka, 2005b).

and a slender shape, respectively. The ridge separations of both scales are $\sim 1.0\,\mu$m. It seems that the cover and ground scales essentially have the same structures, and there is no division of roles in these scales. The wing of the male *M. rhetenor* consists of overlapping, rectangular ground scales, with very small cover scales at their roots. The ridge separations of the ground scales are $\sim 0.7\,\mu$m. The ground scale is pigmented as in *M. didius*. Thus, the iridescence in *M. rhetenor* is caused solely by the ground scales.

In the past, these regularly running ridges were considered as an origin of the iridescence, but the following single-scale experiment clearly denies this expectation. The transmission and reflection patterns obtained for a single cover and ground scale of *M. didius* are shown in Fig. 3.14. For the transmission side, diffraction spots are clearly observed particularly for a cover scale. We can see up to the second order for a cover scale, while only the first order for a ground scale. In the latter case, yellow diffuse scattering is also observed. The diffraction angle calculated from this pattern just agrees with that expected from the separation of the ridges. On the contrary and really surprisingly, the reflection pattern does not show any diffraction spot and a broadly extended, blue pattern is only observed, which extends perpendicular to the ridge direction. The reflection pattern from a ground scale of *M. rhetenor* is particularly slender. Moreover, in a cover scale of *M. didius*, a double pattern with slightly different colors is observed. These

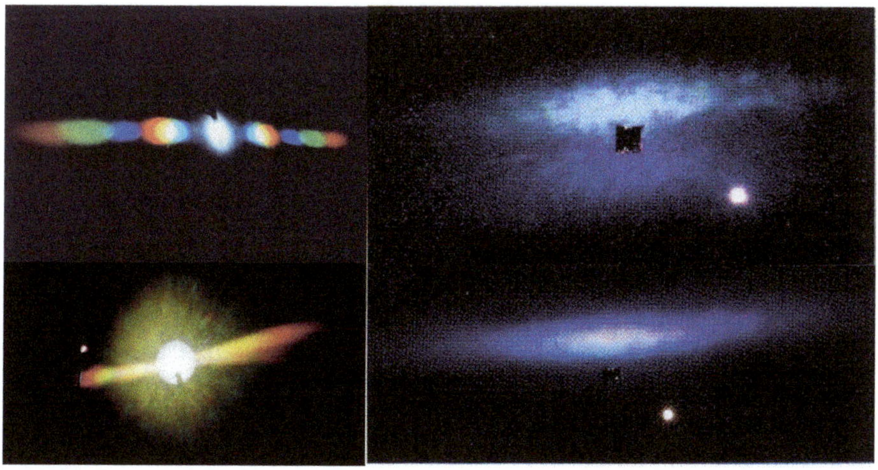

Fig. 3.14 Transmission (left) and reflection (right) patterns from a single cover (upper) and ground (lower) scale of *M. didius* (Yoshioka and Kinoshita, 2004). See Plate 6 in Appendix B.

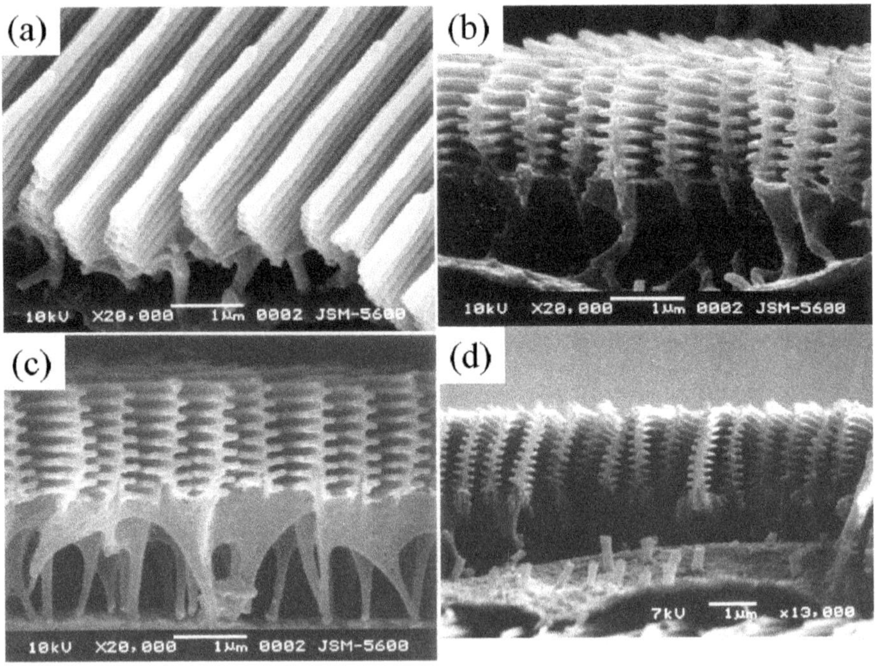

Fig. 3.15 Scanning electron microscopic images of the cross-sections of ground scales of (a, b) *M. didius*, (c) *M. rhetenor*, and (d) *M. sulkowskyi* (Kinoshita *et al.*, 2002a,b).

results are quite unusual in an ordinary sense of optics. That is, the spacings of the ridges work actually as a diffraction grating, while the ridges themselves are not. Thus something mysterious is hidden within a ridge.

In order to investigate the structure of the ridge more in detail, we cut a scale by a pair of scissors and observed the cross-section under a high magnification. First, we show the case of a ground scale of *M. didius*. Obviously, a ridge has a microstructure (Fig. 3.15(a)), which is more vividly observed at the cross-section (Fig. 3.15(b)). The cross-section of a ridge has a size of $\sim 2\,\mu$m height and $\sim 0.3\,\mu$m width. Furthermore, it has a structure like a bookshelf in a library. We call it "shelf structure" hereafter. The vertical separation of the shelves is about $0.2\,\mu$m, which reminds us of blue coloring through light interference from a simple calculation, $0.2\,\mu\text{m} \times 2 \times$ (average refractive index ~ 1.15) $= 0.46\,\mu$m. We have investigated a transmission electron microscopic image and have found the shelf structures, as a whole, are not so regular as atoms in crystal, but not so random as particles in powder. Thus, the moderate regularity is present as is usually the case in living systems. Close inspection of the cross-section shows that a ridge consists of 6–7 shelves of 0.08–0.20 μm in width and 0.05–0.07 μm in thickness, and a stem of 0.05–0.12 μm in width standing at the center. The average thicknesses of the shelf and air are approximately 0.055 and 0.165 μm, respectively. The directions of the stem are slightly distributed and only 4–5 shelves near the bottom seem to grow fully. The left- and right-hand sides of shelves seem to stick alternately, but they are not so regular in periodicity and direction. In *M. sulkowskyi* (Fig. 3.15(c)), the shelf structure is essentially the same as that in a ground scale of *M. didius*, although it seems more regular. In a ground scale of *M. rhetenor*, the shelf structure is more developed than those in the other two, and more than 10 shelves are discernible (Fig. 3.15(d)).

Detailed observation reveals that the shelf is not parallel to a base plane, but is oblique to the plane with the inclination angle of 7–10°. As shown in Fig. 3.15(a), the upper ends of the shelves are easily observed in the image, which are seemingly in a random distribution. If this is true, the height of the ridge will also be distributed when we see at a cross-section. The distribution of the height is not so completely random in principle and should remain within the separation of the shelves, say 0.2 μm. This is more clearly seen from the side view of the scales in a SEM image (Fig. 3.16), where we can see shelves running obliquely with respect to the base plane, while the ends of the layers are randomly distributed on the upper surfaces of the ridges. We have investigated the ridge height distribution at several

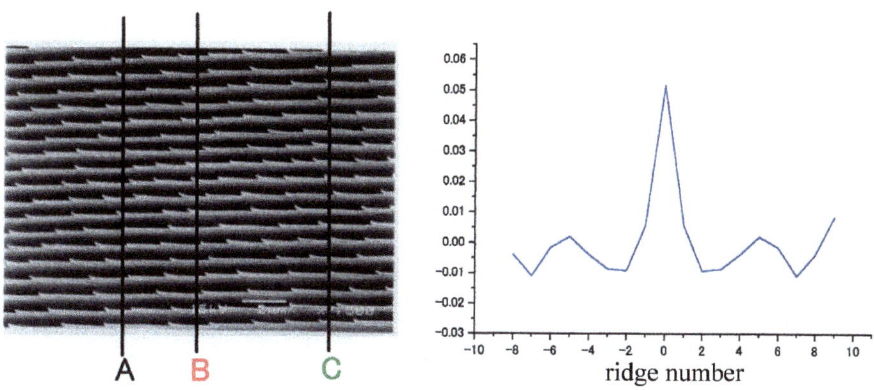

Fig. 3.16 SEM image (left: side view) of a ground scale of *M. didius*. Note that the upper ends of the shelves are distributed randomly. We estimate the height distribution of the ridges at three cross-sections designated as A, B and C, and calculate the spatial correlation function for the height distribution (right) (Kinoshita *et al.*, 2002a).

cross-sections using the distribution of the ends of shelves in the SEM image of *M. didius*. Under the assumption that the ends of shelves take an equal height and that the average length of the end-to-end distance corresponds to the shelf separation of 0.2 μm, we have determined the ridge height distribution and have calculated its spatial correlation function as shown in Fig. 3.16. It is clear that the ridge heights are almost randomly distributed within a range of a shelf separation and the spatial correlation between the neighboring ridges is not clearly discernible at a glance. This fact is quite a common feature among various *Morpho* species and is extremely important to understand the blue iridescence truly, because the random height distribution eventually cancels the interference between the neighboring ridges, and then produces the diffusive reflection as if each ridge scatters the light independently.

Next, we will investigate cover scale of *M. didius*. As shown in Fig. 3.17(a), the ridges in cover scale are distributed regularly but seem to be rather sparse. The shelf structure is similar to the ground scale (Fig. 3.17(b)), but seems not to grow well. The shelves are also running obliquely to the base plane. However, the characteristic of the cover scale lies on the fact that the ridges are attached on a thin base plate of thickness 190–230 nm, which is in contrast with the case of a ground scale, where the ridges are built on the complicated trabeculae. At a glance, it seems that the ground scale is mainly responsible for the structural color in *M. didius* so that a transparent cover scale little affects the structural colors. However, as will be described later, the cover scale plays an important role in the wing coloration and even in the reflectivity of the iridescent wing.

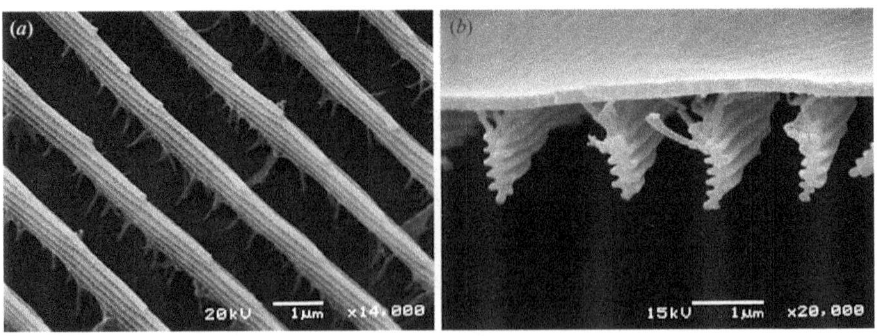

Fig. 3.17 SEM images of (a) a top view from a slightly oblique direction and (b) a cross-section of a cover scale of the male *M. didius*. The cover scale is upside down in (b) (Yoshioka and Kinoshita, 2004).

3.2.5.2 *Optical Measurements*

3.2.5.2.1 Diffuse Reflection and Transmission Measurements

In the following, we will show the optical properties of the wings of the above three species together with *M. adonis* quantitatively. First, we show the reflection and transmission spectra of the wings. Since both the reflection, and transmission patterns are diffuse, we have to use a spectrophotometer equipped with an integrating sphere as shown in Fig. 3.18. Since the optical response of matter should be generally expressed by the sum of transmission, reflection, and absorption of light, we have divided the response of the wing into the above three parts under the assumption that fluorescence process is neglected. This assumption is valid for the *Morpho* butterfly wing because the absorption owes mostly to melanin pigments (Ghiradella, 1998), which is known to be less fluorescent, although biopterin pigment was also present in the *Morpho* wing, which is fluorescent under UV illumination (Kumazawa *et al.*, 1996).

The result for the *Morpho* species is shown in Fig. 3.19. The transmittances of the wings of *M. didius*, *M. rhetenor*, and *M. adonis* are very low below 500 nm and increases toward a near-infrared region, while that of

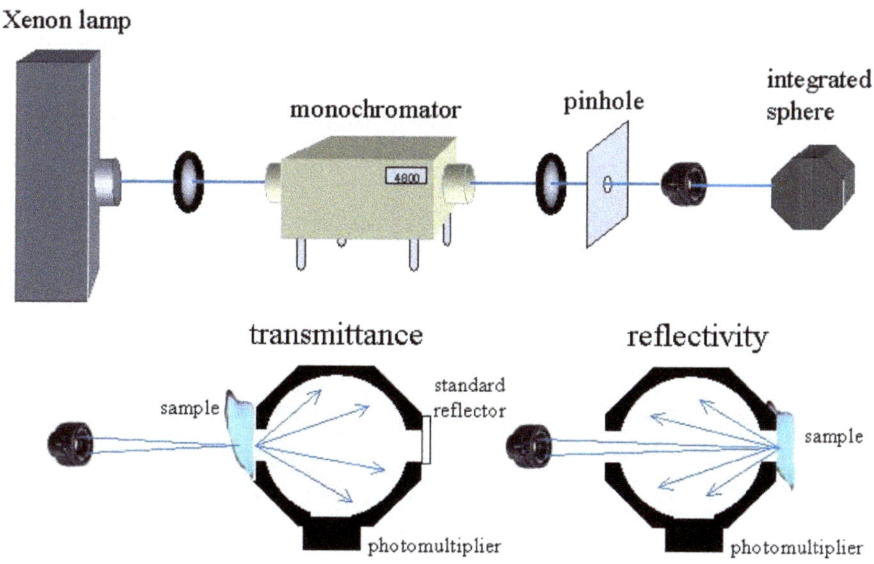

Fig. 3.18 Schematic diagram of the experimental arrangement using an integrating sphere. The absolute transmission and reflection spectra of a diffusely scattering sample are measured in reference to the reflection by a standard reflector.

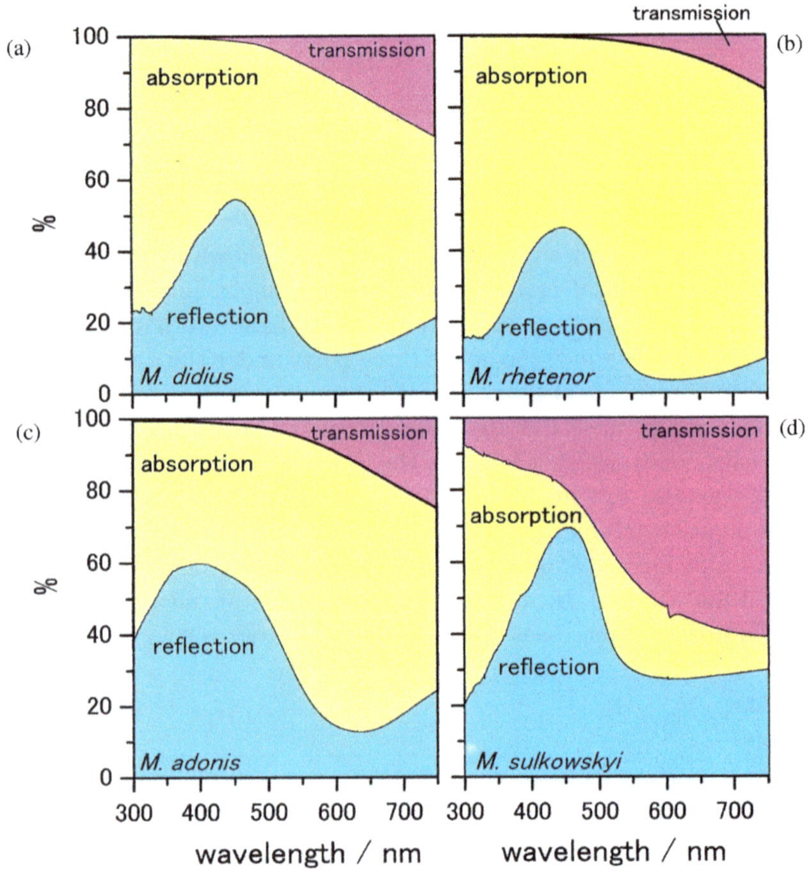

Fig. 3.19 Transmittance and reflectivity of the wings of (a) *M. didius*, (b) *M. rhetenor*, (c) *M. adonis*, and (d) *M. sulkowskyi*, measured by the instrument shown in Fig. 3.18. The rest of the part is assigned to be due to absorption (Kinoshita *et al.*, 2002a).

M. sulkowskyi shows an extraordinarily high value especially above 500 nm. The most remarkable is that the area of the absorption is extraordinary small for *M. sulkowskyi*, while the other three occupy a large part in a visible region of 380–770 nm. Thus, the whitish appearance of *M. sulkowskyi* may come mainly from the lack of pigment in the wing. However, we should not haste the conclusion, because what we see is not the absorption but the reflection from the wing. Thus, if we compare the reflectivities, it is surprising that *M. sulkowskyi* shows the highest reflectivity up to 70% at around 460 nm, which is by far larger than those of 55%, 45% and 60% for

M. didius, *M. rhetenor*, and *M. adonis*, respectively. It is remarkable that strongly blue-reflecting *M. rhetenor* shows the lowest reflectivity.

The important thing to be considered is not the maximum reflectivity at 460 nm, but the reflectivity at a complimentary color around 560 nm. The strongly blue-reflecting *M. rhetenor* hardly reflects the light in a complimentary color region. In fact, it reflects only 3–4% in strongly blue *M. rhetenor* and 10% in the slightly dull-blue wing of *M. didius*, while it amounts to 27% in the whitish *M. sulkowskyi*. The highest reflectivity at 460 nm of *M. sulkowskyi* seems to owe mostly to the high background reflectivity. Considering this, we deduce that the net reflectivity due to the shelf structure reaches essentially a similar value of ∼45% as that of *M. didius* of $55 - 10 = 45\%$ and of *M. sulkowskyi* of $70 - 27 = 43\%$. *M. sulkowskyi* thus adds white background to blue structural color to adjust the saturation of blue color. The white background of *M. sulkowskyi* comes mostly from the scattering within scales at the dorsal and ventral sides, and wing membrane, which will be analyzed more quantitatively later (Sec. 3.2.5.5).

3.2.5.2.2 Angular Dependence of Reflected Light Intensity from the Morpho Wings

Next, we will show the angular dependence of the reflected light intensity from the *Morpho* wings. First, we show the result for the intact wing. Even if the incident light is nearly monochromatic, the reflected light is widely distributed in a plane nearly perpendicular to the ridges and no trace of diffraction spots is seen at a glance. We have measured the angular dependence at various exciting wavelengths using an instrument shown in Fig. 3.20. The results are shown in Fig. 3.21(d). It is immediately noticed that the reflected light under normal incidence is widely distributed, regardless of wavelength, and shows a sharp peak at a normal direction. It is also noticed that toward shorter wavelengths, the distribution has small shoulders at 40–70° and at the wavelengths shorter than 420 nm, the reflection at large angles is more intense than that at longer wavelengths. These results support our observation that the color of the *Morpho* wing does not change largely with the viewing angle except for the observation close to the wing plane.

Since the wing of *M. didius* is covered with two types of scales, we investigated the roles of these two different types of scale through the angular reflectivity. For this purpose, we have removed cover scales using adhesive tape and have measured the angular dependence of reflectivity for the wing covered only with ground scales (Fig. 3.21(b)). It is surprising that the

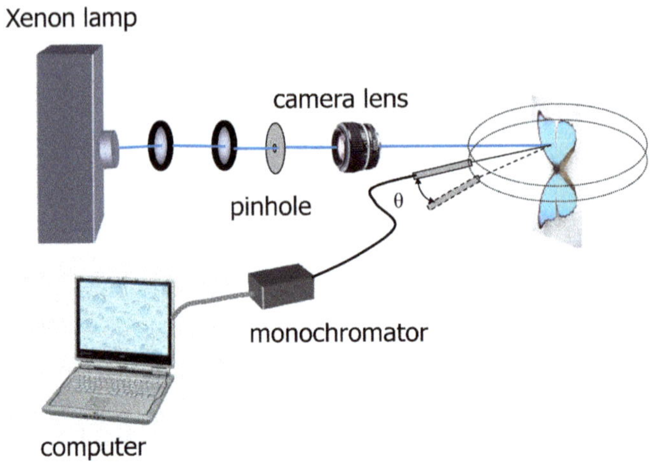

Fig. 3.20 Experimental setup for the angular dependence of light reflection from the *Morpho* wing. The light from a xenon lamp is focused into a pinhole through two lenses, and then onto a wing through a camera lens to avoid chromatic aberration. The reflected light is collected by a movable fiber, which leads the light to a monochromator.

angular dependence in a plane perpendicular to the ridges considerably differs from that covered with cover scales, as shown in Fig. 3.21(d). Namely, the angular dependence is rather widely distributed and the difference in wavelength becomes more conspicuous. It shows a maximum at around the normal direction for longer wavelengths under normal incidence, while it shows a peak at around $\pm 60°$ for shorter wavelengths. It is also noticed that the oscillatory structure is observed in the angular dependence and is particularly prominent at shorter wavelengths. Furthermore, a peak observed at the angle of incidence becomes broad and less conspicuous at longer wavelengths.

Under 30°-inclined incidence, the overall angular profile seems to shift to the angle of incidence. The clear peak is still observed at the angle of incidence. Furthermore, at shorter wavelengths, the intensity is rather larger around 30°, the angle of specular reflection, while at longer wavelengths, the reflection normal to the wing is rather intense. Without cover scales, the overall profile is essentially equivalent to that with cover scales, but the peak at the angle of incidence becomes less prominent as in the case under the normal incidence.

The angular dependence of the reflection parallel to the ridge is confined within a small angular range (Figs. 3.21(a) and 3.21(c)). Actually, the

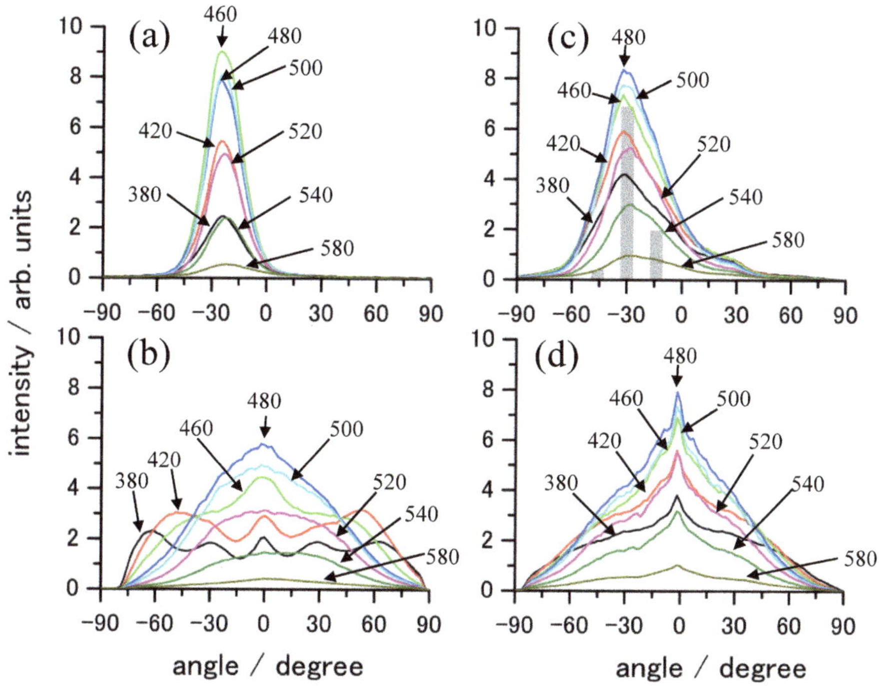

Fig. 3.21 Angular dependence of the reflected light intensity under monochromatic light illumination between 380 and 580 nm with a step of 20 nm (Yoshioka and Kinoshita, 2004). (a) and (c) are obtained under normal incidence and measured within a plane parallel to the ridge for the wing without cover scales and the intact wing of *M. didius*, while (b) and (d) are those measured within a plane perpendicular to the ridge. The histogram shown in (c) is described in Sec. 3.2.5.6.

center of the reflection is inclined from the normal to the wing. This inclination was due to that of the scale with respect to the wing membrane and also of the obliquely running shelves (Mason, 1927a; Tabata *et al.*, 1996a; Vukusic *et al.*, 1999). The difference in the angular reflectivity for the directions perpendicular and parallel to the ridges originates from the diffraction effect due to the unidirectional growth of the ridge. Namely for the direction perpendicular to the ridge, the angular range is roughly determined by the width of the shelf. On the other hand, for that parallel to the ridge, the diffraction is probably based on the exposed portion of the shelf with the length of $\sim 2\,\mu$m, which is determined by the length between the ends of obliquely running shelves. The comparison between the angular dependence for the intact wing and the wing without cover scales reveals that the former is clearly distributed ($\pm 20°$) and asymmetric (Fig. 3.21(c)), while the latter is confined within a small angular range of $\pm 10°$ (Fig. 3.21(a)). Thus, the cover scales in *M. didius* anyway have a function to reduce the anisotropy of the reflection. This hypothesis is further confirmed by comparing the reflection pattern for various *Morpho* species, which do not accompany cover scales. The origins of the asymmetric broadening are due to the combined effect of the thin-film interference at the base plane and the interference at the obliquely running shelves in a ridge within a cover scale (Yoshioka and Kinoshita, 2004), which will be more quantitatively discussed in the later section (Sec. 3.2.5.6).

3.2.5.2.3 Angular Dependence of the Reflected and Transmitted Light Intensity from a Single Scale

In order to investigate the optical properties of each scale, we conducted a single-scale, angle- and wavelength-resolved experiment (Kinoshita *et al.*, 2002b; Yoshioka and Kinoshita, 2006a). The obtained data in the reflection agree basically with those for the intact wing. Under normal incidence, the reflected light toward the normal direction has a peak at 440 nm. With increasing angle of observation, the relative intensity at shorter wavelengths increases and the angular range reaches $\pm 80°$. It is noticeable that the reflection intensity at around 600 nm is quite low, which is in contrast with that of the intact wing of $\sim$30%. In the transmission, the first-order diffraction spots due to the periodic ridge structure are clearly resolved and its peak shifts with the angle of observation. In case of a cover scale of *M. didius*, the diffraction spots up to the second order are clearly seen, while in a ground scale, only the first-order spot is visible with the complex transmission pattern above 500 nm. The wavelength dependence of the diffraction angles is in good agreement with that expected for multi-slit interference. These results are basically in good agreement with those reported previously (Vukusic *et al.*, 1999).

3.2.5.3 *A Simple Model Based on Light Diffraction and Interference*

We are in a position to explain the above experimental results through a simple theoretical analysis (Kinoshita *et al.*, 2002a,b). Although more sophisticated calculations using a Finite Difference Time Domain (FDTD) method and a photonic band calculation have appeared recently (Plattner, 2004), these works mostly pay attention to the regular part of the microstructure, and little attention has been paid to the irregular part, which is quite important for the *Morpho* wing to produce the diffusive light. Our model is too simplified in one sense, yet contains essential points to express the fundamental features of the *Morpho* wing.

Here, we model the scale structure by consisting of many ridges, each having several shelves and a pillar. For the incidence from the upper side, the pillar may hardly play a role in the reflection. Then, we remove the pillar. The height of ridge is distributed in a random way within a range of shelf separation. Under this condition, we consider the diffraction/interference problem, as shown in Fig. 3.22. To extract the essential part of regular shelf structure, we assume that each shelf has an infinitesimal width. This assumption eliminates complex boundary conditions at each surface of the shelf, the refraction inside the shelf, and also the reduction of incident and reflected light at the other shelves. Instead, this model neglects the multiple reflections within the shelves and with neighboring ridges, which

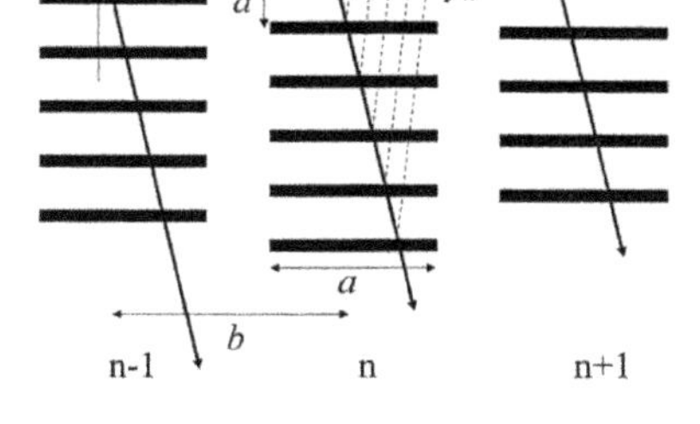

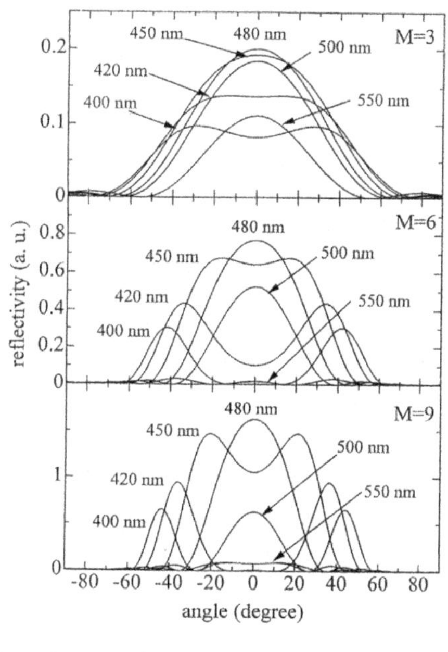

Fig. 3.22 Left: a simple model to calculate the diffraction/interference properties of the ridges consisting of M shelves on a scale. A plane wave is assumed to be incident on N ridges with different heights, without disturbance of the shelf structure. Right: angular dependence of the reflectivity calculated under the assumption that a plane wave incident on three, six, and nine layers each infinitely thin and 300 nm in length with a 235 nm interval is diffracted at each layer under normal incidence (Kinoshita *et al.*, 2002a).

may weaken the interference effect and neglect the effect of coherent back scattering. However, owing to the narrow width of the shelf, the interference effect due to multiple reflections at the shelves within a ridge is considered to be not so large. We also assume that the shelf separation agrees with the optical path length calculated for an actual shelf structure, putting the refractive index of the shelf as 1.60. Thus, in our model, regularly arranged shelves produce the quasi-multilayer interference, while a shelf of size $\sim 0.3\,\mu$m in width and $\sim 2\,\mu$m in length will contribute to anisotropic light diffraction. Furthermore, the random height distribution gives incoherent effect between the ridges. The detailed calculations on the optical responses based on this model will be described in Sec. 9.2.1.

We calculate the angular dependence of the reflectivity under normal incidence according to the relations, Eqs. (9.64) and (9.67). In Fig. 3.22, we show the results calculated for three, six, and nine shelves, which roughly corresponds to the ground scale of *M. didius* for fully expanding shelves, those including intermediately expanding ones, and the whole shelves, respectively. It is evident that with increasing number of shelves, the effect of the interference is enhanced so that more pronounced angular dependence is exhibited. The simulated data for six shelves reproduce fairly well the angle- and wavelength-dependence of the reflection/diffraction from the wing covered with only ground scales in *M. didius* (Fig. 3.21(b)). Particularly, the difference of the angular dependence for various wavelengths is well reproduced. It is concluded that the wavelength difference in the reflection to the normal direction can be explained in terms of constructive and destructive interference at 450–480 and 400–420 nm, respectively, while the peaks of ± 30–50° at 400–420 nm can be explained in terms of constructive interference.

The present model reproduces the angular dependence of the reflection fairly well, which can never be explained using a simple multilayer interference model. The neglect of multiple reflection within a ridge offers a surprisingly good result for normal incidence. This is partly due to the strong diffraction effect and partly due to the irregularity in the actual shelf structure.

However, the above model fails to explain the peculiar features such as a peak at the angle of incidence and the oscillatory structure found at shorter wavelengths. It is also noticeable that the angular range is limited within $\pm 60^\circ$ for the calculation, while it extends over $\pm 80^\circ$ for the actual wing. We consider two possible mechanisms for the former characteristics by introducing the effect of spatial correlation of the ridge height distribution (Sec. 9.2.3) and the coherent backscattering of light (Kinoshita *et al.*, 2002a). We also introduce the effect of alternately sticking shelf structure and the angular distribution of the ridge direction, which will be described in Sec. 9.2.2. Anyway, these extra effects improve the angular dependence described by the model fairly well. Thus, it is concluded that the essential mechanisms of the *Morpho* blue is the combined effect of the quasi-multilayer interference within a ridge and the anisotropic diffraction effect due to the slender-shaped shelf, each of which is further modified by various effects.

3.2.5.4 *High Reflectivity of the Morpho Wing*

In the above model, however, it is difficult to estimate the absolute reflectivity due to the shelf structure, because we eventually neglect the reflection at each shelf. Since the simple geometrical inspection fails to explain when the ridge spacing is smaller than or comparable to the wavelength of light, we will consider a general problem of the light reflectivity. Because of the complexity of the actual shape of the shelves and ridges, here we will simplify the structure by assuming regularly spaced ridges having equal heights.

If we consider the spatial Fourier transformation of the permittivity for periodically arranged lamellae, it is immediately understood that the incident field does not couple directly with the periodic structure, unless the ridge spacing fits that of the wave fronts of the incident light at the interface. In such a case, the incident light couples only with the zeroth-order Fourier component of the dielectric layer, i.e. a layer having a spatially averaged permittivity. Thus, the separate shelves are reduced to a simple multilayer model with two alternate layers having averaged permittivities.

In Fig. 3.23, we show the reflectivity calculated against the ridge separation using a transfer matrix method. In this calculation, we employ the width of a shelf as $0.3\,\mu$m with the refractive indices of shelf and air as 1.6 and 1.0, respectively. It is clear that the high reflectivity is maintained roughly for the intervals up to the wavelength of light. Beyond this, the reflectivity decreases quickly. Thus, the reflectivity of the *Morpho* wing is determined both by the number of layers in a ridge and by the ratio of the shelf width and separation. Since the reflectivity from a multilayer with large difference of the refractive indices easily saturates with increasing number of layers, the most important factor to increase the reflectivity is

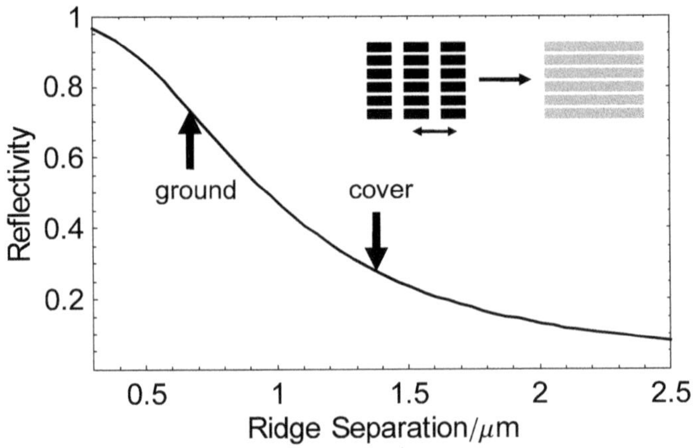

Fig. 3.23 Reflectivity at 480 nm calculated for alternate 11 layers of air and an averaged dielectric layer corresponding to various rides separations. The thickness of air layer is 145 nm, while that of the averaged dielectric layer is 60 nm. The refractive index of air and shelf is set to be 1.0 and 1.6, respectively. The length of the shelf is set to be 300 nm. Arrows indicate the ridge separations corresponding to a ground and cover scale of *M. didius* (Kinoshita and Yoshioka, 2005a).

to reduce the ridge separation. A typical example of this effect appears as the difference in the reflectivity of cover and ground scales in *M. didius*, where the experimentally determined reflectivities for the former and the latter amount to 30% and 55% at 480 nm, respectively. The reflectivity of the cover scale includes that due to thin-film interference at the base plane of the scale and the net reflectivity due to the shelf structure is estimated at 15% (Yoshioka and Kinoshita, 2004). Thus, the present expectations of 25% and 70% roughly agree with the experimental results. However, since the above estimation is based essentially on a continuum model, whose validity has not been confirmed yet, more vigorous calculation will be anyway necessary for quantitative comparison.

3.2.5.5 *Effect of Pigment*

Next, we will investigate the origin of different colors observed in two types of *Morpho* species, *M. didius* and *M. sulkowskyi*. Although both microscopic lamellar structures are similar to each other (see Figs. 3.15(b) and 3.15(c)), the apparent wing colors are completely different. The difference in the optical response shown in Fig. 3.19 will give an answer to this question. Since the shelf structures of both species are very similar, they should equally contribute to the blue reflection around 400–500 nm, the major difference comes from the amount of the absorption and/or background reflection above 530 nm. The absorption in green to red region is due to the presence of pigment, which is believed to be melanin and is distributed in the lower part of the iridescent ground scale of *M. didius*. The pigment is also distributed in the strongly iridescent scales of *M. rhetenor*, while only a negligible amount of pigment is recognized in two kinds of iridescent scale of *M. sulkowskyi*.

Thus, the presence of the pigment beneath the iridescent structure is quite important to exhibit the peculiar colors in the *Morpho* butterflies, because it contributes to the absorption of the complementary color and enhances the contrast of blue coloring. The absorption due to the pigment is effective to reduce the scattering within scale and at wing membrane, and also to reduce the transmitted light from the ventral side. Comparing the reflection measurements of an intact wing and a single scale of *M. sulkowskyi*, and also taking the reflection measurement on the wing without ventral scales, we deduce that the background reflection at around 560 nm in this butterfly mostly comes from the light scattering at the wing membrane and dorsal/ventral scales. Thus, the whitish tone of *M. sulkowskyi* originates mainly from the background reflection due to its own structure.

To investigate the origin of background scattering quantitatively, we have performed the experiment on *M. cypris* (Yoshioka and Kinoshita, 2006b), whose fore and hind wings have a distinctive white stripe pattern on the structurally colored blue wing, as shown in Fig. 3.24. In Fig. 3.25, we show the scanning electron microscopic images on blue and white scales on the dorsal wing. It is clear that the microscopic shelf structures are essentially the same for the blue and white parts, which resembles the ground scale of *M. rhetenor*. Thus, the difference of color is mainly due to the presence of pigment. This butterfly aims to show the white spots clearly by removing the pigment completely from the wing membrane and from the scales at the dorsal and ventral sides. So there is one-to-one correspondence for the white spots on the dorsal and ventral sides (see Fig. 3.24).

In Fig. 3.25, we also show the optical properties of the blue and white scales. It is clear that the white scale is almost lack of absorption and strong background reflection is observed. We have measured the reflection and transmission characteristics of each component separately: blue and white scales on the dorsal side, brown and white scales on the ventral side, and brown and transparent areas of the wing membrane, and have analyzed

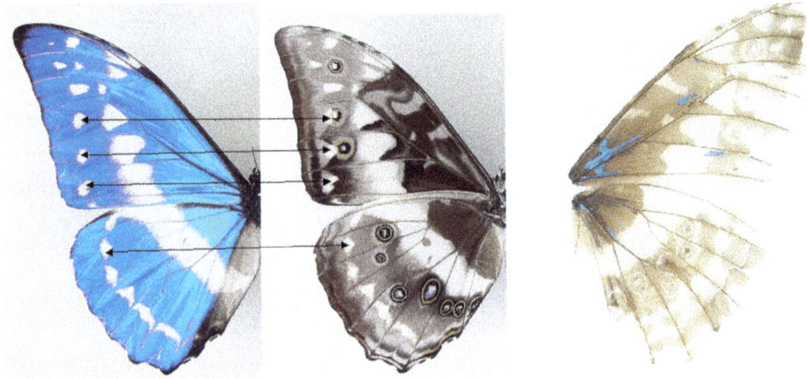

Fig. 3.24 Left: a dorsal wing of *M. cypris* and its corresponding ventral side, where we see one-to-one correspondence between white spots of dorsal and ventral sides. Right: wing membrane of *M. cypris* obtained by removing the scales on both sides (Yoshioka and Kinoshita, 2006b).

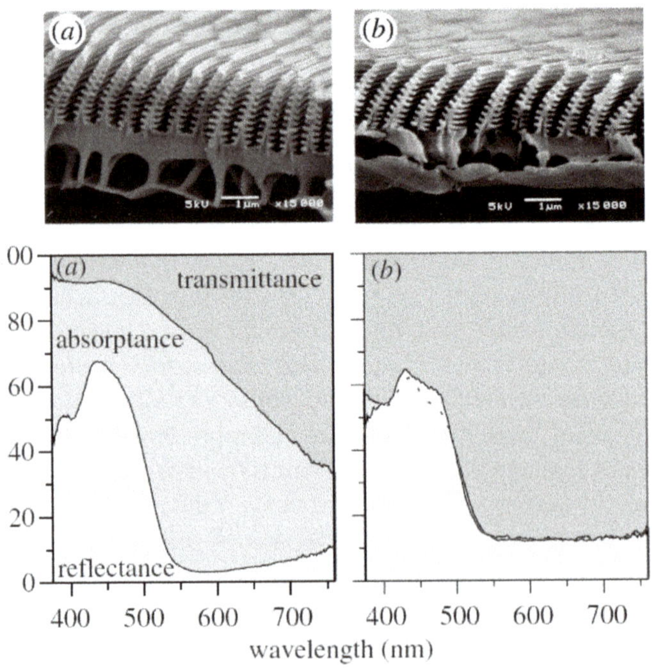

Fig. 3.25 Upper: SEM images of cross-sections of (a) blue and (b) white scales of *M. cypris*. Lower: optical properties of a single (a) blue and (b) white scale on the dorsal side. Percentages of reflectance and transmittance are plotted, while the rest of the part is assigned as absorption (Yoshioka and Kinoshita, 2006b).

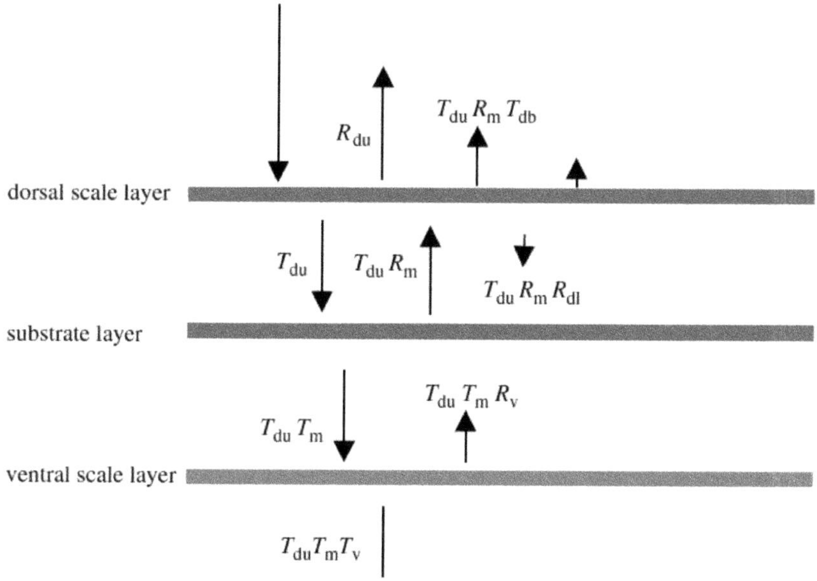

Fig. 3.26 A three-layer model considering multiple reflection between a dorsal scale, wing membrane, and a ventral scale (Yoshioka and Kinoshita, 2006b).

using a three-layer model considering the multiple reflection among them, as shown in Fig. 3.26. The result is shown in Fig. 3.27, where the calculated reflectance is shown for only a dorsal scale, that with a wing membrane, and that with a wing membrane and a ventral scale. The calculations for the blue and white areas can reproduce the data for intact wing very well. Thus, the background reflection comes almost equally from the three components, dorsal and ventral scales, and a wing membrane, with the multiple reflection among them. Among these, at least, a ventral scale shows a usual isotropic scattering, whereas the wing membrane shows a specific reflection and the dorsal scale may show the mixture of anisotropic reflection due to the shelf structure and the scattering due to the trabeculae. Thus, the background scattering bears anyway an isotropic character in scattering direction, which may strongly impress the whitish color in contrast to the blue color, because the blue color is only visible from the particular direction. Another reason is that the perception of eyes obeys a logarithmic detection with respect to the light intensity, known as Weber's law, which strengthens the weak white background and suppresses the strong blue coloring.

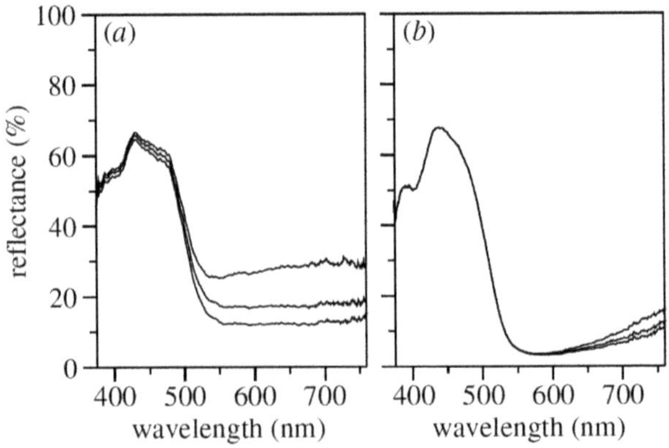

Fig. 3.27 Calculated reflectance at (a) white and (b) blue area. The reflection spectrum is obtained by assuming the reflection from a dorsal scale, that on wing membrane, and scales of both sides with wing membrane, from the bottom to top (Yoshioka and Kinoshita, 2006b).

Thus, the role of the pigment in the iridescent wing is summarized as (1) the enhancement of blue coloring by adding the pigment beneath the iridescent structures, and (2) the adjustment of color saturation by reducing the pigment without sacrificing the elaborated structures.

3.2.5.6 *Role of Cover Scale*

As described above, from a viewpoint of structural colorations, the presence of cover scales on ground scales will be a key role in phylogenetic study (Fig. 3.5). Namely, in ancient groups, whose larval host plant is monocotyledon, the cover scale plays a central role to produce the iridescent colors or does an equivalent role to that of the ground scale, but gradually its role decreases with decreasing size. In a more evolved group such as *M. rhetenor*, the cover scale degrades almost completely to expose the ground scale directly, and in another group, the size of the cover scale is large enough to cover completely the ground scale. Thus, the cover scale in this group seems to attain a special function to display the wing differently. In order to investigate its role quantitatively, we choose *M. didius* as a sample (Yoshioka and Kinoshita, 2004). The cover scale of *M. didius* has a peculiar structure that the ridge is sparsely placed on the base plane (Fig. 3.17), and its reflection shows a double pattern with slightly different colors, light and deep blue (Fig. 3.14). Furthermore, the angular dependence of the reflection from the wing with and without cover scales

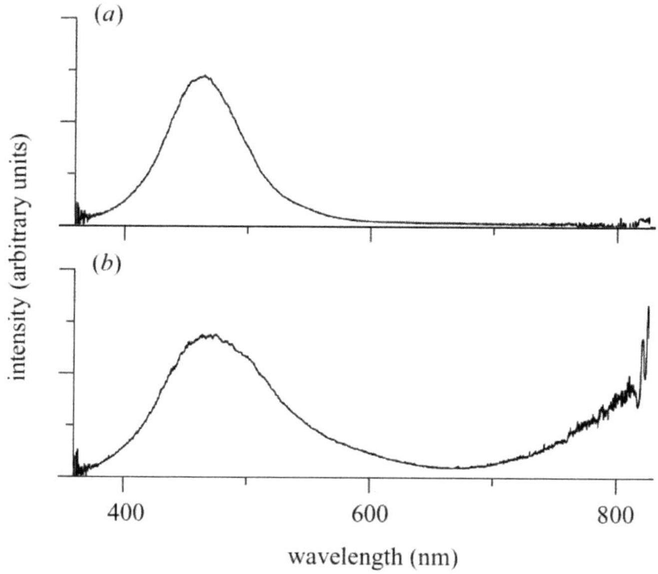

Fig. 3.28 Reflection spectra obtained for (a) deep-blue and (b) light-blue reflection band from a single cover scale of *M. didius* (Yoshioka and Kinoshita, 2004).

is quite different (Fig. 3.21). In the former case, the reflection is somewhat centralized in the direction perpendicular to the ridge direction, while it is broadened in a parallel direction. Thus, it seems to have a function to reduce the anisotropy.

The reflection spectra of light and deep-blue bands are very different as shown in Fig. 3.28. The deep-blue band has only a single sharp peak around 460 nm, while the light-blue band shows a rather broad peak around 460 nm with increasing reflectivity toward the longer wavelengths. Considering the microscopic structure, we deduce that the deep-blue band comes from the shelf structure in a ridge, while the light-blue band does from thin-film interference at the base plane. In fact, the calculation using the thickness of 215 nm with the refractive index of 1.56 gives the second-order interference peak at 447 nm with the maximum reflectivity of 17%, which almost agrees with the experiment. If we assume the total reflectivity as the sum of the two mechanisms, which is determined by the area ratio of 8:2 for the base plane to the ridge, the reflectivity amounts to 31% and is quite in good agreement with the experimental value of 31%. Thus, the two reflection bands consist of the reflection from a basal plane through thin-film interference, and of the reflection of obliquely running shelf structure with the gradient of 8°. It is also noticed that the reflection from the basal plane does not show a specular reflection nor diffraction spots, but is distributed similarly to that from a ridge. The reason for this spatial broadening is not clarified yet.

Anyway, if such a component as cover scale is placed on ground scale of strongly anisotropic reflection, it has a function to extend the reflection angle in the direction of running ridge, and thus to weaken the anisotropic reflection. Simple estimation illustrated in Fig. 3.29 shows that normally incident light should be divided into three directions, normal, 16°, 32°, with the intensity ratio of 0.14, 0.54, and 0.03, respectively, which quantitatively explains the angular dependence obtained experimentally as shown

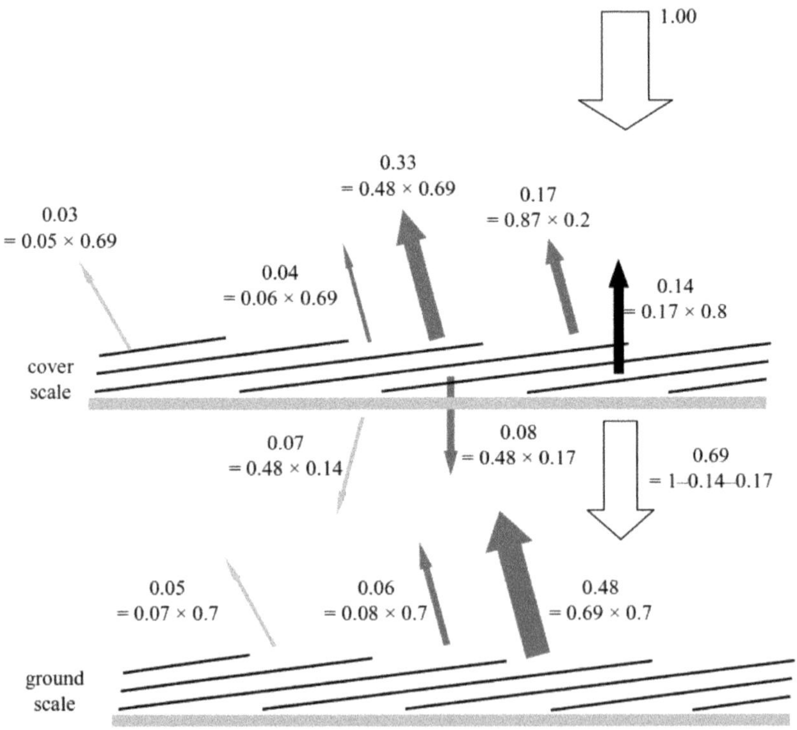

Fig. 3.29 Schematic illustration of the direction and intensity of the reflected and transmitted light in a cross-section of cover and ground scales along ridges under normal incidence. The intensity of the incident light is set as unity and each number indicates the intensity for each direction. In total, the incident light is reflected into three directions with the intensity ratio of 0.14:0.54:0.03 from the normal to the left direction, which is indicated as histogram in Fig. 3.21(c) (Yoshioka and Kinoshita, 2004).

in Fig. 3.21(c). To summarize, the cover scale of *M. didius* has carefully designed microstructures that function as a selective optical diffuser for blue light. Furthermore, the cooperation of the ground and cover scales makes possible the high reflectivity in a wide range of solid angles. This function makes it possible to reduce the gloss of the wing, which is usual the case where the ground scales showing strongly anisotropic reflection are exposed on the wing. The role of the cover scales and the difference of gloss will offer an interesting problem for phylogenetic and ecological studies.

3.2.5.7 *Reproduction of Morpho Blue*

Thus, we have come to the conclusion that the interplay between the regular structure and the irregularity cooperated with the pigment is essential to the blue coloring. In order to confirm it experimentally, we try to reproduce the *Morpho* substrate artificially only by mimicking its principle (Yoshioka *et al.*, 2004; Saito *et al.*, 2004; Kinoshita and Yoshioka, 2005b). First, we have prepared a transparent or black glass plate. One surface is polished roughly to produce the irregular surface using diamond paste of various grain sizes. Then, multilayer consisting of alternate layers of SiO_2 ($n = 1.47$) and Ta_2O_5 ($n = 1.97$) is coated on this rough surface, which assures the regularity. Thus, the coexistence of regularity and irregularity is introduced, in addition to the cooperation with the pigment. The result is shown in Fig. 3.30. The reproduced substrate looks quite similar to the *Morpho* wing, because the strong blue reflection accompanying the diffusive reflection and the color change into violet, when viewed obliquely, are clearly observed. This is in contrast with the case without roughness, because the blue color is only visible when specular reflection occurs, otherwise the substrate looks black. Thus, our hypothesis for the mechanism of the *Morpho* blue has been almost confirmed. However, several problems have still remained. The roughness is a key point to the present substrate. If it is too large, multiple scattering occurs and the substrate looks whitish. On the other hand, if it is too small, specular reflection will occur. Furthermore, anisotropic reflection cannot be introduced in the present method. Thus, controlled randomness is anyway necessary.

Then, we employ the method of nanotechnology to create the controlled randomness. First, electron-beam resist is placed on the substrate, and then the successive treatment is performed such as chemical reaction using electron beam, chemical treatment, dry etching by ion beam, and then

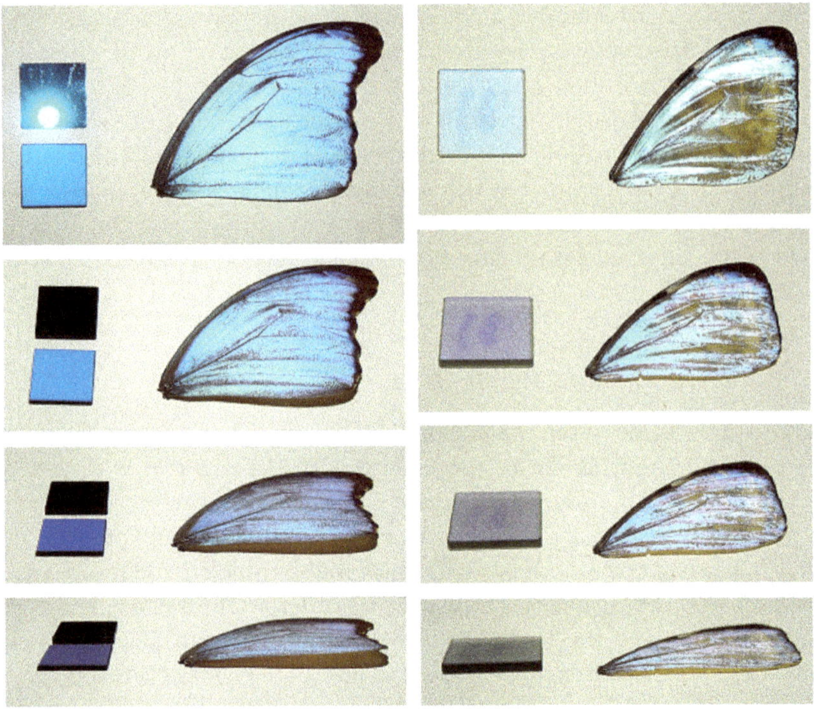

Fig. 3.30 Left: Viewing angle dependence of *M. didius* wing and a multilayer-coated black substrate with (lower) and without (upper) irregular surface, obtained under the illumination of nearly normal to the wing/substrate. Right: that of *M. sulkowskyi* wing and a multilayer-coated transparent substrate with the irregular surface. (Reproduced from Yoshioka *et al.* (2004), with permission.) See Plate 15 in Appendix B.

multilayer coating using 14 layers of SiO_2 and Ta_2O_5. We consider a unit rectangle, whose size is $\sim 2\,\mu$m in length and $0.3\,\mu$m in width. These units are randomly distributed on the substrate taking two values for height. The difference of the heights is set to be $0.11\,\mu$m, which is designed to cancel the specular reflection at 440 nm under normal incidence. The result is shown in Fig. 3.31. The upper photograph shows a SEM image of the substrate and the lower one the photograph of the substrate whose size is $6 \times 6\,\text{mm}^2$. The reflection pattern, shown in the right-hand side of Fig. 3.31, almost completely agrees with that of a *Morpho* wing. Our hypothesis proposed for the mechanism of the *Morpho* blue is thus confirmed experimentally. It is concluded in order to reproduce the *Morpho* blue, we need not mimic the structure itself but only does its principle.

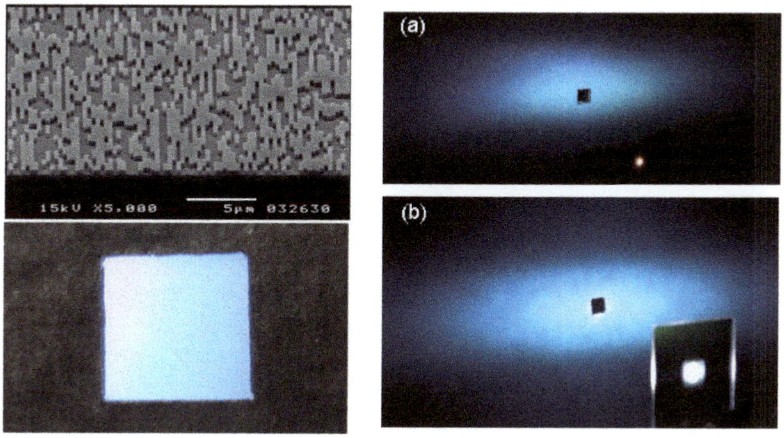

Fig. 3.31 Upper left: SEM image of a glass substrate, whose surface is processed with randomly distributed tips of a size $0.3 \times \sim 2\,\mu m^2$ and a two-valued random height of 0 or $0.11\,\mu m$. Lower left: Multilayer coating consisting of SiO_2 and TiO_2 on the substrate of $6 \times 6\,mm^2$ reveals the brilliancy similar to that of the *Morpho* butterflies. Right: reflection patterns obtained for a *M. didius* wing (a) and the fabricated substrate (b). (Reproduced from Yoshioka *et al.* (2004), with permission.)

3.2.5.8 *Summary of the Morpho Blue*

The physical significance of the above results is summarized in Fig. 3.32: (1) The structural color originates mainly from light interference within shelf structure on the scale through quasi-multilayer interference. (2) The slender shape of a shelf gives strongly anisotropic reflection. (3) The irregularity in the ridge height destroys the interference among neighboring ridges, which results in the diffuse reflection in a wide angular range. (4) High reflectivity is realized owing to the presence of 6–10 layers in a ridge with a large difference in refractive index and with sufficiently small separation between adjacent ridges. (5) The pigment beneath the iridescent structure absorbs the unnecessary complementary color, and enhances the blue structural color, while with reducing pigment, the scattering at dorsal and ventral scales, and wing membrane adds whitish color to the wing. Thus, the coexistence of regularity and irregularity, and the cooperation with the pigment are essential for the structural color in these butterflies.

3.2.6 *Morpho Mimicry*

Another stream of the *Morpho* research has been quickly developed in this decade. That is to mimic the *Morpho* blue to construct the elaborate

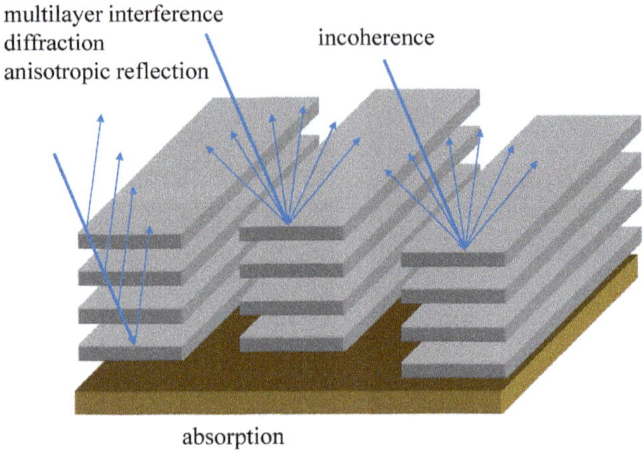

Fig. 3.32 Schematic illustration of factors associated with the structural colors in the *Morpho* butterflies.

structure using recently developed nanotechnology. This work is classified roughly into two categories: one is to mimic the *Morpho* blue coloring irrespectively of mimicking the microstructure, and the other is to mimic its microstructure itself. The former work is related to the vision technology such as painting, ink, display, fiber, cosmetics, and so on, while the latter to the photonic technology to search for new photonic devices and their fabrication methods.

In a textile world, a quite sophisticated fiber was invented through mimicking the scale structure of the *Morpho* butterfly (Iohara *et al.*, 2000). This fiber is made of polyester and has a flattened shape of thickness 15–17 μm, within which 61 layers of nylon 6 and polyester with the thickness of 70–90 nm are incorporated as shown in Fig. 3.33. Because of the multilayered structure, the wavelength-selective reflection and change with viewing angle are attained. Moreover, the flat shape of the fiber makes it possible to align the direction of the multilayer, which increases the effective reflectivity. They have demonstrated it by weaving a wedding dress using this fiber. However, since the polymer materials employed have similar refractive indices ($n = 1.60$ for nylon 6 and $n = 1.55$ for polyester), the wavelength range of high reflectance is limited within a small region. In fact, the wedding dress seems to bear pale blue and is really in harmony with the ceremony, although it differs considerably from that of the *Morpho* butterfly concerning the color impact and luster.

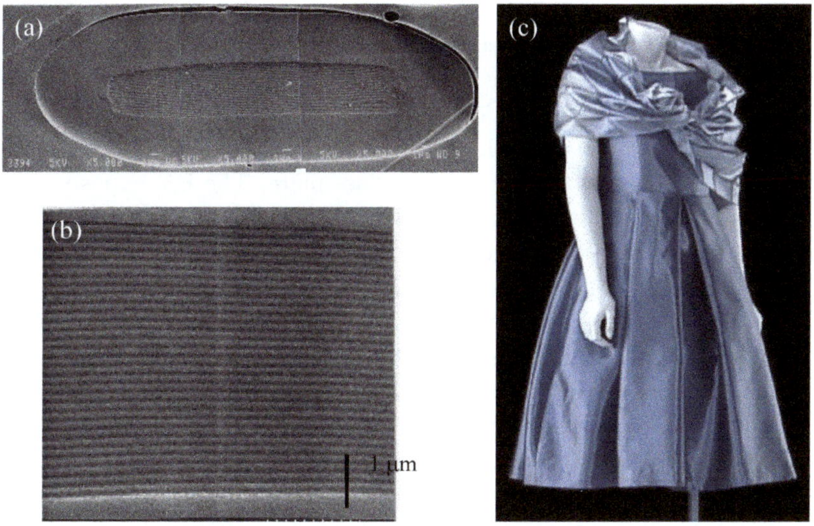

Fig. 3.33 (a) SEM image of the cross-section of a flattened polyester fiber mimicking the *Morpho* butterflies. (b) The fiber includes 61 alternate layers of nylon 6 and polyester with a thickness of 70–90 nm. A dress woven using this fiber is shown in the right-hand side (c). (Courtesy of Teijin Fibers Limited.) See Plate 16 in Appendix B.

Wong *et al.* considered that the wing color of the *Morpho* butterfly came from the grating structure of the ridges. Using the fact that a fine grating reflected only blue color within a visible region in a wide angular range, they fabricated a multigrating structure (Wong *et al.*, 2003). This structure consisted of 2D array of randomly oriented small hexagonal gratings with rectangular grooves having a pitch of 440 nm, a depth of 125 nm, and a duty cycle of 0.5. The orientation of the grating was distributed among three directions. The minimum separation of the gratings was set to be 33 μm. They fabricated this structure using electron beam lithography. The obtained multigrating appeared blue over the viewing angles between 16° and 90°, and bluish-green for the angles less than 16°. However, the diffraction spots inherent to the grating structure like the X-ray Laue pattern were clearly seen owing to a double-slit effect coming from two nearby hexagonal gratings and thus the reflection pattern was quite different from the actual wing.

Saito *et al.* developed our method of fabricating the *Morpho* substrate to be useful for the mass production (Saito *et al.*, 2006). He employed nanocasting lithography (NCL) to obtain the fine pattern replication using various polymer materials. This method is superior to the conventional nanoimprint lithography, in which some issue is molded into the nanomaterial to fabricate the replica, because the former is better for the replication without mold damage and the mixing bubbles are easily removed by vacuum baking. They prepared a quartz master plate as described previously, and UV curable resin was spin coated. UV light was then exposed after the coated surface was covered with a glass plate. The UV exposure made the resin to adhere the glass plate, while the surface of the master quartz plate was covered with antisticking surfactant beforehand. After removing the glass plate with the replicated microstructure from the master plate, 40 nm thick TiO_2 and 75 nm thick SiO_2 layers were alternately deposited up to 14 layers, which was designed to have the maximum reflectivity at 440 nm. The fabricated replica looked very similar to the master plate under a scanning microscopic observation. Furthermore, the reflection pattern was found to be similar to the *Morpho* wing, which was just the same as has been described in Sec. 3.2.5.7. Thus, a low-cost reproduction is attainable, which promises the cost down and mass production of the *Morpho* substrate.

Completely different approach to mimic the microstructure of the *Morpho* butterfly itself was progressed by Matsui *et al.*, who utilized a focused-ion-beam chemical-vapor-deposition (FIB-CVD) technique to reproduce the complete structure of the ridge (Matsui *et al.*, 2000). They succeeded to make a complete three-dimensional structure including overhang structure. This is based on a method that a Ga^+ ion beam is focused on a sample surface with a small focal size of typically 10 nm, where aromatic hydrocarbon precursor gas is injected through a nozzle located near the sample surface. They used phenanthrene ($C_{14}H_{10}$) as a precursor and demonstrated to fabricate a microbeaker with 1.0 μm in diameter, a branch structure with 0.08 μm in diameter, a 0.6 μm diameter microcoil, a 0.25 μm diameter microdrill, and even a 2.75 μm microwine glass. The deposited material was found to be mostly amorphous diamond-like carbon, which promised hard and transparent qualities. They demonstrated to make a ridge structure by this method (Watanabe *et al.*, 2005a,b). They employed the distance of lamellae so adjusted to agree with that of *Morpho* butterfly and even to reproduce the alternate structure of the lamella (Fig. 3.34). The time to fabricate a structure with 2.60 μm height, 0.26 μm in width, 20 μm in length, and 0.23 μm pitch of the shelf was reported to be 55 min. The replica illuminated by white light is found to actually glitter blue to violet under microscopic observation.

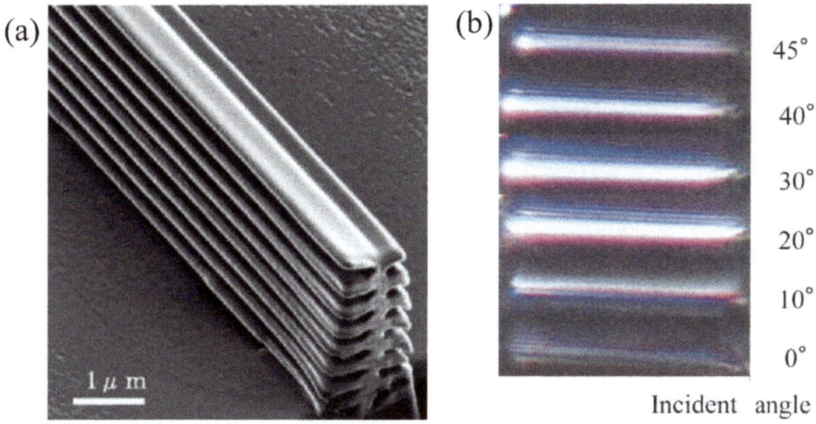

Fig. 3.34 "Shelf" structure of the *Morpho* butterfly fabricated by focused-ion-beam chemical-vapor-deposition method. (a) SEM image of the structure and (b) optical microscopic image observed under 0–45° incidence angle of white light. (Reproduced from Watanabe *et al.* (2005b) with permission.) See Plate 16 in Appendix B.

The replication of the *Morpho* scale was also reported (Cook *et al.*, 2003; Silver *et al.*, 2005; Huang *et al.*, 2006). Huang *et al.* (2006) aimed to mimic the photonic structure of butterfly scale to find out new photonic devices. They employed atomic layer deposition (ALD) technique to replicate the wing structure of *M. peleides* using Al_2O_3 under a temperature of 100°C. They found that the wing color was changed from blue to green, yellow, and pink according to the thickness of the Al_2O_3 coating of 10–40 nm. After the coating, the sample was annealed at 800°C for 3 h to crystallize the coating and to burn out the organic wing. Then, the completely hollow replica was obtained, which consisted of robust polycrystalline Al_2O_3 shell structure with a crystal grain size smaller than 3 nm. They made transmission electron microscopic observation and also performed optical measurement for the shell structure by manipulating scales side-by-side to cover an area of 1 mm^2. The reflection spectrum had a clear peak at 420 nm, which was slightly shorter than that of the original wing itself. They also reported that the hollow lamellar structure became a waveguide constituting 2D photonic crystal with a unit cell of $50 \times 60\,nm^2$ based on the repeating shelves and regular trabeculae between them.

Thus, the recent rapidly developing nanotechnology has proved that even the most elaborate structure in nature comes true. Therefore, the future of the study on structural colors is extremely promising, because the optical and physical investigations on their mechanisms will be immediately confirmed by fabricating the reproduction directly.

3.3 Overview of the Structural Coloration in Butterflies and Moths

In the present section, we will overview the world of butterflies and moths from a viewpoint of elaborate structures decorated on their scales. Although more than 100 papers have been devoted to the structural colors in the butterflies/moths wings during the past century and the number is definitely growing particularly in this decade, yet there are very few papers that truly analyze the optical properties and properly describe their physical mechanisms. Moreover, really surprising, the systematic researches are extremely few.

Here, we will again refer to the excellent works performed by Ghiradella, who extensively investigated the morphological characteristics of scales in conjunction with their functions, and further proposed rational classification for the scales specialized to the structural colorations (Ghiradella, 1998, 2005). The result is summarized in her well-known illustration shown in Fig. 3.35. She classified the specialized scales into three categories (Ghiradella, 1998), according to the location of the elaborate structures: (1) ridge, (2) flat between ridges, and (3) lumen at the lower side of a scale (see Fig. 3.35).

First, we show the ridge-iridescence type. As described in Sec. 3.1.2, ridges run straight along the longer side of scale on its obverse surface, keeping constant separation of ~1 μm with neighboring ridges. In a normal scale, the ridges are slightly decorated by the kind of slender lamella, overlapping partly with each other and growing obliquely by an angle of ~10° from the base plane. Below the ridges, trabeculae are present to support the ridges on the base plane. Between ridges, crossribs keep a spacing possibly as a mechanical supporter. The rest of the part is open window, which makes it possible to connect the inner lumen with the outer space. Sometimes, the structure of ridge is specialized by increasing its height, broadening the width of the lamellae, and constituting partially stacking multilayer, which produces quasi-multilayer interference as described in the *Morpho* butterflies (type A). As a modification of this type, the central pillar of the ridge sometimes takes the form of a Christmas tree (see Fig. 3.39). Sometime, shelves sticking out from the pillar becomes small so that the multilayer interference seems to no longer hold, and thus the volume diffraction effect

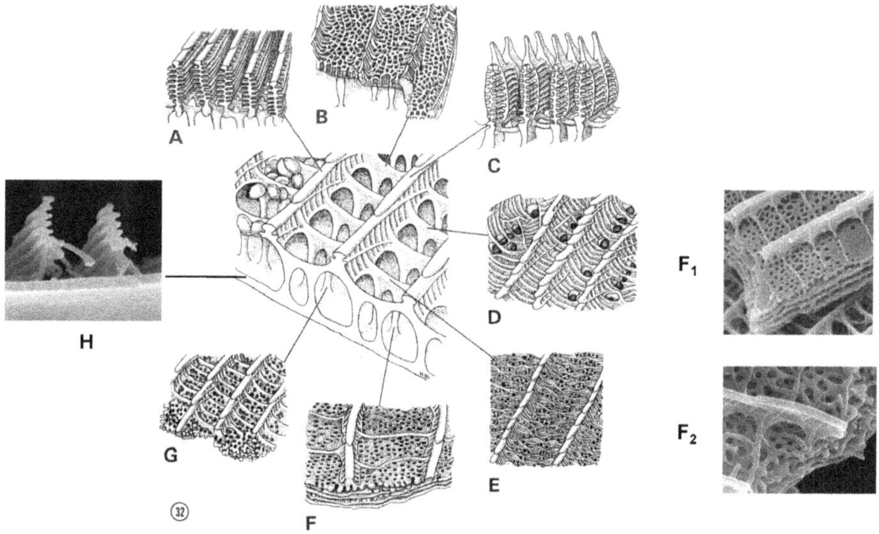

Fig. 3.35 Elaborate structures created at various sites in a lepidopteran scale: (1) Ridge-iridescence types A and C, (2) flat-iridescence types B, D, and E, and (3) lumen-iridescence type F and G. We have further added H type into base-iridescence type, and have divided the F type into two, F_1 and F_2 types. In the figure of H, ridge-lamellae (A type) are also seen in addition to a thin base film (H type) (Ghiradella, 1998). (Reprinted with permission of Wiley-Liss Inc., a subsidiary of John Wiley & Sons, Inc.)

will come out. Anyway, in this type, microribs, running perpendicular to the lamellae, seem to little contribute to the optical response. However, when the gradient of the lamellae becomes steep, both lamellae and microribs will contribute to the optical responses as two independent periodic structures. In the extreme case, the lamellae stand nearly perpendicular to the base plane, and the periodic microribs now contribute to optical interference. Ghiradella exemplified the scale of *Eryphanis aesacus* for this type (type C) (Ghiradella, 1998).

Next, we show a flat-iridescence type. Ghiradella showed three types as examples of this type. One is the case where the flat is elaborated into randomly distributed alveoli (type B), and *Papilio zalmoxis* is exemplified (see Fig. 3.37). The second one is that the flat is covered with rather regularly running microribs that are perpendicular to the ridges, which make the flat look like a wide plane with pleats (type D). The example is seen in *Diachrysia balluca*. The third one is the modification of type B, whose flat is filled with the plates-and-pores pattern (type E), and is characteristic of pierid androconical scale. The extreme case of this type is that the obverse and reverse planes of the scale are fused together to make a thin film, on which ridges are regularly running (type H, which I dare to add into the Ghiradella's illustration). This type of scale sometimes produces thin-film interference, which has been already explained as an optical diffuser in case of a cover scale in *Morpho didius* (Yoshioka and Kinoshita, 2004). The stiffness of the scale in the last case is clearly weak so that even under a scanning electron microscope, the scale is sometimes turned up by a function of electron beam.

The last one is lumen-iridescence type, in which the lumen located below the ridges is filled with elaborated structure. She exemplified two cases for this type. One is that the lumen is filled with body-lamellar structure, which will produce the multilayer interference. The body-lamella sometimes takes the form of a true or curved multilayer (see Fig. 3.47), or that bearing randomly distributed holes (pepper-pot perforation) (type F_1: *Arhopara japonica japonica*, see Fig. 3.41), and in other cases, the multilayer is completely replaced by the multilayered network of the fibrous materials (type F_2, *Celastrina argiolus ladonides*, see Fig. 3.41). Thus, simple multilayer interference no longer holds and the sophisticated treatment that takes into account both regular multilayer and irregular structure is anyway necessary. The second type is that the lumen is filled with three-dimensional lattice to form a photonic crystal (type G). This type is widely distributed in lepidopteran species and *Teinopalpus imperialis* is exemplified (see Plate 9 in Appendix B). The photonic crystal in butterflies is not a complete crystal, but the irregular structure is always accompanied, in which the domain structure of a small size is distributed with the orientation of the crystal changing from domain to domain (see Fig. 3.42). Anyway, the trabeculae to support the lumen of the scale disappear in lumen-iridescence type and the elaborate structure itself maintains mechanically the shape of the scale.

Unfortunately, the theoretical approaches to understand the optical effect from such a wide variety of elaborations have not been established at the present stage, except for the numerical approach using finite difference time domain (FDTD) method. However, if one wants to understand the physical mechanism of such elaborations truly, it is necessary to build the simplest model to explain their characteristic optical response. Thus, the physical interpretation of the above elaboration such as type B, E, and F_2 are still behind the veil.

In addition, the macroscopic structure is also important to offer particular visual effect on the butterfly wing. One of such effects occurs on the order of the scale size. In fact, the scales are sometimes curved in the longitudinal direction, which produces the special angular effect and also polarization effect, as in the scale of *Urania riphenus* (now called *Chrysiridia ripheus*), which will be described later (Sec. 3.3.6). The optical interaction between the cover and ground scales will also produce the important effect, e.g. to make the matting wing as in *Morpho didius*. Thus, the size of the cover scale compared to the iridescent ground scale plays a decisive role. It is also reported that the cover–ground interaction enhances the efficiency of the light scattering, which makes the wing more whitish as in pierid butterfly (Stavenga *et al.*, 2006a).

Finally, the distribution of the pigment definitely affects the visual effect of the wing. In case of *Morpho sulkowskyi*, the lack of melanin pigment clearly shows the pearl-like whitish wing, while the presence of melanin pigment at the lower side of the ground scale enhances the blue coloring of *Morpho rhetenor* by absorbing the light of complimentary colors. In this case, the pigment distribution of the wing membrane and ventral scales should be considered in addition to the iridescent scale on the dorsal side. In fact, in *Morpho cypris*, these components are proved to contribute evenly to produce the white spots by light scattering. In the following, let us look at the structural colors in various butterfly wings and investigate their origins from various viewpoints such as structural elaborations, macroscopic effect, and chromatic effect.

3.3.1 *Papilioninae (Papilionidae)*

This subfamily Papilioninae is currently called swallowtail and cattleheart, and is familiar for its large wing, strong flight power, and variously colored wing. A well-known "old world swallowtail" (*Papilio machaon*) belongs to this group, which is distributed world-wide in the Northern Hemisphere. It is quite natural that the butterflies in this group bear the structural colorations. To date, 50 papers have been devoted to the structural colorations of 65 species in this group.

The most quoted butterfly in this group is "green-banded peacock swallowtail" ("emerald swallowtail"), *Papilio palinurus*, which inhabits in the forests in Southeast Asia (Malaysia, Indonesia, Philippines) as shown in Fig. 3.36 and Plate 7 (Appendix B) (Huxley, 1975; Ghiradella, 1985, 1999; Vukusic and Sambles, 2000; Vukusic *et al.*, 2000, 2001a; Vukusic, 2003; Vukusic and Sambles, 2003; Kinoshita and Yoshioka, 2005b; Vukusic, 2006). Its green band in the dorsal wing emits the color not so strongly but somewhat mysteriously, which has attracted scientist's attention. Its wing span

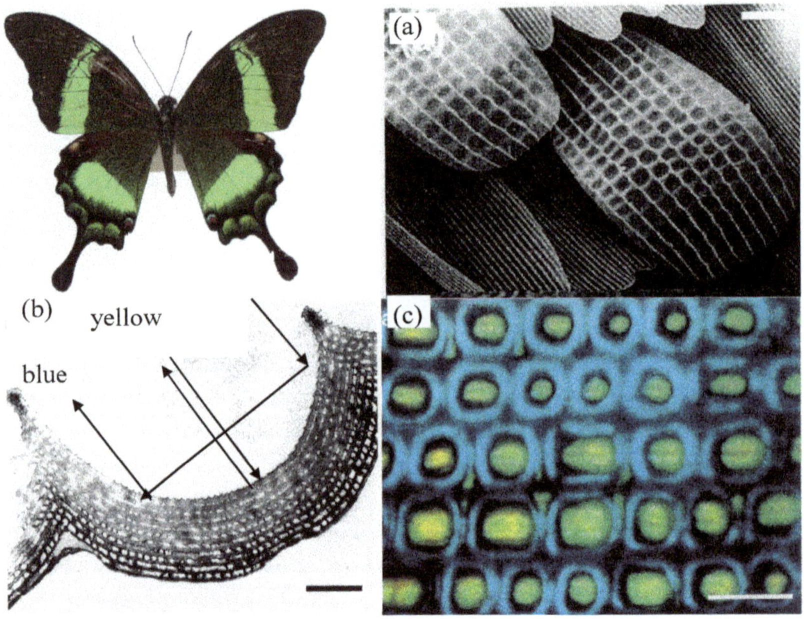

Fig. 3.36 Color mixing mechanism due to concavities in a scale of *Papilio palinurus*. Scale bars are (a) 10 μm, (b) 1 μm, and (c) 10 μm. (Reproduced from Vukusic *et al.* (2001a) with permission. Specimen of the butterfly is in the possession of The Nature and Human Activities, Hyogo, Japan and the photograph was taken by Dr. M. Kambe.) See Plate 9 in Appendix B.

is typically 90–100 mm. The main habitat is primary forest and the larval host plant is the rue family, Rutaceae.

The first quantitative investigations using the electron microscope and optical measurement were performed by Huxley (1975). He found that the scale of *P. palinurus* have a peculiar shape with the wide ridge–ridge spacing and with almost circular, shallow pits distributing between the ridges, which were similar to the scale of *P. karna*, showing greenish blue patch in the hind wing. Beneath the ridges, the multilayer consisting of 8–11 cuticle-air layers were distributed, which were curved along the circular pits. He considered its peculiar structure to serve as the reflection of light over a large solid angle. Thus, this butterfly is clearly classified into lumen-iridescence of type F_1. He measured the reflection spectra and analyzed using multilayer interference. He employed the refractive index of cuticle layer of 1.58 and the incidence-angle (θ)-dependent refractive index for air layer as $1+\theta/300$, with thicknesses of both layers as 108 and 94 nm, respectively, and found good agreement involving the angular dependence. He also noticed that the uppermost lamella was covered with periodic microribs of 50–60 nm height and 170 nm periodicity, which might serve as the broad-band impedance transformer similar to corneal ripple.

Its peculiar pit structure was analyzed later by Vukusic *et al.* (2000, 2001a). They investigated the scale of *P. palinurus* and found that the concavity (pit in Huxley's paper (1975)) was a hemispherical shape having a diameter of 4–6 μm with a depth of 0.5–3 μm. Below a round concavity, the curved multilayer consisting of about 20 alternate layers of cuticle and air was present. They deduced the optical mixing mechanism for this green-color production (see Fig. 3.36). Namely, when white light was incident on this structure, yellow light hitting directly on the bottom of a concavity was selectively reflected, while blue light hitting on the side of the concavity was reflected, hit again on the opposite side of the concavity wall and was then reflected back to the reverse direction. Thus only yellow and blue colors were selectively retro-reflected, which produced green color in our eyes. Because of the oblique reflection of blue color at the wall of the concavity, the color change with the polarization direction of incident light was also observed. They analyzed the reflection spectrum using a multilayer interference model including the imaginary part of the refractive index for a sum of single- and double-reflection paths, and found a good agreement with the experiments. They compared this result with the similar *Papilio* butterflies, *P. ulysses* ("blue mountain swallowtail," see Plate 9 in Appendix B) and *P. blumei* ("green swallowtail"), where the concavities were found to be shallow, which produced the usual multilayer reflection, while the diffusivity of light in these butterflies was realized by their concaved layers.

A large African butterfly "blue swallowtail," *Papilio zalmoxis*, is the second quoted butterfly (Mason, 1927a; Huxley, 1976; Ghiradella, 1994, 1985, 1998; Prum *et al.*, 2006), which inhabits in tropical forest canopy in Central Zaire to Nigeria and Liberia (Fig. 3.37 and Plate 9 in Appendix B). The whole wing bears pale blue with the wing span of 140–170 mm. The blue color of this butterfly wing is eventually due to the lack of saturation and iridescence. Mason commented on this butterfly that the seat of its iridescence was beneath the surface layer (Mason, 1927a). Huxley made a detailed investigation on its scale and found that the scale consisted of a thick layer of close-packed, air-filled vertical tube like alveoli (type B), whose thickness was 1.9 μm, internal diameter 200–243 nm and wall thickness 83–126 nm (Huxley, 1976). It was also found that the fluorescent dye, e.g. kynurenine, was distributed as a film across the upper ends of the alveoli. In fact, it

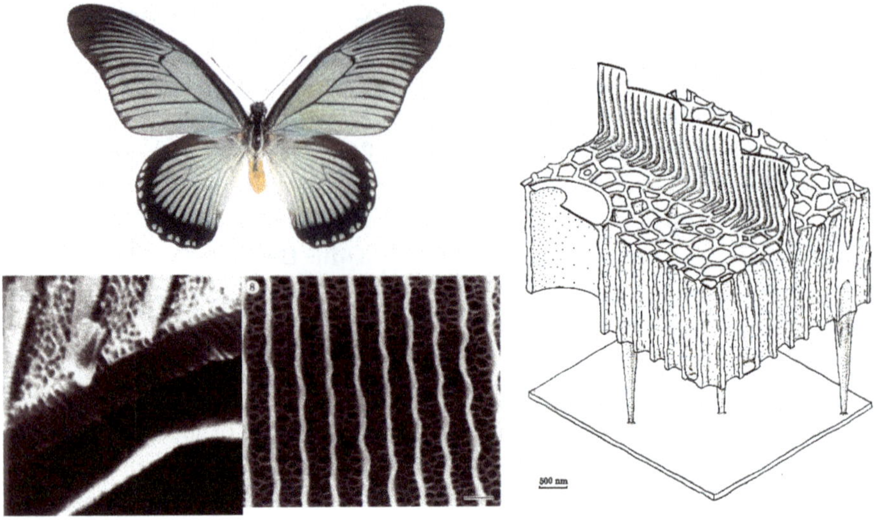

Fig. 3.37 Air-filled tube structure like alveoli found in *Papilio zalmoxis*. (Reproduced from Huxley (1976) with permission of Royal Society Publishing. Specimen of the butterfly is in the possession of The Nature and Human Activities, Hyogo, Japan and the photograph was taken by Dr. M. Kambe.) See Plate 9 in Appendix B.

emitted blue fluorescence under ultraviolet excitation. He concluded that the alveoli was a major cause of blue coloring due to Tyndall scattering, which was modified by the reflection due to the thin-film interference at the base plane, and also by the strong pigmentary absorption. However, he showed contradictory data in his paper that after extraction of pigments in methanol–acetic acid solution, the reflection spectrum, originally peaked at 470 nm, almost lost the structure.

Recently, Prum *et al.* performed a series of experiment, in which electron microscopic image of the complex structures producing the structural colors was Fourier transformed (FT). They analyzed the **k**-distribution to deduce the reflection spectrum using the average refractive index (Prum *et al.*, 2006). They found the deduced reflection spectrum differed considerably from the observed spectrum peaked at 320 and 500 nm. They concluded that the incoherent Tyndall scattering did not occur in this butterfly and the blue coloring came from the fluorescent pigmentary color.

In relation to alveoli structure, Vukusic and Hooper have recently reported that the emission efficiency is affected by the presence of alveoli in *Papilio nireus* (see Plate 9 in Appendix B), whose alveoli layer has 2 μm thickness, 240 nm diameter, and 340 nm spacing (Vukusic and Hooper,

2005). They have performed the photonic band calculation for air cylinders with triangular symmetry and have found that the band gap coincides with the fluorescence band involved in the alveoli. Thus, the absence of density of states in two-dimensional plane enforces the fluorescence emission out of the plane. They have examined the effect of the structure on the fluorescence decay and have found 30% increase of the lifetime. Although this idea seems to be quite fresh and extremely interesting, further studies are clearly needed concerning the irregularity found in the alveoli and also the experimental comparison for the fluorescence measurement.

"Emerald-patched cattleheart," *Parides sesostris*, is the third runner (Ghiradella, 1985, 1991, 1994, 1999; Vukusic and Sambles, 2003; Vukusic, 2003; Ghiradella, 2005; Vukusic, 2005; Prum *et al.*, 2006) in the subfamily Papilioninae, which is distributed around Central and South America (see Plate 9 in Appendix B). It shows a strong green patch in the fore wing. Ghiradella reported that the scale of this butterfly showed spectacularly regular lattice that filled the lower part of the scale from the electron microscopic observation (type G) as shown in Fig. 3.38 (Ghiradella, 1994, 1991, 1985, 2005). The structure was composed of crystallites with a face-centered cubic lattice of granules, which produced a green diffraction color. It was also noticed that very peculiar ridges are standing regularly on the lattice. Vukusic reported from the minute observation that the three-dimensional lattices were divided into irregular but distinct domains of a few microns, which made it possible to offer a constant-color effect by spatial averaging (Vukusic, 2005). Thus, the cooperation of regular and irregular structures

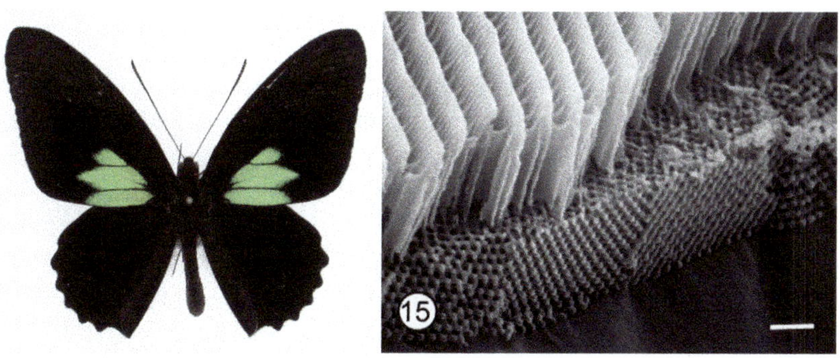

Fig. 3.38 SEM of a fractured green scale of *Parides sesostris*. Scale bar is 1 μm. (Reproduced from Ghiradella (2005) with permission. Specimen of the butterfly is in the possession of The Nature and Human Activities, Hyogo, Japan and the photograph was taken by Dr. M. Kambe.) See Plate 9 in Appendix B.

again appears explicitly as in the *Morpho* butterflies. In the *Morpho* butterflies, the regularity comes from the ridge-shelf structure giving rise to a quasi-multilayer interference, while the irregularity in the height of the ridges cancels the interference between the neighboring ridges, which gives a diffusive reflection through the diffraction effect (Kinoshita *et al.*, 2002a,b; Kinoshita and Yoshioka, 2005b). In *P. sesostris*, the photonic crystal is the regular structure, while variously oriented domain structure corresponds to the irregular structure. Although the domain size of a few microns certainly contributes to the diffraction, the distribution of the crystal orientation seems more direct. Thus in this butterfly, the diffusive light is produced macroscopically using geometrical optics. In this sense, the optical effect of the peculiar shape of ridge should be further discussed.

Blackness is another important factor to the butterfly wing, as it clearly increases the contrast of the colored wing patterns. The blackness means the light of all the wavelengths are absorbed effectively. However, it will be soon understood that if we want to increase the absorption, we have to increase the imaginary part of the refractive index, which inevitably decreases the penetration depth of the light at the surface and increases the reflection. This is the case for pure metal like aluminum or silver, which looks rather whitish. Thus, the particular mechanism is necessary. Vukusic *et al.* found the elaborate mechanism in the wing of *Papilio ulysses* (see Plate 9 in Appendix B) (Vukusic *et al.*, 2004a; Vukusic, 2005). They chose this butterfly because it shows two different types of blackness, one is deep mat black in appearance, whereas the other more lustrous black. They made the measurement of the absorption with and without immersing index-matching liquid, bromoform, and also investigated using a transmission electron microscope. They found that the reflectivity was less than 2% with the light absorption of 90% and 95% for lustrous and mat black scale. It was also found that the immersion of bromoform decreased the absorption by 40%, which definitely showed that the blackness was deeply connected with the microstructure. The electron microscopic image showed the periodic tapered ridge of 2–3 μm pitch with an aperiodic lattice-work of struts and walls in the lumen of the scale. The difference of the two scales was that the density of the lattice was higher for the matt black region. They analyzed that these structures were to increase the optical path length by multiple scattering at the surface, where the pigments were diffusely distributed. The tapered shape of the ridge was used to decrease the abrupt change in the refractive index at the surface. Furthermore, they found at the surface of each ridge, sub-wavelength impedance-matching elements, which

Table 3.2 Types of iridescent scales in various Papilioninae species.

Name	Type[a]	Name	Type[a]
Battus philenor	D	*Papilio ulysses*	F1
Graphium sarpedon	D,A	*Papilio ulysses autocylus*	F1
Papilio antimachus	B	*Papilio zalmoxis*	B
Papilio blumei	F1	*Parides erithalion polyzelus*	A
Papilio karna karna	F1	*Parides sesostris*	G
Papilio lorquinianus	F1	*Teinopalpus imperialis*	G
Papilio nireus	B	*Troides brookiana*	A
Papilio oribazus	B	*Troides magellanus*	A
Papilio palinurus	F1	*Troides priamus hecuba*	A,F1
Papilio palinurus daedalus	F1	*Troides priamus urvillianus*	A,F1
Papilio paris	F1		

[a]Classification according to Ghiradella's classification with some modifications as shown in Fig. 3.35. See Appendix A.

might reduce the backscattering of light and the scattering for nonnormal incidence. The blackness will become one of the most important topics in butterfly optics, and quantitative measurements and analyses are strongly needed, in which the optical effects due to their hierarchical structures with various spatial forms should be taken into account.

Finally, various Papilioninae species are summarized in Table. 3.2 with the iridescence-related structures reported in the past. We categorize the iridescent scales into several types, which were employed in the Ghiradella's classification with some modifications. It was clear that the structural colorations in Papilioninae species are originated from various types of scale decorations ranging over (1) ridge-lamella (A), (2) alveoli (B), (3) body-lamella with concavities of various depths (F_1), and (4) photonic crystal with domains (G). For the ridge-lamella case, two types have been reported. One is the case of the hind wing of *Troides magellanus*, where the lamellae are so steeply slanted by 36° to the vertical that they become multilayered stacking only when viewed and illuminated at near grazing-angle incidence. Lawrence *et al.* analyzed this structure using a bigrating model originating from the periodicities in the x and y directions, where the lamellae are assumed to be slanted in the xy plane (Lawrence *et al.*, 2002). The other is found in *Trogonoptera brookianus*, in which adjacent ridges are in perfect register and form an unusual ridge-lamella structure (Ghiradella, 1985; Wickham *et al.*, 2006). However, optical measurement on this peculiar scale has not appeared yet. Thus, the Papilioninae group offers various iridescent gadgets to decorate their wing color as a patchwork.

3.3.2 *Pierinae and Coliadinae (Pieridae)*

The family Pieridae involves about 1000 species in the world, within which ~50 species have been investigated and discussed in relation to the structural aspects of their scales. The wing color of this group is usually orange, yellow, white, and black, and is no longer iridescent to our eyes. However, the male is sometimes structurally colored in ultraviolet (UV) region. Further, peculiar granules are usually observed within the lumen of the scale, which have been attracting the scientist's interest.

Although UV reflective character of the male butterfly in this group was known before (Nekrutenko, 1965), the exact microscopic structure was clarified using electron microscope in 1972. Ghiradella *et al.* investigated the UV reflection characteristics of "little yellow," *Eurema lisa*, as a sample (Ghiradella *et al.*, 1972). This butterfly is distributed in the southeast part of North America with the wing span of 3.2–4.4 cm. They first observed the butterfly under the ultraviolet illumination and found either side of the wing was lights-up, from which they deduced that the layer structure was inclined by 20° to the plane of the wing. Under scanning electron microscope, striking microstructures revealed out, which resembled the ridge-lamella in the *Morpho* butterflies (Fig. 3.39). The ridge was a Christmas-tree type and had seven to eight lamella symmetrically sticking from the tree, which slanted by 10° from the base plane. At the central part of the tree, the sinuses (holes) were periodically arranged, which lined out of phase with the sticking lamellar structure. The thicknesses of lamellae and air were found to be 55 and 82.6 nm, which assured the "ideal multilayer" in this system, by assuming the refractive indices of 1.60 and 1.00. Within a lumen of the scale, the crossribs were decorated with a lot of ellipsoidal beads.

Later, Allyn and Downey investigated the microscopic and optical properties of several pierid species, and found that most of male pierids showed UV reflection, but the male *Aphrissa boisduvali*, *Phoebis sennae*, and *Colias philodice* lacked the UV reflecting scales (Allyn and Downey, 1977). They found the UV reflecting scales had over twice as many ridges per unit width and the number of lamellae depended on the species. They also confirmed that the ridge sinuses were in an apposite position to the ridge shelves. The reflectance showed a peak near 360 nm with a steep increase toward the longer wavelength, depending on the species. Thus, UV reflective scale of this group essentially belongs to type A, but the optical importance of the ridge sinuses is not clarified yet. It is speculated that symmetrically sticking shelves and a ridge sinus in anti-phase will work as destructive

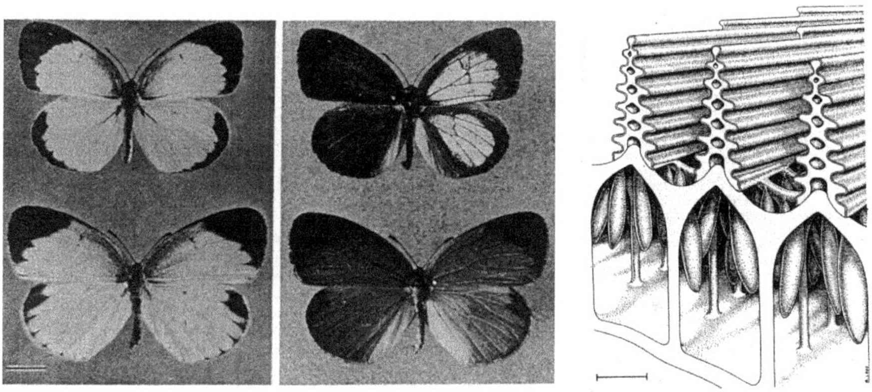

(a) (b)

Fig. 3.39 Left: photographs obtained under visible (a) and ultraviolet (b) illuminations for male (upper) and female (lower) specimens of *Eurema lisa*. Right: schematic view of a cross-section of the ultraviolet iridescent scale of *Eurema lisa*. We can see well-developed ridge-lamellae with anti-phase sinuses and also ellipsoidal granules below the ridges. Scale bar is 0.5 μm. (From Ghiradella *et al.* (1972). Reprinted with permission from AAAS.)

interference for the reflection to the normal direction when it is illuminated normally, thus such a structure will be useful to extend the direction of the reflection in the plane perpendicular to the ridge direction. This is the same principle as the alternately sticking shelves found in *Morpho* butterflies, where either shelf having anti-phase position to the other functions to cancel the reflection to the normal direction (see Sec. 9.2.2).

Ghiradella performed extensive studies on the development of UV-reflecting scale using "alfalfa butterfly," *Colias eurytheme*, which is distributed in Northern America as shown in Fig. 3.40 (Ghiradella, 1978, 1991, 1994, 1998). The ridges of this species are similar to the other, with the center-to-center ridge distance of 0.9 μm, and the thickness of air and lamella of 84.8 and 51 nm, respectively. The pupal stage lasts about 7 days and the scale formation continues throughout this period. Ridge formation was found to occur before the third day, when bundles of microfilaments were arranged under upper and lower surfaces of a flattened scale. The bundles in the lower part were larger than those in the upper part. The ridge formation occurred between the upper bundles. Cuticulin deposition was going on at the time of early ridge formation along the entire surface of the scale. The ridges continued to heighten and became narrower. The first lamellae appeared as buckling of the cuticulin at the top of the ridge

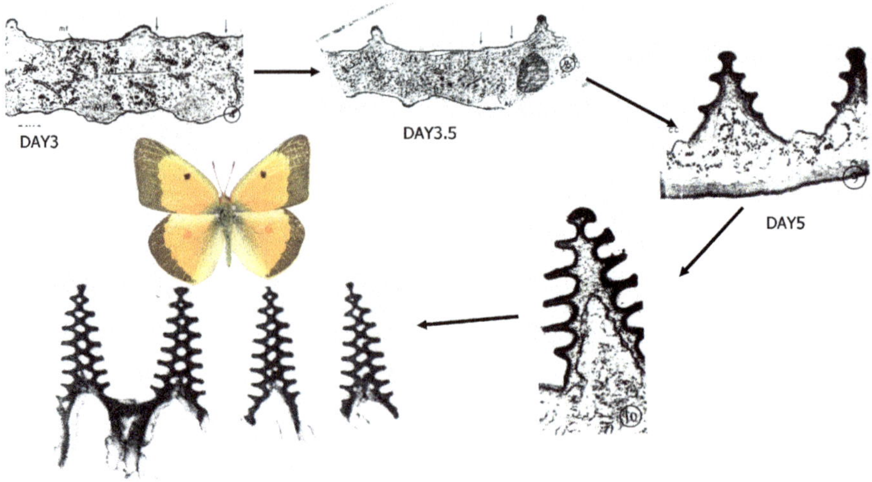

Fig. 3.40 Developmental study of ultraviolet-reflecting scale of *Colias eurytheme*. The cross-sections of a developing scale at various stages after pupation are shown. (From Ghiradella (1978). Reprinted with permission of Wiley-Liss, Inc., a subsidiary of John Wiley & Sons, Inc. Specimen of the butterfly is in the possession of The Nature and Human Activities, Hyogo, Japan and the photograph was taken by Dr. M. Kambe.)

and the others appeared more basally. Then, the protein epicuticle began to fill in the lamellae quickly. The crossribs formed later than the ridges. The plasma membrane was close to the ridge at first, then retreated during the later stage except for close-contact region, which becomes a window in the later stage. Finally, the trabeculae and ellipsoidal beads (pigment granules) were laid down. She proposed a model of elastic buckling, sinusoidal modulation, for the formation of the ridge-lamella system. According to this model, the lamellae could be formed as a result of either compression or stretch in a longitudinal direction of the scale due to the contractile element of microfilaments. Thus, the elastic stress due to the function of the microfilaments caused the buckling on the surface of the nascent ridge. However, she also mentioned that once the lamellae begun to fold, an additional mechanism, perhaps an interaction between adjacent parts of a fold, was necessary to cause the observed final form of the ridge-lamella. Unfortunately, the researches following her observation have not been reported to date.

Next, we will pay attention to ellipsoidal beads observed in the lumen of the pierid scale. As for this, it is still controversial. The crossribs, and partly trabecular braces, of this species are densely decorated with the ellipsoidal beads, sometimes called pigment granules or pterinosomes (Allyn and Downey, 1977). Stavenga *et al.* performed an electron microscopic observation and optical measurement on "small white," *Pieris rapae rapae* and *Pieris rapae crucivora* (Stavenga *et al.*, 2004). The beads had a dimension of 100–500 nm. To be surprised, the sectioned beads appeared hollow with 5–8 nm thick membrane. They measured the reflection spectra for the white and black parts of the wing, and observed the time course of the reflection spectra during the evaporation process after giving a drop of liquid xylene to the wing. The reflection spectrum of the male *P. r. crucivora* remarkably changed in the visible region above 420 nm, while that of the female, which was UV reflective, did not change much. The absorption below 420 nm was probably due to the presence of pterin pigment. They deduced that its marked whiteness came solely from the structural origin due to the presence of the beads and not from the pigment, since only the male carried the elliptical beads. Thus, it was concluded that the beads made the wings brighter at wavelengths where the pigmentary absorption was negligible.

On the other hand, Rutowski *et al.* insisted through the experiment on the change of the reflectance due to the pterin extraction that the combination of the pigmentary and structural mechanisms should be considered, since the beads carried the pigment (Rutowski *et al.*, 2005). They compared the increase of UV reflectance for two incident angles in which UV reflection was displayed or not displayed, and found that it was generally larger than that after extraction of pterin by wetting of 70% isopropyl alcohol and 1% NH_4OH solution. Thus, they insisted that the simple addition of the reflectance was no longer applicable to this system. However, both papers involve the ambiguities in their analyses. The shape of the bead and its arrangement should be more properly considered, because they will offer special optical effect in combination with the pigmentary distribution. Thus, the further studies are strongly expected.

3.3.3 *Lycaenidae*

The family Lycaenidae is the largest and the most diverse group in all butterfly families, and comprises more than 6000 species. The Lycaenidae is currently divided into five subfmilies: Riodininae (metalmarks), Poritiinae, Miletinae (harvesters), Lycaeninae (hairstreaks, blues, cuppers,...), and Curetinae (sunbeams). However, works on structural colors have been confined only within the subfamilies of Lycaeninae and Riodininae, and about 50 papers have been devoted to ~120 species.

Tilley and Eliot performed the surveying work through scanning electron microscopy on the scale structures of 54 species in the Lycaenidae including 7 species of the Riodininae (Tilley and Eliot, 2002). Here, we will follow their work to classify the iridescent scales from the structural aspects. They divided the structurally colored scales into four types according to their structures. One is *Morpho*-type scales, as they described, and some species in the Lycaeninae (tribe Aphnaeini, and the subtribes Deudorigina and Hypolycaenina in the tribe Eumaeini) and in the Riodininae show this type. This type definitely belongs to type A according to the Ghiradella's classification (Ghiradella, 1998), and is characterized by the well-developing ridges with periodic modulation standing on the hollow interior held by trabeculae. The ridges are standing closely enough, between which the deep dips are present. The characteristics of this type in the Lycaeninae are peculiar, because the structures mostly contributing to structural colors come from microribs, which are normally arranged vertically to the scale plane, and not from shelf (lamellae) structure as is usually the case for the *Morpho* butterflies. The microribs incline by less than 40° from the vertical direction so that the iridescent color is observed when the wing is tilted. In some cases, the microribs run at 20° (*Rapala rhoecus*, Lycaeninae) to the scale plane, and some at 30–45° (*Hypolycaena erylus*, Lycaeninae), depending on the species. In *Artipe eryx* (Lycaeninae), the microribs smoothly run up the ridge, forming seven to eight layers along the side of ridge with keeping in parallel to the base plane. On the other hand, the type A in the Riodininae is the same as the *Morpho* butterflies and the shelf structure is clearly observable.

The second type is *Urania*-type, which belongs to types F and G in Ghiradella's classification with the subclassifications F_1 and F_2. The name of *Urania* comes after the Mason's observation that the scale had a layered structure in the lower side of the scale as in *Urania riphenus* (now called *Chrysiridia ripheus*, Uraniidae, Geometroidea). This type is the major cause of the structural colors in these groups, and is widely distributed throughout the whole subfamilies. *Urania*-type bears either multilayer or photonic-crystal structures in the interior of the scale. The multilayer structure is more or less modified by pepper-pot perforation or fibrous network, which anyway belongs to F_1 and F_2 types (see Fig. 3.41).

In a scale of *Chliaria othona* (Lycaeninae), only a single layer with pepper-pot perforation is present. However, this butterfly develops the microrib on the ridge to form type A structure. Furthermore, apical scales

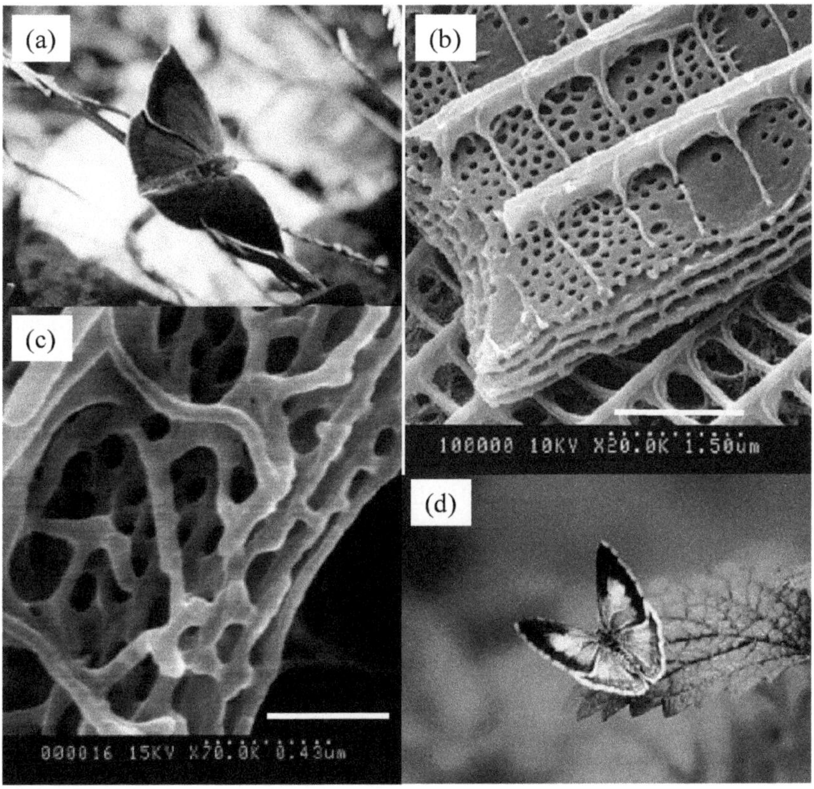

Fig. 3.41 Scanning electron microscopic images of iridescent scales of licaenid butterflies. (a,b) *Arhopaia japonica japonica* (Type F_1), and (c,d) *Celastrina argiolus ladonides* (Type F_2) (Kinoshita and Yoshioka, 2005b).

do not possess even a pepper-pot structure and only *Morpho*-type structure develops. Thus, this species locate at an intermediate position between the *Morpho*- and *Urania*-type. In the blue iridescent scales of *Anthene lycaenina* (Lycaeninae), the upper surface of the interior is almost completely closed by a membrane to make a "bag," which shows holes here and there, into which we can see pepper-pot structure inside. The scales of *Eumaeus minijas* (Lycaeninae) show the "satin" type surface between ridges, say type D. The scales of *Callophrys rubi* and *C. avis* (Lycaeninae) have no layered structure but possess crystallites, thus they belong to type G, which will be explained later.

The third type is the piliform and plume scales, which is observed in *Tomares ballus* (Lycaeninae) (Ghiradella, 2005). The scales of this species

are of a piliform with the pepper-pot structures inside. The fourth type is the undifferentiated colored scales. They have no significant internal structure, yet show the iridescence, and thus the further investigations are
necessary.

From the structural and phylogenetic considerations, Tilley and Eliot speculated that the old type of iridescent scales were of a *Urania*-type and the transmutation to *Morpho*-type took place in the past (Tilley and Eliot, 2002). During the evolutional process, the loss of the lamellae structure and the recreation of the trabeculae should be attained. *Chliaria othona* makes it possible to imagine how the transformation actually progressed during the evolutional step.

In the following, we will concentrate on a few species to investigate their microscopic structures in detail. First, we select *Callophrys rubi* (Lycaeninae, see Fig. 3.42 and Plate 9 in Appendix B) as an example (Onslow, 1920b, 1923; Schmidt, 1943; Morris, 1975; Ghiradella and Radican, 1976; Ghiradella, 1984; Srinivasarao, 1999; Tilley and Eliot, 2002; Vukusic, 2005; Kinoshita and Yoshioka, 2005b). This small butterfly distributed widespread in Europe to Asia and North Africa was first received attention by Onslow (1920b, 1923). He noticed that the scale of this butterfly was covered with irregular polygonal dark patches when observed under transmitted illumination (see Fig. 3.42). Each polygonal area was rimmed by bright lines. Under the reflected light, the scale showed green spangles, which corresponded to the dark areas in transmission. He speculated that modulated thickness of pigmentary scale caused this phenomenon. He also performed the developmental study and no particular difference was observed between immature and fully developed scales.

Schmidt called this type of scale with irregular patches "mosaic scale," and investigated the similar scale of *Teinopalpus imperialis* (Papilioninae, see Plate 9 in Appendix B) under an optical microscope (Schmidt, 1943). He speculated the cause of structural color in the scale as the lattice consisting of obliquely running thin lamellae. He compared it with the scale in *C. rubi* and deduced that the similar architecture was an origin of the green coloring. The microstructure of this mosaic scale was later investigated through an electron microscope by Morris (1975). He observed that the polygonal grain, seen as mosaic patch, was $5.4\,\mu$m in mean diameter and each consisted of a simple cubic network with hollow spheres at lattice points and the lattice constant of $0.257\,\mu$m. Ghiradella and Radigan performed an electron microscopic observation of the scale and found that the

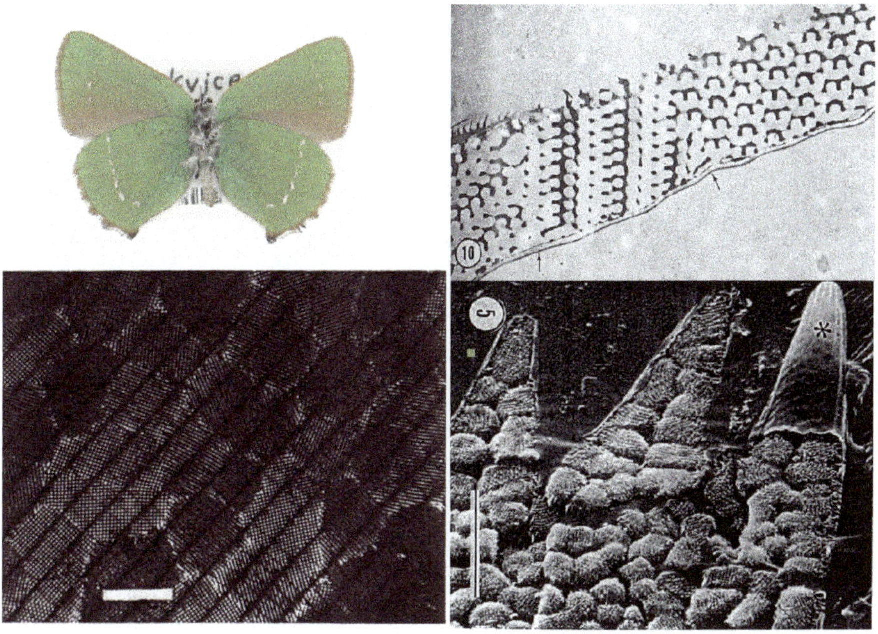

Fig. 3.42 Lower left: TEM of green-iridescent scale from *Callophrys rubi* showing many domains of regular structure (scale bar 5 μm). (Reproduced from Morris (1975) with permission of the Royal Entomological Society.) Upper right: Longitudinal section of the iridescent cell of *C. rubi* (upper left), showing photonic crystal with three domains. (From Ghiradella and Radican (1976).) Lower right: SEM view of the interior of a tip of cover scale on the *Mitoura grynea* wing, showing many small domains of lattice structures (scale bar 10 μm). (From Ghiradella (1989). Reprinted with permission of Wiley-Liss, Inc., a subsidiary of John Wiley & Sons, Inc.) (Specimen of the butterfly is in the possession of The Nature and Human Activities, Hyogo, Japan and the photograph was taken by Dr. M. Kambe.) See Plate 9 in Appendix B.

hollow spheres were arranged in a close-packed cubic (fcc) (see Fig. 3.42) (Ghiradella and Radican, 1976). They considered that the lattice structure had its origin in a trabecular network.

Thus, the elaborated structure in the interior of the scale in *C. rubi* is now well known as a typical example of biological photonic crystal of an air-hole type. Similar structures are found in *Callophrys siva* (see Plate 9 in Appendix B) (Allyn and Downey, 1976; Stavenga *et al.*, 2004), *C. avis* (Tilley and Eliot, 2002), *Mitoura grynea* (Lycaeninae) (Ghiradella, 1991, 1994, 1989, 1998; Kinoshita and Yoshioka, 2005b; Prum *et al.*, 2006) and *Thecla damo* (*Pseudolycaena damo*, Lycaeninae, see Plate 9 in Appendix B) (Ghiradella, 1991). Although the optical measurement of this scale has not been performed yet, it seems quite attractive to consider how such air-hole type crystal displays the peculiar optical properties and also how the domain structure contributes to the diffusion of light.

Ghiradella reported the developmental studies on two lycaenid species with different elaborations in the lower part of the scale (Ghiradella, 1989). One was *Mitoura grynea*, which showed a crystalline structure (Fig. 3.43), while the other was *Celastrina ladon* (Lycaeninae) showing the lamellar structure consisting of fibrous network. She performed the electron microscopic observation during pupal periods of these butterflies. At first, the scales of these two species gave the similar architectures often seen in ordinary scales, but after the eighth day in *M. grynea* and the seventh day in *C. ladon*, the internal specialization began. In *M. grynea*, the combination of the membrane with cuticle formed a unit, enclosed network structure due to smooth endoplasmic reticulum as a nucleus and formed quite regular arrangement of a close-packed lattice structure (see Fig. 3.43). The enclosing membrane–cuticle units were connected with each other and also with the outer space, and thus essentially had an extracellular character. The formation process of the lattices continued for 2 days and the structure became firmly fixed at that position as the cell died back and disappeared.

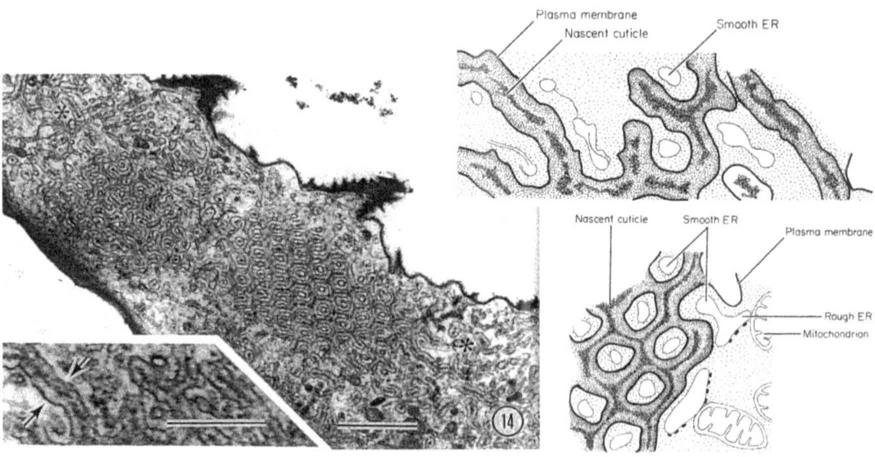

Fig. 3.43 Transverse section (left) of a developing scale 9 days after pupation of *Mitoura grynea*, showing two crystallites in a cell (scale bar $1\,\mu$m). Each crystallite consists of membrane-cuticle units (lower right), which at first appear as nascent cuticle enclosing smooth endoplasmic reticulum as a nucleus (upper right: schematic illustration of the inset of the left photograph (scale bar $0.5\,\mu$m)) (Ghiradella, 1989). (Reprinted with permission of Wiley-Liss, Inc., a subsidiary of John Wiley & Sons, Inc.)

In case of *C. ladon*, the stack of endoplasmic reticulum began from the lower end of the scale cell. In addition to this, electronically dense and pale vesicles appeared. The fibrils and tubes appeared to be lined up into stack, and were closely associated with the plasma membrane.

This experiment shows an important implication how such elaborated structures are formed in the developmental processes. As she described (Ghiradella, 1989), at the present stage, these processes seem to adopt different ways of organization, yet these species are phylogenetically very near and the intermediate species having lamellar structure with regular holes actually exists. Nevertheless, similar experiments have not appeared to date. In addition to the biological pursuit, the adoption of nonbiological model system is quite important to derive the physical essence of this phenomenon. Because such self-assembled system and order formation known so far are extremely limited such as crystal growth, convection, chemical oscillation, and so on, whose spatial scales are extraordinarily larger as compared with the intracellular phenomena.

3.3.4 *Nymphalidae*

The family Nymphalidae is the second largest group in butterflies and about 4500 species are now belonging to it, though the classification have been changing frequently. This family is now divided into 10 subfamilies, among which the butterflies belonging to the Morphinae are most extensively studied: actually 70% of the past investigations are turned to this small groups. The investigation on this group, however, have been fully described above, here we will concentrate on the other species.

The most noteworthy researches on this group are those concerning the pupae of the Danainae, Heliconiinae, and Nymphalinae. The pupae of these groups show silvery or golden appearance in a whole, or distribute silvery/golden spots.

First, I will show the experiment performed in my laboratory. Fujii (one of my students) investigated golden pupae of *Idea leuconoe* (Danainae) (see Fig. 3.44), which was kindly supplied from Itami City Museum of Insects. At first, the color change during the pupal period is shown. Just after the pupation, the color of the pupa was rather yellow. After 20 h, the color changed from shiningly greenish, reddish and finally golden 48 h later. Two days before the emergence of the adult, the wing pattern became clearly visible, and 1 day before, the golden color disappeared and greenish color appeared.

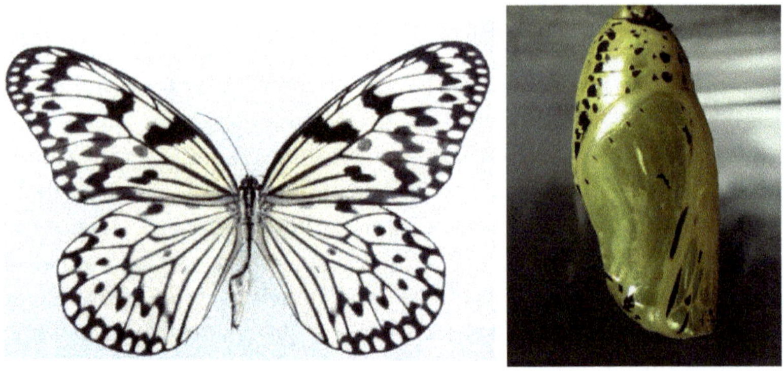

Fig. 3.44 *Idea leuconoe* (Danainae) and its golden pupa (2 days after pupation) (S. Kinoshita *et al.*, *J. Jpn. Soc. Colour Mater.* **75**, 493 (2003)). See Plate 11 in Appendix B.

When we see an excised cuticle piece of the pupa, it shows a golden appearance at first, and then loses the brilliancy and becomes yellow, while the addition of water restores the golden brilliancy. He measured the reflection spectrum at the surface of the living pupa every 1 h during the pupal period from 1 to 60 h after the pupation. The results are summarized in Fig. 3.45. The reflection spectra are roughly divided into three periods: (1) 1–18 h, (2) 19–48 h and (3) 49–60 h. In the first period, the reflectivity is rather low with 3–15% with a clear rise at ~510 nm, which makes the pupa yellowish. In the second period, the reflectivity changes drastically and rather stepwise from shorter to longer wavelength, that is, at first around 440 nm and then ~530 nm, ~600 nm, ~700 nm in succession. This successive increases of the reflectivity make the reflection at longer wavelengths flat and high with the reflectivity of 35–45%. The peak positions and the total reflection curve seem to depend on the individuals, but the tendency is essentially the same. It seems that the suppression of the reflectivity is clearly observed below 510 nm, which make the total appearance of this pupa golden. We notice that the dip in the reflectivity at 453 and 485 nm, which just corresponds to the absorption spectrum of carotenoid such as β-carotene. This implies that the suppression of the reflectivity at shorter wavelengths comes from the light absorption by carotenoids (Rothschild *et al.*, 1978). In the last period of 48 h and later, the reflection spectrum hardly changes and thus the arrangement of the inner systems is completed.

TEM image at 24 h (beginning of the reflectivity increase) and 48 h (completion of the reflectivity increase) after the pupation (Fig. 3.45) show

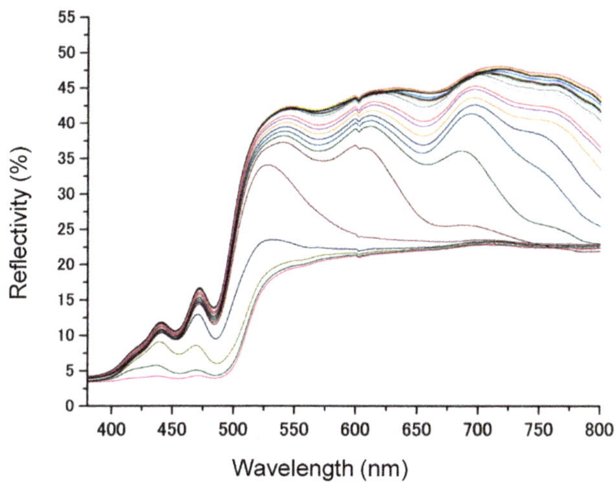

Fig. 3.45 TEM image of the cross-section of a pupa surface of *Idea leuconoe* 48 h after the pupation. We can see 230 layers within the thickness of 40 μm. Reflection spectra from the pupa measured every hour between 19 and 48 h after pupation (from bottom to top). The reflectivity is increased rather stepwise and finally covers the visible region above 500 nm. Notice two small dips appearing at 453 and 485 nm are originated from the absorption due to carotenoids (S. Kinoshita *et al.*, *J. Jpn. Soc. Colour Mater.* **75**, 493 (2002)).

the layered structures at the surface, but the thickness and form are very different. In the former, the thickness is 20 μm and ~150 layers seem to be tightly packed, while in the latter, thickness grows drastically to 40 μm with the number of the layers of 230. Thus about 80 layers are increased with the thickness of 20 μm. Newly growing layers seem to be rather sparse, and dense and lucent regions are clearly observable. It is also noticed that the interval of the layers increases toward the inside. The depth dependence of the layer interval shows that the thickness distribution is rather constant with the average of 125 nm within 150 layers, while newly increased layers have a clear increase from 150 to 300 nm. Thus, the multilayer interference with increasing interval is the cause of this phenomenon with the suppression below 510 nm. However, in the present image, we cannot find the stepwise change in the layer thickness, probably owing to the low resolution of this observation and also to the characteristic shrinkage of the sample.

Thus, in order to explain the reflectivity of the pupal surface and also of the temporal evolution quantitatively, we must take into account the contributions of (1) the surface rigid layers that give the initial reflection spectrum, (2) the multilayer interference due to variable interval thickness, (3) the absorption due to carotenoid, and possibly (4) backscattering due to the muddy body flow. Finally, we compare the reflectivity of pupa with metallic gold. The agreement on the overall spectral profile is pretty good, though its reflectivity is considerably low.

Golden and silvery surface of the butterfly pupa have attracted the physicist's and biologist's attention since more than 100 years ago. Lord Rayleigh described, referring to Poulton's paper (1887), that nonmetallic substances revealed its metal-like appearance in the pupa of *Vanessa urticae* (*Aglais urticae*, Nymphalinae), which disappeared when drying, while it restored when wetting again (Rayleigh, 1919). The liquid with higher refractive indices such as benzene involving a small amount of carbon disulfide made the surface very dark, while alcohol and water recover the golden appearance. He considered that light interference within the alternate layers of chitin and liquid was the major cause of this appearance. He thought that by drying, the liquid films became thinner and finally solid films contacted with each other, which destroyed the interference.

The detailed observation and exact conclusions were experimentally attained later using an electron microscope. Neville investigated various insect species ranging from the pupae of butterflies to beetles' elytra, the latter of which will be described in the following chapter, and showed qualitatively that alternate layers of substances with high and low refractive indices generated the silvery and golden appearance (Neville, 1977). He took up three butterfly pupae of *Heliconius erato* (Heliconiinae), *Aglais urticae* (Nymphalinae), and *Araschnia levana* (Danaiinae). He found that the metallic appearance came from the broadband reflection due to the multilayer whose intervals were distributed. The multilayers in these pupae were located in an endocuticle region and the golden appearance was somehow associated with the pigment absorbing the shorter wavelengths. He also commented that the pigment in *H. erato* resulted from the tanning polymerization, while *Danaus plexippus* (Danaiinae) was due to a nontanning pigment, which showed metallic golden spots even in cast pupal skins.

Later, Steinbrecht conducted a detailed investigations on the pupae of *Euploea core*, *E. midamus*, *Amauris ochlea*, *A. niavius*, and *Danaus chrysippus* (Danainae), and *Heliconius charitonius* (Heliconiinae) (Steinbrecht, 1985; Steinbrecht *et al.*, 1985; Steinbreht, 2005). He stated that shining golden luster of *Euploea* and silvery appearance of *Amarius* pupae, which covered almost the entire body surface, came from multilayered structure in an endocuticle region. From an electron microscopic observation, he found that the layer consisted of a characteristic pattern of alternating dense and lucent layers. The thickness of dense layer was rather constant, while that of lucent layer changed systematically, first increasing and then decreasing again from the outside to inside. He also found that the thickness of the endocuticle in nonreflecting region and the summated thickness of the dense layer in reflecting region almost coincided with each other, implying that lucent layers were a mere addition to the normal endocuticle.

Detailed observation on the multilayer revealed that 10–20 nm diameter dense filaments were distributed in lucent layer, which connected the adjacent dense layers with an angle of 45°. There were pore canals within a dense layer, which connected the lucent layers continuously. The media involving in lucent layers was fluid-like, with the refractive index close to that of water, which are presumably made of aqueous solution or highly plastic gels.

Twenty to twenty-four hours after pupation of *A. ochlea*, the first bluish reflection came out, when the outer endocuticle with helicoidal pattern appeared and the cuticle began to be deposited. The lucent layers generated initially were markedly thin. In 24–28 h, the reflection color changed via greenish to golden. At this stage, 70 double layers were observed with increasing thickness of lucent layer. The deposition continued till 48 h and the final number of a pair of lucent and dense layer became 270, with a total thickness of 57 μm. Apolysis (the separation of the cuticle from the epidermis) occurred at 2–4 days after the pupation, and at 1 day before the emergence of the adult, the multilayer and outer endocuticle were completely decomposed.

During the pupal period, he made the reflection measurement and found the spectral change with time (Steinbrecht *et al.*, 1985). He employed the refractive index of the dense layer as 1.58 from the observation of the Becke line for the golden cuticle with that of the lucent layer assumed as 1.37, and reproduced the change in the reflection spectra fairly well. He discussed that too many numbers of layers might be due to the absence of the stop command for creating the layers. He also insisted that the golden appearance was not due to the pigmentation, but was purely of a physical origin, deduced from the comparison of the transmission and reflection spectra.

The deposition of cuticle layer began with that of the innermost three to four lamellae (deposition zone), which showed helicoidal pattern that was common in nonreflecting cuticle (Steinbrecht, 1985, 2005). The cuticle deposition took place in a usual way, in which tighter packing of the microfibres occurred followed by water expulsion and strong cross-linking. He proposed a hypothesis that the cross-linking was somehow interrupted at regular intervals, which eventually determined the thickness of the dense layer, and between the layers, expelled water was retained to form the clear layer and the microfibres glued alternately to the adjacent dense layer, which corresponded to the connecting filaments and determined the thickness of the lucent layer. The mechanism proposed by Steinbrecht seems quite attractive and will give a basic scenario to create regular multilayer structure in the biological system. Further experimental confirmation is strongly expected.

Finally, let us briefly survey the studies on structural colors in the family Nymphalidae. Ghiradella offered an interesting topics from the structural aspect of the butterfly scale (Ghiradella, 1984, 1994, 1998, 1999). Owl butterflies (Brassolini, Morphinae) show enormous eyespots on the ventral side of the hindwing and show weakly blue brilliancy on the dorsal side. The scale structure of this group shows well-developed microribs. In *Eryphanis aesacus*, the lamella/microrib system is "rocked-back" so that the microribs become reflective element as shown in her classification of type C. In *Caligo beltrao*, *C. memnon*, *C. uranus* (see Plate 9 in Appendix B), and *C. illioneus*, ridge-lamellae and microribs tilt in the opposite directions to form the double reflective elements. Furthermore, in *C. memnon* and *C. illioneus*, satiny scales (type D) are also seen. Although this group shows such a variety of scales, the optical measurements are quite a few, and further the quantitative explanation to such effects has not appeared yet.

The other topic is concerned with the corneal ripple on the compound eye of *Vanessa* sp. (*Aglais* sp.) (Bernhard *et al.*, 1965; Parker, 2002, 2005b), which will be described together with moth-eye structure in the next subsection (Sec. 3.3.7). The other works, to which I cannot refer, are concerned with the polarized iridescent pattern in *Heliconius cydno* and *H. melpomene* (Heliconiinae) (Sweeney *et al.*, 2003), the measurement of reflectivity for *Heliconius melpomene* (Stavenga *et al.*, 2004), *Cynandra opis* (Limeritidinae) (Brink and Lee, 1999), and several species including *Callicore aegina* and *Caligo martia* (Morphinae) (Wickham *et al.*, 2006).

3.3.5 *Hesperiidae (Hesperioidea)*

This group of butterflies is featured by their characteristic forms, flight styles, and antennas, and usually called skippers. Only a very small number of species are known to show the structural color, and hence the papers devoted to this group are very few. Ghiradella reported that iridescent bristles of *Astraptes azul* (Pyrginae) showed the multilayered structure (Ghiradella, 1994), while those of *Urbanus proteus* (Pyrginae) showed a lattice structure as in *Callophrys rubi*, which seemed not so regular as the latter. In the scales of *Phocides lilea* (Pyrginae), the internal structures were entirely missing, which shows a fused appearance of the upper and lower layers of scales (type H), and often appears as a satiny sheen. In *Elbella polyzona* (Pyrrhopyginae), the inner layer are completely missing, allowing the fragile envelopes to mat together, which produces a frosty white appearance.

3.3.6 *Moths*

Among 130,000 moth species, probably the most notable and striking moths are those belonging to the genera *Alcides*, *Chrysiridia*, and *Urania*, which are categorized into the subfamily Uraniinae (Uraniidae, Geometroidea). These species are well known for their brilliantly colored wings (see Plate 11 in Appendix B), diurnal character, and their migratory flight. *Alcides* inhabits in Australia, while *Chrysiridia* in Madagascar and East Africa, and *Urania* in neotropical. Although most of the ordinary moths have their wings broadly outspread at rest, *Chrysiridia* shows the wing held vertically over the back at night.

The moths in this group attracted great attention of scientists in 1920s, who performed the surveying work on the structurally colored animals. Onslow investigated the scales of *Urania fulgens* (Urania swallowtail) and found a scale consisting of an upper membrane and a lower pigmented membrane, with thin striae (ridges) separated considerably (Onslow, 1923). Süffert conducted the detailed work on *Chrysiridia croesus* (East African sunset moth) (Süffert, 1924). He placed this moth as a typical example of the structure-based coloring, and called *Urania*-type. He performed various experiments on its scale such as microscopic observation, optical properties such as transmission and reflection, application of local pressure, penetration of various liquids, polarization dependence, and so on. He found, from the theoretical considerations, the iridescent nature and polarization characteristics of the scale were deeply connected with the strongly curved multilayer system, located at the lower part of the scale, which consisted of alternate layers of chitin and air. He concluded that the multilayer interference was the major cause of the iridescence. Mason also performed similar experiments on the scale of *C. ripheus* (Madagascan sunset moth) and characterized the scale as the multiple thin films with the superposition of the vaned or meshed structure (ridge and crossrib) placed on the outer surface of the upper lamina (Mason, 1927a).

These authors tried various basic methods to investigate the cause of the structural color in this moth, e.g. (1) color change with increasing angle of incidence such that green changed into blue to purple, while blue changed into red to orange, (2) birefringence in the transmission with the optic axes along the ridge and perpendicular to it, (3) application of pressure changed the color as in (1), (4) penetration of liquid with the refractive index of 1.5–1.6 destroyed the color. All these experimental results strongly suggested the presence of multilayered structure with the substance of refractive index 1.5–1.6 and air.

The essential structure of this moth's scale was later confirmed through electron microscopic investigation by Lippert and Gentil, who investigated the two species of Uraniinae with different genera, *C. ripheus* and *U. leilus* (Lippert and Gentil, 1959). They reported that the scale of *C. ripheus* had a size of 60 μm in width and 300 μm in length, and that strongly dotted lamellar structure lay in the lower part of the scale. These dotted structures were distributed over a whole area of the lamellae and related with small supports between the layers to make the layer distance constant (later Ghiradella described with respect to the scale of *U. fulgens* that the spacers were distinctly bead-like with no hint of rod structure (Ghiradella, 1985)). The lamellae consisted of maximum seven dense layers and decreased as approaching the edge of the scale. The thickness of the dense layer lay between 0.11 and 0.15 μm, which corresponded to the maximum reflection at 550 nm. He also investigated the polarization effect for the transmitted light through a scale and concluded that the strongly curved scale with the multilayer generated the apparent birefringence (Gentil, 1941).

Thus, the scale structures of the genera of *Urania* and *Chrysiridia* have been assigned to the multilayer interference and have been treated as a typical case among butterfly/moth scales. Recently, Vukusic took up this scale and categorized into his classification of Type IIa, within which the scales comprising multilayering (body-lamellae) were classified. He added his opinion from the observation of the multilayer system with sparsely distributed thin ridges, that the ridges formed impedance matching against outer space to reduce the broadband reflection from the topmost surface, and also functioned as diffraction elements to diffusely spread the reflected color in angle (Vukusic, 2003, 2005). However, actually, the essential features of the iridescence of this moth does not simply lie in the multilayered structure within a scale, but lies rather in the interscale interaction, which plays a decisive role as is described in the following.

Yoshioka and the present author (Yoshioka and Kinoshita, 2006a, 2007) have investigated the wing of *C. ripheus* in detail by an optical microscope and have found the peculiar pattern, which differs remarkably from that of the *Morpho* butterfly. Under the epi-illumination, it is found that the neighboring scales constitute narrow reflective band running laterally, which form separate parallel bands over a wing (Fig. 3.46). It is also noticed that pale color with different hue covered the area between these bands. For example, in the red-purplish area of the ventral hind wing, the narrow bands are seen purplish, while broad yellow band covers the rest of the area. Since the sizes of the scale of 0.25 mm in length and 0.1 mm in width are too small to discriminate, the mixture of the colors is expected to our eye. Thus, this moth exhibits a kind of color mixing in the wing.

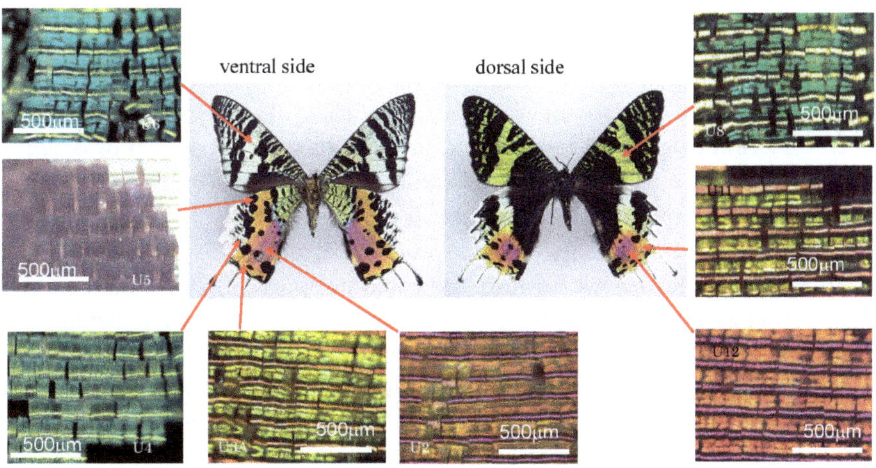

Fig. 3.46 Madagascan sunset moth, *Chrysiridia ripheus*, and the enlarged views of the scales at variously colored positions. Note narrow reflective bands are running in parallel with each other (Yoshioka and Kinoshita, 2007).

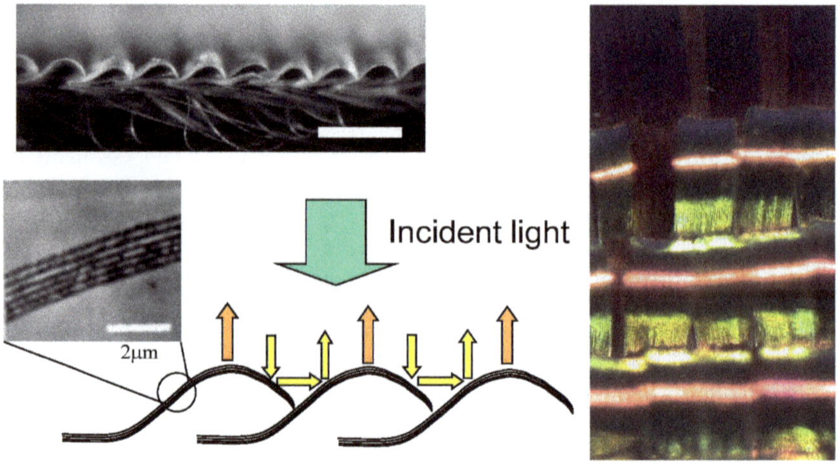

Fig. 3.47 Left: SEM view (upper) of the longitudinal section of strongly curved scales of *Chrysirida ripheus* and the schematic illustration (lower) showing the principle of interscale color mixing. TEM of the longitudinal cross-section is shown in the inset. Right: enlarged view of the scale under epi-illumination (Yoshioka and Kinoshita, 2007).

The optical and electron microscopic observations of the longitudinal cross-section show that the distal part of the scale is strongly curved to make a hemisphere (Fig. 3.47), while in the proximal part they are rather flat, where the hemisphere part of an adjacent scale overlaps. Owing to this structure, the two scales form a valley-like deep groove between them. Below these specialized scales (cover scales), flat ground scales are arranged. Under a scanning electron microscope, the surface of a cover scale is smooth with very thin ridges of 0.4 μm width running sparsely, separated by 3.5 μm. Below the surface, the multilayered structure is clearly seen under a transmission electron microscope. The multilayer is also curved along the surface of the scale, and the number of the layers are six at the top of the scale with decreasing number to two toward the proximal part. The thickness of the dense layer is 170 ± 20 nm, while the lucent layer is somewhat distributed from 100 to 150 nm. Thus, it is strongly suggested that the narrow purplish band observed above comes solely from a specular reflection at the top of the curved scale, while the broad yellow band from the retro-reflection at the side of the hemispheres (Fig. 3.47).

In order to confirm this hypothesis, the reflection spectrum is measured under the polarized light. This is because the reflectivity is strongly affected by the polarization under oblique incidence. The reflection spectrum consists of a small peak around 400 nm, which is essentially independent of polarization, and a broad peak around 700 nm, which is largely dependent on the polarization. Since the polarization-dependent reflection solely comes from the oblique incidence for retro-reflection, we calculate the difference of the reflectivity measured experimentally and obtain the reflectivity only for the retro-reflection, which shows a broad peak at 650 nm. Considering the condition of multilayer interference, we conclude that the former peak at 400 nm comes from the second-order ($m = 2$) peak of 394 nm under the normal incidence, while the latter broad peak at 650 nm is the sum of the first-order ($m = 1$) peak under normal incidence (787 nm) and the first-order ($m = 1$) peak under $\sim 45°$ incidence (653 nm). This analysis shows the color mixing is of an interscale phenomenon due to a specialized form of the overlapping curved scales, and also attained by using the sum of the first- and second-order multilayer reflection. Further this color mixing is strongly polarization-sensitive even under the illumination of unpolarized light, because it employs the difference of the reflectivity for naturally existing s- and p-polarized light. From more quantitative considerations, the irregularity of the planarity of the multilayer due to the presence of particles and bumps on the layer should be considered, which dulls the square-like spectral shape usually obtained for multilayer interference for large difference of the refractive indices. Furthermore, it is found that the irregularity of the interval of the multilayer makes the tail of the reflection band smooth.

Thus, the Uraniinae group exhibits surprisingly sophisticated structure that produces variously colored wing. On the other hand, most of the moths are believed to be rather quiet and plain in color, since they mostly show dark brown wings. Thus, the researches on moth wings are extremely limited although remarkable species inhabit even in Japan. I have searched brilliant species in my collection of Japanese moths and have found that the following species are noteworthy to be investigated. One is the family Zygaenidae (smoky moth) in which *Eterusia aedea*, *Erasmia pulchella*, and *Illiberis consimilis*, emitting glossy dark blue at the bases of the wing, body, and antenna. The others are rather distributed among various families: Glossy black wings are seen in *Adela nobilis*, *Nemophora albiantennella*, and *N. aurifera* of the Adelidae (fairy moths), and in *Glyphipterix nigromarginata* of the Glyphipterigidae (glyphipterid moths). Silverly wing and spots are seen in the species of Cosmopterigidae (cosmopterigid moths), *Pyralis regalis* of the Pyralidae (pyralid moths), *Tarsolepis japonica* and

Spatalia doerriesi of the Notodontidae (prominents), species in the Plusiinae (Noctuidae: owlet or noctuid moths), *Aletia pryeri* of the Hadeninae (Noctuidae), and *Rhodinia fugax* of the Saturniidae (giant silk moth and royal moth). Glossy green wings are apparent in *Lepidotarphius perornatella* of the Glyphipterigidae. Golden brilliance is seen in *Trichoplusia intermixta* of the Plusiinae (Noctuidae).

The species closely related to the last example in the Plusiinae have been frequently reported by Ghiradella as those showing satiny scales (Ghiradella, 1984, 1991, 1998), and have been analyzed in terms of diffraction grating with multiple components and structures by Brink *et al.* (1995) and Brink and Lee (1996). Silvery areas in the larvae of *Antheraea pernyi*, and gold and sliver spots in those of *A. militta* (Saturniidae) were also reported (Neville, 1977). The golden cocoon of Indonesian *Cricula triphenestarata* is expected as a wild silk in the local industry. Thus, the position of moth in the researches of structural colorations will be soon raised.

3.3.7 *Moth Eye*

Bernhard and Miller found that the corneal surfaces of the facets of some lepidopteran compound eyes were covered with congruent cone-shaped protuberances (see Figs. 3.48 and 3.49), which they called corneal nipples (Bernhard and Miller, 1962). The nipples are present only on the corneal chitin and cover its surface completely. They are around 200 nm in amplitude and are arranged closely in a hexagonal close-packing with the separation of the adjacent nipples of 200 nm. The shape of the nipple somewhat resembles a hanging bell. The size and shape are, however, dependent on the species.

They thought such a regular and minute structure would function as antireflection effect at the corneal surface (Bernhard and Miller, 1962; Bernhard *et al.*, 1965, 1968). This was because the interface between the corneal substance with high refractive index of typically 1.57 and that of the outer space of 1.0, would give the reflection loss at the surface and on the other hand, it would cause the glare seen from outside. Thus, the nipple structure anyway contributed to the impedance transformer by gradually changing the refractive indices at the surface.

To confirm this hypothesis, they performed the transmission and reflection measurements in a microwave region by enlarging the system size to 10^5, in which the nipple size of 1.0–6.0 cm with the wavelength of the microwave of 0.7–1.5 cm were employed (Bernhard *et al.*, 1965, 1968). They

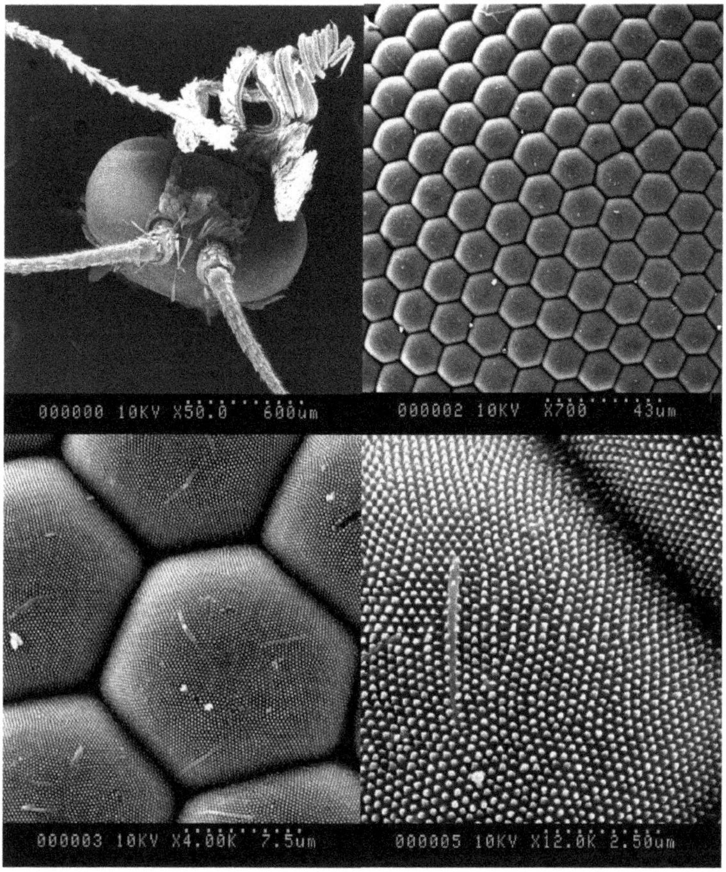

Fig. 3.48 SEM views of moth eye, *Agrotera nemoralis*, with enlarging magnification. (Courtesy of Dr. S. Yoshioka.)

modeled the nipples by choosing the center-to-center distance of d, the amplitude d, and the base diameter of the nipple of $0.6d$, each placed on the curved plane with the curvature of the radius $100d$. The materials of the model nipples were composed of 75% paraffin and 25% beeswax to accommodate a dielectric constant of 2.25, which corresponded to the refractive index of 1.5. They found a slight increase of the transmission and a reduction of the reflection in a wide wavelength range with this system. Furthermore, they made the optical spectroscopic measurement on the reflection properties by comparing a grasshopper's eye without nipples and that of a night moth with nipples. They found a reduction of the reflection in the

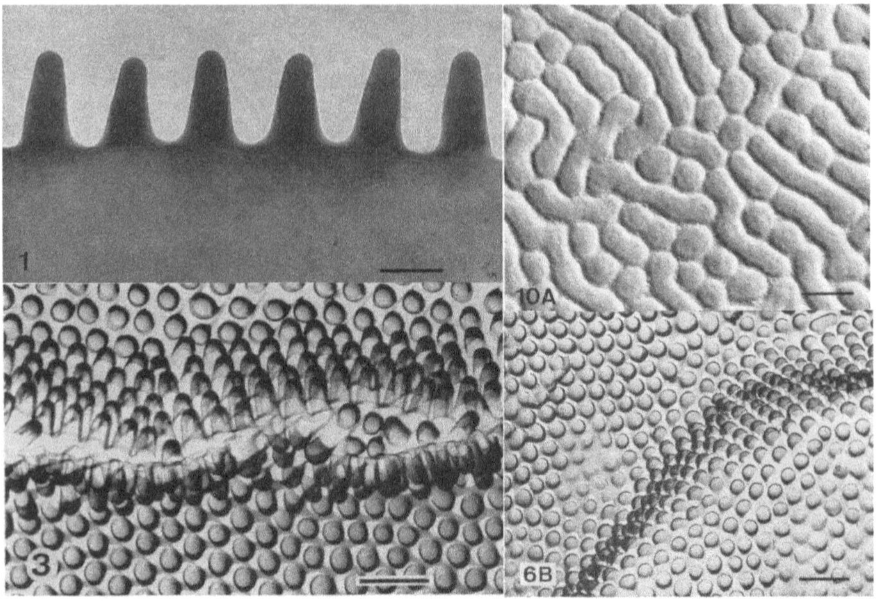

Fig. 3.49 Upper left: electron microscopic image of a row of well-developed corneal nipples of a moth, *Deilephila elphenor* (Sphingidae). Scale bar is 200 nm. Lower left: replica of facet surface of a night moth, *Cerapterix graminis* (Noctuidae). The replica is made by evaporating 5 nm thick platinum together with a supporting film of carbon. The replica is freed after dissolving the cornea with sodium hypochlorite or sulphuric acid (see Bernhard *et al.* (1970)). Upper right: replica of corneal surface in a fly *Lucilia caesar* (Diptera), showing mingled small-amplitude nipples. Lower right: replica of corneal surface of a night moth, *Acronycta* sp. (Noctuidae), showing intermediate-amplitude nipple arrays. Except for upper-left figure, scale bars are 500 nm. (With kind permission from Springer Science + Business Media: Bernhard *et al.* (1970).)

moth structure around the wavelength of 550 nm, which differed considerably from the calculation using a model of continuously changing refractive index. They considered that the shape of the nipple was important for the wavelength dependence. Thus, corneal nipple, which will be widely called moth-eye in technological fields, was considered to be the biological adaptation to reduce the reflection by changing the discontinuous change of the refractive index at the surface into the gradual change, which was already studied about 80 years ago by Lord Rayleigh (1879).

The distribution of corneal nipple was surveyed extensively by Bernhard *et al.* who investigated 361 species belonging to most of the insect orders (Bernhard *et al.*, 1970). They found that the sizes and shapes of nipples were variously distributed and classified them according to the amplitude of the nipples into group I–III. Group I included the species without nipples or with those having the amplitude lower than 50 nm, group II with those of 50–200 nm, and group III with those higher than 200 nm (see Fig. 3.49). Thus, the medium- and well-developed nipples anyway belonged to groups II and III. They found 56% of the investigated species assumed corneal ripples, which were, however, restricted within only four orders, Thysanura (silverfish), Diptera (flies), Trichoptera (caddisflies), and Lepidoptera (moths and butterflies). It was quite interesting that the distantly related silverfish to the other three assumed the corneal nipples, which stimulated the scientist's imagination in later years. For example, Bernhart *et al.* speculated that the distribution in silverfish implied the regression, rather than progression, from the well-developed nipples with the common ancestors among these four species (Bernhard *et al.*, 1970). The regression was incomplete for these orders, while the most of the other orders did completely.

The hypothesis that the nipples appeared as the phylogenetic product to accommodate nocturnal activity in moths to increase the transmission would be puzzled by the fact that only Papilionidae among various families of diurnal butterflies lack the nipples. In Lepidoptera, 95% of the species belonging to the families Pierididae, Lycaenidae, and Nympharidae show the nipples, while the nipples are present in only 30% for Hesperiidae and none for Papilionidae. In moth species, this ratio becomes 75%. The distribution in butterflies in three families was later investigated by Stavenga *et al.* on 19 diurnal butterflies, which showed similar tendency that only small arrays of about 30 nm amplitude were found in Papilionidae, while in Pieridae and Nymphalidae, they ranged from 90 to 230 nm with only one exception of 40 nm for *Charaxes fulvescens* (Stavenga *et al.*, 2006b).

The ontogenetic process of this ultrastructure was investigated in the same era. Gemne performed a detailed observation using optical and electron microscopes on pupae of *Manduca sexta* (tobacco hornworm moth) (Gemne, 1966, 1971). After pupation, the process of the corneagenous cell surface became slender and were separated from each other, which in 5 days after pupation, became microvilli of 400–500 nm in length with 70–100 nm width separated by 170 nm. The microvilli were arranged in a hexagonal way, which were finally separated by 200 nm, though their bases were less regular with sometimes bifurcated. At the tip of the microvilli, a patched lamellar element appeared to constitute a dome. After 5.5–6 days, the dome assumed the shape of somewhat higher cupole with the amplitude of 150 nm, and after 6–8 days the increase in height of the cupole was observed, which finally amounted to 200–250 nm. The corneal material was then deposited to fill the evaginations created by evaginating epicorneal membrane above the tip of the microvilli, and the consolidation occurred. Thus, the hexagonal array of well-developed corneal ripples were completed.

Similar moth-eye structures seem to be widely distributed among insect world. Yoshida *et al.* reported that similar structure was found in the wing of *Cephonodes hylas* (hawkmoth), which lost almost all the wing scales at the time of emergence (Yoshida, 2005; Yoshida *et al.*, 1996, 1997). The resultant transparent wing assumed the nipples with the height of 250 nm with the separation of 200 nm in a hexagonal packing. They compared the reflectivity before and after crushing the protuberances, and found 70% reduction of reflectivity was observed. Watson and Watson performed atomic force microscopic (AFM) investigation on the wings of cicada (*Pflatoda claripennis*) and termite (Rhinotermitidae sp.), and found the nipple with the height of 225–250 nm and spacing of 200 nm for the former, while somewhat rough protuberances spaced 700–1000 nm with the height of 150–250 nm for the latter (Watson and Watson, 2004). They speculated the structure for the former would presumably offer the protection against the predators due to the antireflection effect, and the self-cleaning effect against the small particles including water drop due to the structurally hydrophobic nature, in addition to the mechanical rigidity. On the other hand, that for the latter would only contribute to the mechanical rigidity. Parker (2005b) and Parker *et al.* (1998a) investigated the compound eyes of Eocence fly in Baltic amber (45 Ma). They found that the corneal surface comprised of a series of parallel ridges with approximate sine-wave profile with the periodicity of 240 nm and the depth of 146 nm. They also found similar structures in two extant Diptera species, *Neurotexis primula* and *Zalea minor*. They measured the reflectivity of a model substrate having similar surface fabricated by a photoresist method, and found considerable reduction of the reflectivity was obtained for both *s*- and *p*-polarizations.

The application of moth-eye principle appeared immediately after the above optical and physiological investigations were published. Clapham and Hutley (1973) and Wilson and Hutley (1982) attempted to fabricate the moth-eye structure by a photoresist method. They utilized two collimated Kr laser beam of 351 nm to intersect each other to generate interference fringe of straight lines with the spacing of 210 nm, then rotated the substrate by 90° to impose two such pattern at right angles, which made it possible to create sinusoidal modulated pattern. From the condition of zeroth-order diffraction grating, the wavelength of light λ should satisfy $\lambda > d$ or $2d$

for normal and grazing incidence. On the other hand, to function the antireflection effectively the nipple height h should roughly satisfy $\lambda < 2.5h$, thus the antireflection was satisfied in the limited region, which was confirmed experimentally (see Sec. 9.3.2).

After then, moth-eye type antireflector has been practically applied as a new optical tool in rapidly growing field of nanophotonics, which will take the place of conventional antireflectors such as monolayer and multilayer coating, and material of graded refractive index. The characteristics of the moth-eye structure as an optical element and the calculation of the reflectivity will be described in Sec. 9.3.2.

Chapter 4

Beetles and Other Insects

4.1 Overview

Insects live everywhere on the Earth except undersea. The number of their species is said to be now over 1 million and yet about 3000 species are newly recorded every year. This exceptionally large group, compared with the other groups of animals such as 8000 species of mammals, 10 000 species of birds, 18 000 species of fish, and 110 000 species of Mollusca, the second largest group, will show us the diversity of structural colors.

Bodies, hairs, and wings of insects are variously colored owing to pigmentations and structural colors. The surfaces of the bodies and wings are the places, where structural colors most frequently reveal their appearances. Before entering into the detail, let us see the surface structure of insect body briefly. Since a lot of excellent reviews and books concerning the insect surface structure have appeared up to now, here I will only describe its essential parts according to an excellent work done by Neville (1975).

The surface of an insect is generally covered by a cuticle, which becomes a barrier for various inner organisms against the outer world, and has a function to maintain homeostasis in continuously varying environments. The cuticle is composed of various complex chemicals including proteins, lipids, polyphenols, chitins, and so on, which complicatedly intertwine with each other to form a solid material. In general, the outer part of the insect body is divided into three parts: epicuticle, procuticle, and epidermis, from the outer to inner direction.

Epicuticle is defined as a layer lacking chitins (Richards, 1951; Neville, 1975) (Fig. 4.1). It constitutes the outermost layer of the insect body, which is generally composed of four layers such as cement layer (tectocuticle (Richards, 1951)), wax layer (lipid cuticle), outer epicuticle (sometimes

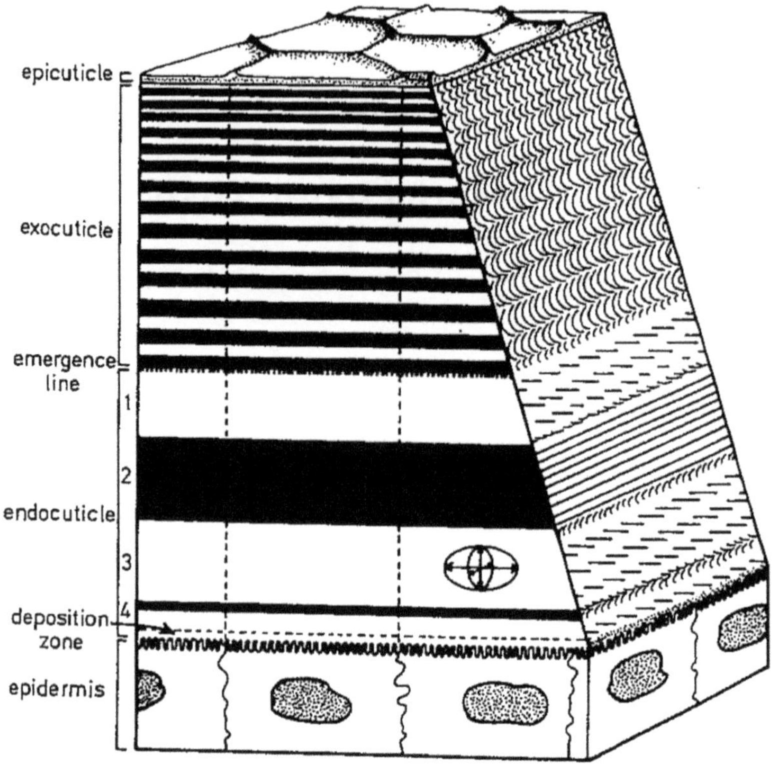

Fig. 4.1 Schematic view of insect cuticle based on adult *Tenebrio*. (Reproduced from Neville (1975) with kind permission of Springer Science and Business Media.)

called cuticlin), and inner epicuticle (dense layer or protein epicuticle). It is common that these four layers are not fully furnished. The latter two are secreted in sequence directly from the epidermal cells before ecdysis. In particular, the outer epicuticle is known as the first layer to be secreted, while the former two are secreted just before or after ecdysis.

Cement layer is secreted through a pipe called dermal gland that perpendicularly breaks through the surface layers, and covers the underlying wax layer to protect from the outer space. It consists of various materials, dependent on the species, such as protein with polyphenol, tanned protein with lipid, wax stabilized with shellac, and so on.

Wax layer consists of lipids of a labile nature and has a function of water control by covering the whole body. It is secreted through thin pipes called wax canals of typically 6 nm diameter. A large number of wax canals fuse into a small number of thicker pipes called pore canals, which are originally projections of epidermal cells, and take the form of twisted ribbons within the procuticle. Within a wax canal, wax filaments are present, which bundle into a large size at their roots and are connected with those in a pore canal. The role of these filaments is believed to be a sort of wick to smoothly transport labile lipid over the outer epicuticle. The filament is believed to have an esterase activity, and thus the synthesis of the wax is performed within the canals.

The outer epicuticle is the first cuticle layer to be secreted and consists of a trilaminar membrane up to 18 nm. At first, the layer appears as patches, which fuse into a layer. The function of the outer epicuticle is to resist the enzymes of the moulting fluid. The inner epicuticle of typically 0.5–2.0 μm thickness consists of tanned protein with polyphenol or lipid, which usually shows laminations. The optical properties of this layer are isotropic, while those of the outer epicuticle are anisotropic.

Procuticle is secreted after the epicuticle and is defined as a layer that involves chitin. This layer is further divided into exocuticle and endocuticle. The exocuticle is secreted before ecdysis and becomes tanned and harder after ecdysis. Thus, the color of the exocuticle is usually brown or black, which is the result of tanning or melanin formation. The exocuticle contains cuticle microfibrils of ~3 nm diameter, which are composed of unidirectionally grown chitin crystallites imbedded into the protein matrix. These microfibrils are arranged uniformly in a plane parallel to the surface, and their directions gradually change with depth from a surface, thus causing the anticlockwise helicoidal structure throughout the exocuticle. The pitch of the helix lies in the visible wavelength range in scarabaeids, which causes the structural colors in these insects. The endocuticle appears as the growth and digestive layer, which also contains bundles of chitin microfibrils. Between the endocuticle and epidermis, cuticle deposition zone of 1 μm thickness is located.

Chitin consists of polysaccharide chains of poly-*N*-acetylglucosamine with varying lengths (Neville, 1975). The molecular chains of chitin are associated together in a highly ordered manner to form a crystallite of 2.8 nm diameter (Neville *et al.*, 1976). Chitin usually exists in three polymorphic forms, α-, β-, and ψ-chitin, depending on the arrangement of the molecular chains in crystallites. In an arthropod cuticle, only α-chitin appears, where the molecular chains are arranged in an antiparallel form. The crystallites are packed in a hexagonal or pseudohexagonal array, but the extent and degree of perfection vary considerably (Neville *et al.*, 1976). Neville estimated that the size of a chitin chain in a crystallite had a dimension of $0.476 \times 0.476\,\text{nm}^2$ in cross section and that a crystallite involved ~20 chains with the dimensions of $2.8 \times 2.8\,\text{nm}^2$. Although chitin has never been found free of proteins, the information concerning the bonding with protein has not been obtained.

The cuticle of newly ecdysed insect is soft and pale pink, but soon it darkens and becomes harder, which is known as tanning and sclerotization. However, these two processes should be considered to be different concerning the chemical reactions for cross-linking, because albino can form hard cuticle. In spite of that, they are closely related with each other. Various mechanisms have been proposed so far. All these include small molecular tanning agent such as *p*-benzoquinone or *N*-acetyldopamine, amine derivative of tyrosine produced during ecdysis, biphenyl linking, and also the protein groups such as sulfydryl, *N*-terminal amino, and ϵ-amino group of lysine. Anyway, these small molecular groups containing movable electrons cross-link the protein molecules, causing tanned protein network. Chitin crystallites become a part of the network.

The tanning pigment is sometimes called sclerotin, but its molecular structure is not unique. It may be defined as a pigment formed during the cuticular hardening, and involving quinonoid and phenolic characters. On the other hand, melanin is another pigment not having a unique chemical structure, which is defined as a dark and insoluble pigment produced by oxidation of polyhydric phenols. Melanins are sometimes classified into several classes such as eumelanins (black, indole-type), phaeomelanins (yellow, red, or brown) and allomelanins (black, catechol-type). Melanin is often bound to protein to form melanoproteins, possibly linked through thioether-type or peptide bonding. Although the melanin formation is sometimes included into the sclerotization, these two processes should also be treated as two independent processes and actually, they sometimes take place in different time courses.

4.2 Beetles

4.2.1 *Scarabaeid Beetles*

It was about 100 years ago that a strange parabolic pattern was observed in the section of the cuticle layer of Dipropodes, millipede, which was first reported by Silvestri in 1903 (see Bouligand, 1965). This pattern was found to be periodically arranged under the cuticle surface with the total thickness

(a) (b)

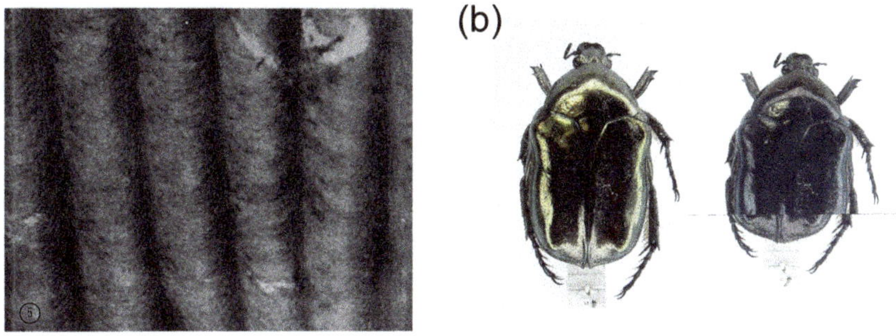

Fig. 4.2 (a) Electron micrograph of pronotum of a mealworm beetle *Tenebrio molitor* pupa, showing a parabolic pattern, which was obtained by cutting obliquely the cuticle, and staining with potassium permanganate and lead citrate. (Reprinted from Neville and Luke (1969) with permission from Elsevier.) (b) *Protaetia pryeri* photographed without (left) and with (right) a right-circular polarizer (Kinoshita and Yoshioka, 2005b). See Plate 10 in Appendix B.

over 100 μm. Later, similar patterns were reported by Biedermann, Hass, and Schmidt at the surface layers of crustaceans and insects (see Bouligand, 1965). The electron microscopic observation revealed that the pattern consisted many curved lines like scratches scraped by a brush, which were periodically arranged parallel to the surface, as shown in Fig. 4.2. Between the parabolic patterns, linear linings were present, which were also composed of an ensemble of somewhat linear scratches. The most remarkable feature of this pattern was that it was present only when an obliquely cut section was observed. Otherwise, only the periodic lamellar structure was observed.

The first analysis of this pattern was made by Bouligand, who considered that the pattern came from fibrous elements uniformly arranged in a plane parallel to the surface (Bouligand, 1965). Further, the direction of the elements gradually changed as a function of the distance from the surface. He explained using geometrical analysis that the origin of the parabolic pattern was due to a gradual change of the preferred direction when one observed the section cut obliquely, as will be described in Sec 9.5.1. Bouligand (1967) noticed that the biological helicoidal structure giving a parabolic pattern was closely related to cholesteric liquid crystal, which was known since the first discovery by Reinitzer in 1888. Botanist Reinitzer mixed cholesterol with benzoic acid and synthesized their ester as a crystal form, which was later found as an anisotropic liquid material and named as liquid crystal by Lehmann in the next year.

Cholesteric liquid crystal takes a layered structure, within which molecules take a preferred direction that gradually changes from layer to layer and takes a form of helicoidal structure, whose axis is normally perpendicular to the surface. The periodicity of the helicoidal structure is determined by the pitch of the helix, P, and if the orientation of the molecules in a plane is isotropic with respect to the direction like a nematic liquid crystal, a half of the helicoidal pitch $P/2$ determines the periodicity of the pattern and also the optical interference. Thus, the selective reflection of light occurs when the wavelength of light λ becomes close to the helicoidal pitch P. As will be shown in Sec. 9.5.2, the peak wavelength of the reflection band is $\lambda_p = (n_\parallel + n_\perp)P/2 = nP$, while its bandwidth becomes $\Delta\lambda = |n_\parallel - n_\perp|P = \delta nP$, where $n_\parallel$ and $n_\perp$ are the refractive indices corresponding to two optic axes in a uniaxial system, and are taken as those parallel and perpendicular to the surface. n and δn are the average refractive index and the birefringence, respectively.

It is remarkable that selective reflection normally occurs for circularly polarized light. In such a helicoidal material, two states of circular polarizations, right- and left-circular ones, become a convenient basis for the optical field (cf. Sec. 2.1.3). The two polarizations are generally defined as follows: when one sees an approaching light beam in front of him and the trajectory of the amplitude of the oscillating electric field at a certain position draws a left- and right-rotating circle, we call them left- and right-circular polarizations.

The important characteristics of cholesteric liquid crystal with an anticlockwise helix defined similarly are that if the left-circularly polarized light is incident normal to this material, only the light with left-circular polarization is reflected owing to the selective reflection (for mathematical treatment, see Sec. 9.5.2). On the contrary, if it is reflected at an ordinary mirror, right-circular polarized light should be reflected. When linearly polarized light, having the wavelength near selective reflection band, penetrates a cholesteric liquid crystal, the direction of the linear polarization will change. This phenomenon is called optical rotation or optical activity, and is widely used to investigate the molecular structure in biomolecules. Remarkable feature is that the direction of rotation changes when the wavelength of light crosses the reflection band.

Cholesteric liquid crystal is usually synthesized by adding chiral dopant to nematic liquid crystal, and is sometimes called *chiral nematic* liquid crystal. The cholesteric liquid crystal is now frequently used as a film thermometer, because the helicoidal pitch is highly sensitive to temperature, and also used as a colored display owing to its selective reflection within a visible region.

Neville noticed exactly the same idea concerning the molecular architecture of the insect cuticle in the year of 1967. Neville and Luke (1969) studied a cuticle structure in locusts and found that the circadian clock controlled two types of orientation of fibrous elements in the cuticles: helicoidal and preferred orientations. They investigated various locusts, cockroaches, crickets, stick insects, giant water bugs, and mealworm beetles, and found that in some cases, the exocuticle was lamellated throughout their thickness by the parabolic patterns, and in other cases, the endocuticle was distributed by such a pattern. In the endocuticle of the locust, a helicoidal structure giving the parabolic pattern and a layer of the preferred direction is periodically arranged, and formed a lamellar structure. The width of each layer was found to be controlled by the circadian clock: in daytime, the preferred layer was deposited, while at night, the helicoidal layer was deposited. Kenchington and Flower (1969) also reported the parabolic pattern in a praying mantis and interpreted it according to Bouligand's model. They estimated the rotation angle of fibrous direction in each successive layer as 18°.

The first observation of the anomalous reflection from beetle's elytron was done by Michelson (1911), who investigated the reflection of light from the elytron of a scarabaeid beetle, *Plusiotis resplendens* (see Plate 10 in Appendix B), using a Babinet compensator. He found that the reflected light was circularly polarized even at normal incidence. The degree of the circular polarization was maximum at blue, decreased gradually toward a yellow region, completely depolarized in orange-yellow, and appeared again in the opposite sense in the red end. He considered the origin of this circular polarization as due to a screw structure probably of a molecular dimension. He assigned that the circular reflection came from the selective absorption for only one of the two polarizations. He also considered that the depolarization in the orange-yellow region, where the two circularly polarized components equally contributed to the reflection, was due to the inhomogeneity within the elytron. Anyway, he described, probably with astonishment, that its absorption coefficient was even comparable to that of metal, which was then distributed within a surface layer of 1 μm thickness.

This observation was succeeded by many researches. Neville *et al.* performed experiments to understand the essential properties of this phenomenon. Neville and Caveney reported the detailed descriptions on the structural and optical properties in various species of scarabaeid beetles including *Potosia*, *Cetonia*, *Lomaptera*, *Heterorrhina*, *Smaragdesthes*, *Anoplognathus*, *Pelidnota*, *Plusiotis*, *Anomala*, *Phanaeus*, *Onitis*, *Goniocanthon*, *Melolontha*, and *Gnorimus* (Neville and Caveney, 1969). They found that most of the species except *Heterorrhina* and *Smaragdesthes* show the circularly polarized reflection. They found that (1) the color of the elytra of these species changed with changing viewing angle; (2) a combination of circular and linear polarizers capable of passing the light with different circular polarization made the elytra completely black (see Fig. 4.2 and Plate 10 in Appendix B), and (3) the transmitted light through the excised surface layer showed an optical rotation.

They scraped the cuticle surface minutely and assigned that the unusual optical properties were located at the outer region of exocuticle, where a regular lamellar pattern of 4–15 μm thickness with the apparent pitch of 153–198 nm was observed. On the contrary, the inner region of the exocuticle was sclerotized, and it showed a dark color with irregular lamellar structure. In *Potosia* and *Lomaptera*, dark pigmentary layer was present between the inner exocuticle and endocuticle. The lamellar spacing in the outer exocuticle normally changed systematically to cause the reflection in a wide wavelength range. The excised cuticle layer showed a clear optical rotation for the transmitted light, the direction of which changed at a particular wavelength such as 620 nm for *Potosia speciosissima* and 650 nm for *Phanaeus festivus*, indicating the selective reflection that occurred around these wavelengths. They found that the overall features were very similar to those of a well-known cholesteric liquid crystal such as cholesteryl cinnamate. From the investigation of the inner region of exocuticle of *Anoplognathus viridaeneus*, they found the microfibrils spaced about 6.5 nm apart in the direction normal to the cuticle surface. Since the helicoidal pitch was found to be $165 \times 2 = 330$ nm, the number of the layers was estimated as $330/6.5 = 50$ and the rotation angle between adjacent layers was $360/50 = 7.2°$.

Caveney (1971) compared two species of scarabaeid beetles, *Plusiotis optima* and *P. resplendens* (see Plate 10 in Appendix B), the latter of which was investigated by Michelson (1911). The former showed silvery color, while the latter did golden luster like brass. The thicknesses of the optically active layers for these beetles were found to be 16 and 22 μm, respectively. Remarkable was the layer of the latter species that showed a sandwiched structure (Fig. 4.3), within which 5 μm of upper and 15 μm of lower layer showed clear parabolic pattern, while a sandwiched layer of $1.81 \pm 0.04\,\mu$m did not show similar pattern: the middle layer showed

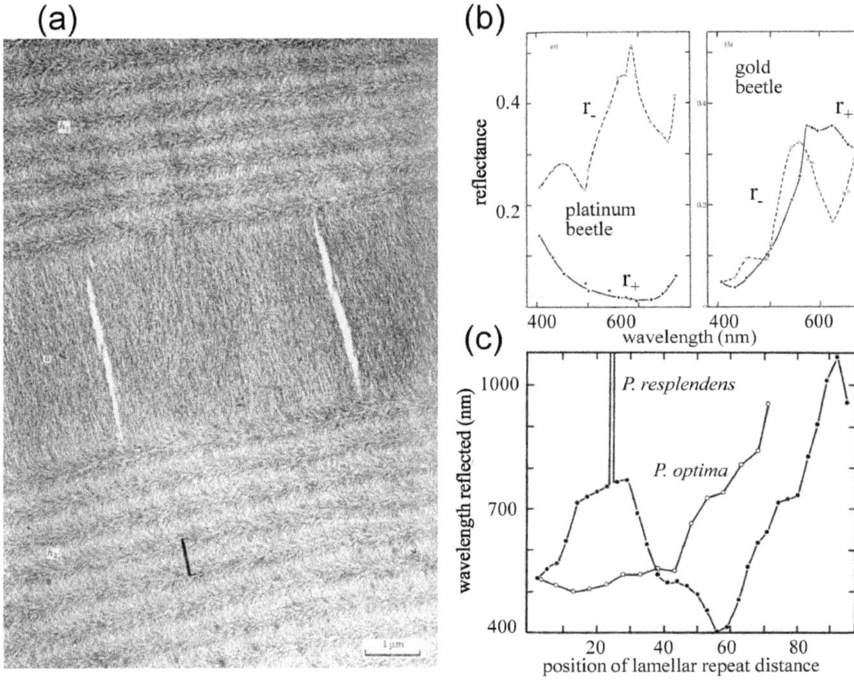

Fig. 4.3 (a) Oblique section of the reflecting layer of *Plusiotis resplendens*, (b) reflectivities for two circular polarizations for the elytra of platinum beetle, *P. optima*, and gold beetle *P. resplendens*, and (c) spacing of the lamella against the lamellar position for *P. optima* (open circles) and *P. reslendens* (solid circles). The position of unidirectional layer is indicated as vertical lines. r_+: right-circular polarization, r_-: left-circular polarization. (Reproduced from Caveney (1971) with permission of Royal Society Publishing.)

unidirectional molecular architecture, which was known as in locusts. The reflection spectra for circular polarizations are very peculiar that for left-circular one, it shows a peak around a green region of 560 nm, while for right-circular one, it shows a peak in an orange region around 575–624 nm (Fig. 4.3). Both reflections show rather high reflectivity of 0.32–0.35, which amounts to the total reflectivity of 0.6–0.7, while in *P. optima*, only the light with left-circular polarization is reflected with the maximum reflectivity of 0.5 (Fig. 4.3). The reflection of both circular polarizations and the wavelength shift of the reflection maximum confirms Michelson's observation. That is, in the green region the left-circular polarization overwhelms the reflection, and around 590 nm, where these two balances, depolarization occurs, while at the red-color side, the right-circular polarization is overwhelming.

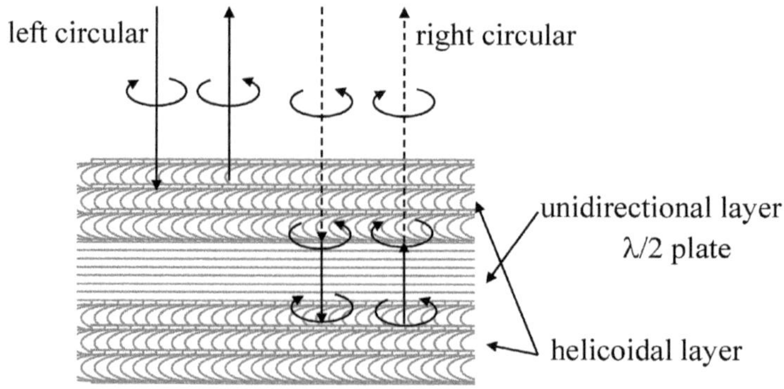

Fig. 4.4 Principle of the reflection from the elytron of *Plusiotis resplendens.* Light with left-circular polarization is reflected directly at the first helicoidal layer, while that with right-circular polarization is converted to left-circular one and reflected at the second helicoidal layer, which is then converted to right-circular polarization and transmits the first layer.

The origin of this peculiar phenomenon comes from the presence of the sandwiched layer, which shows optical anisotropy. Caveney (1971) calculated the phase change between different axes of the anisotropy using the measured refractive indices, and found that the layer had a function of a perfect $\lambda/2$ retardation plate around 590 nm, which worked fairly well in a region of 550–650 nm. What happens when a $\lambda/2$ retardation plate is inserted between two cholesteric liquid crystals of anticlockwise helical system? Figure 4.4 shows the phenomenon schematically. When the light with left-circular polarization is incident normally on the material, the light with left-circular polarization is completely reflected, while that with right-circular polarization penetrates without any loss. The $\lambda/2$ retardation plate changes the right-circular polarization into a left-circular one when transmitted. Then, the converted left-circular light is selectively reflected at the second cholesteric layer, and the reflected light is again left-circular. The $\lambda/2$ retardation plate again changes the left-circular polarization into right-circular one, which penetrates the first layer from the back side without loss and emerges out. Thus, both the left-circularly polarized light originally reflected and the right-circularly polarized light converted in this way interfere with each other and forms the reflected light.

Caveney (1971) measured the anisotropic refractive indices of these layers and found that $n_{\parallel} = 1.603$ and $n_{\perp} = 1.700$ with $\delta n = 0.097$ for a helicoidal layer and $n_{\parallel} = 1.535$ and $n_{\perp} = 1.701$ with $\delta n = 0.166$ for a unidirectional layer, where $n_{\parallel}$ and $n_{\perp}$ were the refractive indices parallel

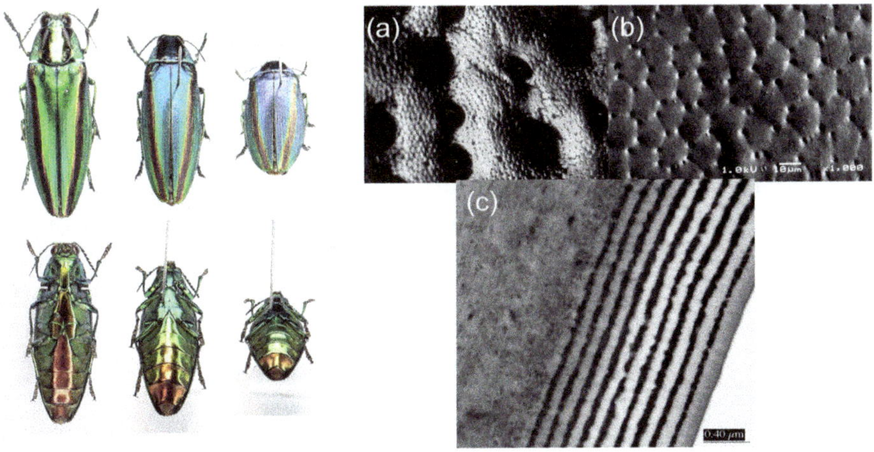

Fig. 4.5 Left: Color changes at the dorsal and ventral sides of a jewel beetle, *Chrysochroa fulguidissima*. Right: (a) Optical and (b) scanning electron microscopic views, and (c) the cross section of the elytron. Scale bars in (b) and (c) are 10 μm and 0.4 μm, respectively. (Courtesy of Prof. T. Hariyama for TEM image, and the others are reproduced from S. Kinoshita, *Cell Technol* **22**: 1113 (2003).) See Plate 11 in Appendix B.

and perpendicular to the cuticle layer. These values were quite in contrast to those of ordinary scarabaeid beetles, such as $n_{\parallel} = 1.580$ and $n_{\perp} = 1.598$ with $\delta n = 0.018$ for *Potosia speciosissima*. He chemically extracted the exocuticle layer using NH_4OH solution and found high content of uric acid (volume fraction of 0.6–0.7) was distributed in this layer. Uric acid crystal is known to have a high refractive index of $n = 1.89$ with high birefringence. In order to make the uric acid incorporation to generate large anisotropy, the highly ordered orientation of the crystal is anyway needed, which has not been clarified yet. Thus, quite interesting works concerning the beetle's elytra were done during 1960–1970s by these authors. However, it is rather strange that successive studies have not appeared in these 30 years.

4.2.2 *Jewel Beetles, Leaf Beetles, and Tiger Beetles*

Elytra of beetles are conspicuous in their brilliant and lustrous reflections and have attracted people's eyes for a long period. In East Asia, more than a thousand years ago, the wings of jewel beetles were used for the decoration of craft work. Buddhist miniature temple decorated by the wings of jewel beetles, which was thought to be made in the 7th century, is one of the most famous architectures in Japan.

In contrast to the general attention, the scientific approaches to elucidate the color-producing mechanisms in these beetles are extremely few. In the first half of the 20th century, scientists solely observed the elytra of beetles by giving a pressure, soaking into a liquid, scraping the surface, and observing the section under an optical microscope. Mason (1927b) investigated a blister beetle, *Lytta cesicatoria*, and found that a color-producing layer was located in a very thin cuticle layer. The colors were not uniform, but was variously dispersed owing to the modulation made by the surface polygonal patterns. He also confirmed that the application of hydrogen peroxide destroyed the pigment without damaging the iridescent coloring. He exemplified Cicindelidae, Carabidae, Buprestidae, Scarabacidae, Tenebrionidae, and Chrysomelidae, as giving essentially similar optical behaviors. He also gave an example of *Plusitotis gloriosa* (see Plate 10 in Appendix B), which showed an exceptionally thick layer of 8–10 μm as an iridescent color-producing seat. The mechanism of the latter beetle has been already explained in Sec. 4.2.1 as due to the optically active helicoidal structure. Thus, most of the iridescent colors in beetles are something different from the helicoidal one.

Before explaining their mechanisms, we will show simple experiments to investigate the iridescent colors of this type using a jewel beetle, *Chrysochroa fulguidissima*, as a sample. In Fig. 4.5, we show a specimen's dorsal and ventral sides. Both sides are lustrous and brilliant, and are glaring when one sees under a specific reflection. The overall view shows brilliant green with copper brown stripes in the dorsal side, while in the ventral side, it is lustrous copper brown. When we change the viewing angle from the normal to the oblique to the elytron, it is easily noticed that the dorsal color gradually changes from green to dark blue, while in the ventral side, it changes from copper to green. Thus, the iridescent feature is clearly seen. It is further noticed that its color change is rather continuous and isotropic, which is in contrast to that of *Morpho* butterfly, where a rather abrupt change of the color from blue to violet is observed and the reflection of light is strongly anisotropic. Under a polarization analyzer, not to pass the left-circular polarization, one will find no change in appearance; thus, the helicoidal structure possibly existing below the elytral surface does not contribute to this color.

When the jewel beetle is soaked in a liquid, for example, ethyl alcohol, a slight color change is observed, i.e. from green to greenish blue. However, this change mostly originated from the change in the angle of refraction at a cuticle interface, i.e. changing from a smaller refraction angle at the air–cuticle interface to a larger refraction angle at the liquid–cuticle interface. Except for that, essentially no change in the iridescent feature is observed.

If we observe the elytron under an optical microscope, many depressions of $\sim$50 μm in diameter are found, which are randomly distributed and cover the whole elytron (Fig. 4.5(a)). Thus, the surface of the elytron is more or less modulated. Under high magnification, it is found that the surface is completely covered with polygonal patterns of irregular pentagons or hexagons with a typical size of 10 μm (Fig. 4.5(b)). No sign of ultrastructures giving optical interference is observed. Thus, for further studies, we need electron microscopic observations.

The first electron microscopic observation of this type of beetle was reported by Durrer and Villinger (1972), who examined a buprestid beetle, *Euchroma gigantea*, as a sample. They found that the elytral surface of this beetle was covered with yellow powder and many depressions of 1–2 μm in diameter. The electron microscopic observation of the transverse cross section of the elytron showed that five layers of electron-dense regions were regularly arranged near the surface, below which irregular fibrous region was located. The regular electron-dense layers were interrupted here and there by electron-lucent regions. They deduced that the electron-dense layers were composed of melanin, which was also distributed below the regular layers. The uppermost electron-lucent layer had the thickness of 0.2–0.4 μm, which followed the electron-dense layers of 0.09 μm thickness for copper brown region and 0.06 μm for green region, while the electron-lucent layers between the dense layers were rather constant with 0.06 μm thickness. He calculated the interference condition including the uppermost layer and confirmed the relation with the apparent color using refractive indices of the dense and lucent layers of 2 and 1.5, respectively. He also noticed that the arch-shaped modulation at the surface changed the incidence angle to the multilayered system and contributed to various colorings of this beetle.

Mossakowski (1979) investigated the iridescent color of a tiger beetle, *Cicindela campestris*, and of a kind of jewel beetle, *Chrysochroa vittata*, using a transmission electron microscope in addition to a macro- and microspectrophotometer. He noticed that the surface of the tiger beetle was strongly alveolated, which disturbed the viewing-angle dependence of the color change, while that of the jewel beetle was only slightly modulated, and remarkable iridescence was observed. He found the regular layers in both species, which were similar to that observed by Durrer and Villinger. In *C. campestris*, the electron-dense and lucent layers showed nearly the same thickness in a green region and the lucent layer was thinner in a red region, while in *C. vittata*, both thicknesses did not change with the color

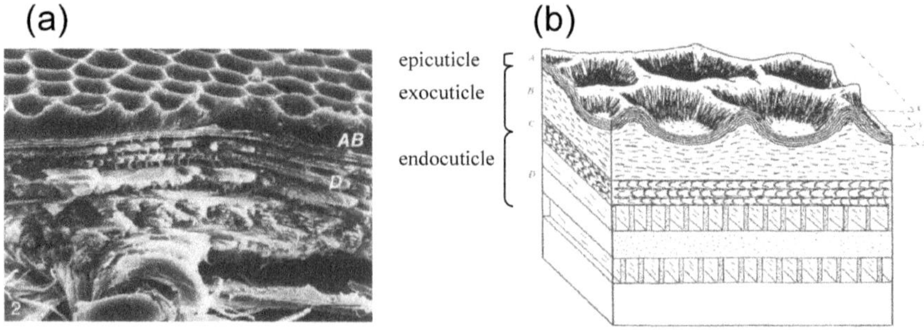

Fig. 4.6 (a) Cross section of cuticular layer of the elytron of *Cicindela splendida* and (b) its schematic illustration. (Reproduced from Schultz and Rankin (1985a) with permission of the Company of Biologists.)

and only the number of layers changed from 9–10 layers in a green region to 7–8 layers in a red region. He assigned the position of the layered region in the inner exocuticle and calculated the reflectivity using various values of the refractive indices between 1.33–1.56.

More complete investigations were reported by Schultz and Rankin (1985a). They investigated four species of tiger beetles, *Cicindela formosa*, *C. splendida*, *C. scutellaris*, and *C. repanda*, using optical measurement, chemical treatment, and electron microscopic observation. Particularly, they performed the last two methods thoroughly. The results are visualized by their illustration shown in Fig. 4.6, in which the surface of the elytral cuticle is divided into three regions: epicuticle, exocuticle, and endocuticle. The epicuticle region shows a thickness of 1–2 μm, and strongly alveolated with polygonal ridges and hollows with the depth of 3.5 μm. Just below the alveoli, 4–9 electron-dense bands are located, which are composed of vertically aligned granules with the thickness of 0.03–0.1 μm. Between the layers, electron-lucent regions with the thickness of 0.06–0.125 μm are present. The total thickness of the layers is large in the ridge region and small in the bottom region. The uppermost region of the epicuticle is covered with electron-lucent layer of 20–30 nm thickness, which is possibly assigned as a cuticlin layer. Below the epicuticle, outer exocuticle is present with the thickness of 2–3 μm, which is composed of 25–35 electron-dense lamellae of 0.055–0.1 μm interval, and may constitute imperfect helicoidal structure. Both epicuticle and outer exocuticle show dark brown and are thought to be melanized. Below it, the inner exocuticle of peculiar helicoidal pattern and the endocuticle of plywood structures appear. It is also found that dermal gland canals of 0.8 μm diameter break through the layers to an upper

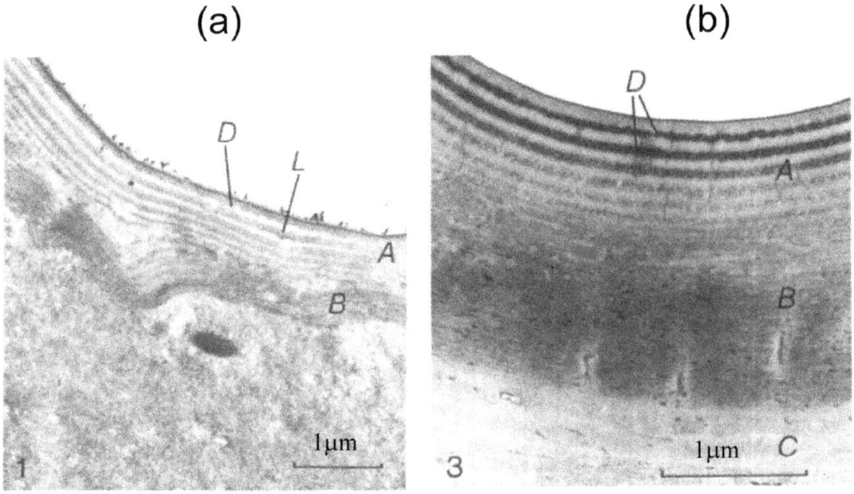

Fig. 4.7 Cross sections of developing elytral cuticle of *Cicindela scutellaris* (a) 10 days after pupation and (b) 12 h after ecdysis. A: epicuticle, B: outer exocuticle, C: inner exocuticle, and D and L: presumptive electron-dense and lucent layers. (Reproduced from Schultz and Rankin (1985b) with permission of the Company of Biologists.)

end, while pore canals of 50–70 nm diameter interrupt the regular layers in the epicuticle.

They investigated the location of the layered structure by chemical treatments (KOH for dissolving the epicuticle, chitosan method to detect chitin, and H_2SO_4 for dissolving endocuticle), and confirmed that the regular layers were located in the non-chitinous epicuticle. They further investigated the effect of H_2O_2, a bleaching agent for melanin, and observed the shift of the iridescent color, which suggested that the dense layers were composed of melanin granules.

Schultz and Rankin (1985b) also performed a developmental study on *C. scutellaris* and *C. splendida* (Fig. 4.7). They observed the color change before and after ecdysis: The color was initially pale white or creamy with slight violet iridescence 4 days before ecdysis, then became golden straw with a distinctive violet hue 8 h after ecdysis, predominantly blue iridescent with black backing within 8–12 h, and entirely blue and pigmented 24 h later. After then, the iridescent hue turned green 5 days later, and the mechanical hardening of the elytron occurred 14 days later.

The electron microscopic observation during these periods revealed that the presumptive epicuticle and the part of exocuticle were already deposited 4 days before ecdysis. The number of the layers was fully prepared for the later growth, but the presumptive electron-dense bands were seen to be rather electron-lucent, while the presumptive electron-lucent layer were slightly electron-dense. Thus, the electron-microgram looked like a negative film. During post-staining using uranyl acetate/lead citrate, extremely narrow, fine grained, electron-dense bands of 15–40 nm thickness appeared within the presumptive electron-dense bands. 12 h after ecdysis, the outer exocuticle was fully formed, but the plywood structure was not constructed. The epicuticle became 80% of the thickness and the electron-dense bands were clearly observed without post-staining and their densities were stronger toward the outer side. From these observations, they concluded that the change in the electron density was entirely due to the chemical change and possibly due to the *in situ* melanoprotein formation, which also accounted for the change in the optical thickness, and hence the interference color, during the development.

They summarized the formation process for the color-producing structure in terms of the electron-dense melanin granules, which appeared within the electron-lucent inner epicuticle region (Schultz, 1986). When the granules were distributed randomly within the epicuticle region, the black and non-reflective pattern was obtained. If the granules were deposited in layers, the iridescent color would be produced; in both cases the melanin also distributed in the exocuticle region to avoid unwanted reflection. On the other hand, when the melanin granules and layered structure were completely lacking, white colour would appear.

In contrast to the early works, recent researches on this type of beetle are considerably restricted. Parker raised a problem whether mathematically nonideal structures were actually appropriate to the needs of the host animal or not, and exemplified the chirped and regular layers in two beetle species, *Anoplognathus parvulus* and *Calloodes grayianus* (Parker *et al.*, 1998b). The former was well known to show optically active reflection as investigated by Neville and Caveney (1969) and Caveney (1971), while the latter had not been investigated yet. He found that the latter showed 12 electron-dense layers of 50 nm in thickness, which were located between the electron-lucent layers of 200 nm in thickness. Thus, the multilayer was strongly nonideal. In contrast to the high and broadband reflectivity of the former elytron, the reflectivity of the latter amounts to only 18% in a green region. He considered that *C. grayianus* showed a wrinkle surface of 0.5–1.0 μm thickness on the elytron, which reduced the reflectivity and diffused the reflection angles. *C. grayianus* would benefit from the low reflectivity as a camouflage effect to resemble the optical property of a leaf, where this beetle usually settled.

Kurachi *et al.* (2002) investigated a leaf beetle, *Plateumaris sericea*, which showed the regional color variations from blue to red (450–750 nm). Through electron microscopic observations, they concluded that the interval of the layered structure, which consisted of only three electron-dense layers, contributed to the specific coloring. The thickness of the dense layer varied from 60–100 nm for blue specimen, to 70–160 nm for bronze one, and to 100–140 nm for red one. They simulated the reflection spectrum using the refractive indices of 1.73 and 1.40 for electron-dense and lucent layers, and obtained good agreement with the measured spectrum for each color group. They also discussed the contribution of layered exocuticle and concluded that only the epicuticle contributed to the reflection spectrum. Hariyama reported the structural color of a jewel beetle, *Chrysochroa fulguidissima* (Hariyama *et al.*, 2005). The reflection spectra for green and copper brown regions of the elytra showed a maximum at 550 nm and 750–950 nm, respectively. Since the electron-dense layer was found to be somewhat irregular in a strictly optical sense, they used a densitometer to convert the change in the electron density to that in the refractive index, and found that the interface of the multilayer was not optically discontinuous but rather gently modulated like a sinusoidal curve. They calculated the reflectivity under this model and obtained fairly good agreement.

Before closing the present section, I would like to describe some comments particularly on the optical response of iridescent beetles bearing multilayer system. Although the multilayer is probably the simplest system to cause the structural coloration, yet various problems remain unsolved:

(1) What on earth is the actual constituents of electron-dense regions in the multilayer and how should we evaluate the refractive indices to calculate the interference phenomenon? Many authors strongly suggested that the electron-dense region is composed of melanin granules, which are formed through *in situ* chemical reactions. However, direct evidence has not been obtained to support this idea. If this is true, the calculation of the multilayer interference should include the complex refractive indices.

The inclusion of the imaginary part into the calculation of multilayer reflection is easily fulfilled within a framework of the conventional calculation. In Sec. 2.3, we show an example of such calculations. The inclusion of the imaginary part in the refractive index considerably reduces the background transmission and diffuses the side-lob structure without any large reduction of the maximum reflectivity. This calculation clearly shows that the inclusion of the imaginary part largely contributes to enhance the contrast of the structural colors with only slight sacrifice of the maximum reflectivity.

(2) The effect of the surface modulation like alveoli on the multilayer interference has not been properly evaluated. Although almost all the authors commented on the presence of the surface modulation, which should contribute to the broadband reflection and also to the diffusion of the light reflection owing to the distribution of the incidence angle, no quantification has appeared yet.

(3) The irregularity of the multilayer should be properly evaluated. Hariyama tried to evaluate the density distribution of the electron-dense layer by a densitometer and found the sinusoidal distribution for the electron density rather than discontinuous layer as has been assumed conventionally (Hariyama *et al.*, 2005). This finding seems very informative if we consider the dense layers to be composed of the ensemble of melanin granules, which inevitably diffuses the shape of the interface as in the antireflection effect due to corneal ripples. It is also noticed that the layer thickness and interval are not so strictly controlled as in manufactured products, and further they are spatially inhomogeneous. These irregular effects on the optical response should be more thoroughly investigated, since they are essential to express the true appearances that the animals are displaying.

4.2.3 *Scale-Bearing Beetles: Weevils*

Among various scale-bearing beetles, weevils offer quite an interesting object to the studies on structural colors. In 1911, Michelson investigated a weevil called diamond beetle, probably *Estimus imperialis*, which possessed brilliant and exquisitely colored scales on the elytron. Since the color distinguished with the liquid immersed from a tear of the scale and also since the reflection color was not widely distributed, he considered that the color of this beetle was due to the fine striations on the interior surface of the scale with an unsymmetrical saw-tooth shape, which was intended to enhance a particular diffraction order. Mallock (1911) observed the polygonal pattern in the transmission view of a scale and investigated the effect of liquid immersion. On the other hand, Onslow (1920b) found a well-defined crossed appearance like the strings of a tennis racquet when seen at the cross section of the scale of *E. imperialis*, and opposed to Michelson's specially designed grating. He also described that a variety of color saturations among species might come from the irregularity of the periodicity. Mason (1927a) also investigated the scale of *E. imperialis* and found the indication of lamellar structure within a scale.

Using an electron microscope, Ghiradella (1984) found a very peculiar structure within a scale of a local weevil, *Polydrusus sericeus*. The scale of this beetle possessed a similar lattice structure to that found in a scale of butterfly, *Callophyrys rubi*, i.e. photonic-type scale reported by Morris (1975). Later, she described that it was a three-dimensional lattice that diffracted light to produce a sparkling green color very similar to that of sunlit leaves (Ghiradella, 1998). Later, Parker *et al.* (2003) reported that *Pachyrhynchus argus* found in the forests in northeastern Queensland, Australia, showed a metallic coloration visible from any direction owing to a photonic crystal analogous to opal (Fig. 4.8). They reported that within

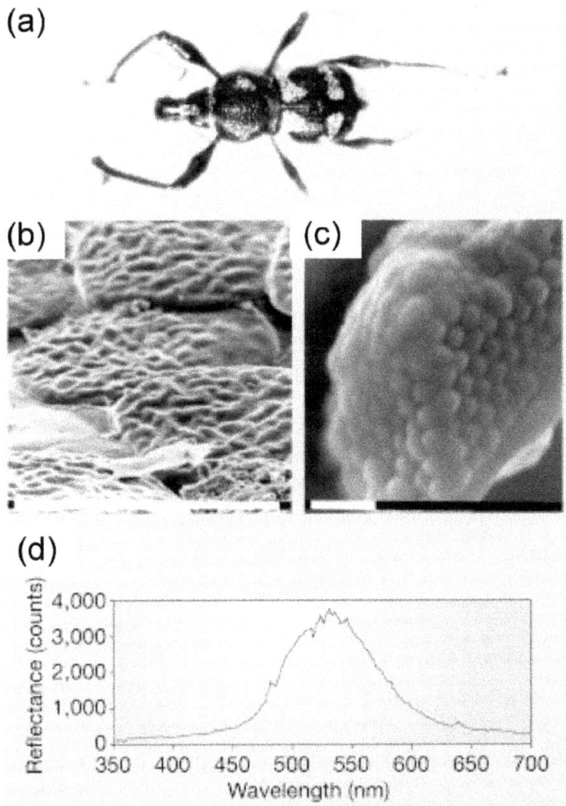

Fig. 4.8 (a) A weevil *Pachyrhynchus argus*, and (b, c) its scales and (d) its reflection spectrum obtained under 20° inclined incidence to the normal of the scale surface. (Reprinted by permission from Macmillan Publishers Ltd: Parker *et al.* (2003), copyright (2003).)

a weevil's scale, many transparent spheres of 250 nm diameter constituted a photonic crystal of hexagonal close-packing and selectively reflected 530 nm light. They estimated the wavelengths giving the maximum reflectivity using the refractive indices of the sphere and medium as 1.56 and 1.33. Further, they considered that the domain structure found within a scale contributed to the omnidirectional color.

4.2.4 *Color-Changing Beetles*

Hinton (1973) visited Venezuela with a friend in 1971 and one evening he caught a large male dynastid, *Dynastes hercules*, that flew into a strip lighting by the petrol station. He kept the beetle in a sock and returned to England. In due course, he noticed that the elytra of this beetle changed into greenish yellow and back again to black. He began to study its mechanism (Hinton, 1973, 1976). The color change was dependent strongly on the humidity. It took 30 s to 2 min to change completely. At night, it became black, while in daytime, it did yellow. If it were yellow at night, its color would be more easily seen by predators, while in daytime, the yellow color would hide the beetle when it was feeding on fruit. The change was local, meaning that when the elytron was locally dried, the color only at that place was changed, and basically autonomous because the color of the detached elytron from a live beetle behaved exactly in the same manner as that of the attached one. However, there was a case where a beetle would become yellow when exposed to conditions that should have made it stay black. Thus, the color was at least partly under central nervous control.

In his illustration, the epicuticle is transparent and has a thickness of 3 μm. Below it, yellowish spongy layer lies with the thickness of 5 μm and further beneath it, the color turns black. Thus, Hinton speculated that under dried circumstances, the yellow spongy structure was filled with air, which made the layer optically inhomogeneous. Thus, yellow light would be scattered. Under wet conditions, the refractive index matching took place and the spongy layer was made optically homogeneous. The light passed through this layer without disturbance and was absorbed by the black layer below it. He confirmed this hypothesis by placing a drop of liquid nitrogen on the black elytra and found a yellow spot appearing immediately. This was explained in terms that the frozen water would scatter the light again. Similar type of color change was observed in *D. granti*, *D. hyllus*, and *D. tityrus*, in which both elytra and pronotum changed the color. On the other hand, African and Oriental species of *Dynastes* and *D. neptunus* did not change the color.

Hinton (1976, 1973) also investigated the color change known in tortoise beetle, *Aspidomorpha*, *Eurypepla*, *Deloyala*, and *Metriona*. The elytra and pronotum of *Aspidomorpha tecta* usually showed bright gold color, which concealed itself by mimicking a raindrop shining in the sun. However, when it was somehow disturbed, it turned reddish copper within 2 min. He made an electron microscopic observation of the section of the elytra and found that 44×2 layers with changing intervals were present in the inner part of the cuticle (endocuticle) directly above the layer of the epidermal cells. He assumed that the epidermal cells directly moved the water from the materials to the layer to increase the interval. Later, Neville (1977) investigated the same tortoise beetle, and reported an experiment of pH change on the dead cuticle. He found that the color was greenish gold at pH 6.2, while at pH 5.0, 8.0, and 9.0, it became reddish gold, particularly at the corners of the elytra. Thus, he concluded that the extracellular pH change produced by the epidermis could alter protein hydration, causing swelling or shrinking, and then changed the lamellar spacing.

4.3 Damselflies and Dragonflies

Damselflies and dragonflies both belong to the order Odonata having the characteristics of four large slender wings, two large compound eyes, tough mouthparts, thin neck, and slender bodies. The Odonata is generally classified into three suborders: Zygoptera with forewing and hindwing of nearly equal size, Anisoptera with those of a different size, and Anisozygoptera which shows the intermediate characters of the former two and possesses the ancient character found in fossils. Damselflies are called for those belonging to Zygoptera, while dragonflies are those to Anisoptera. The number of the species in Anisozygoptera is extremely limited and only two species are extant, which live limitedly in Japan and Himalayas.

The studies on the structural colors in these insects have mainly focused on their wings and bodies. In 1926, Mason described the blue body color of a familiar bluet damselfly, *Enallagmae*, as due to Tyndall blue, which was caused by light scattering by fine particles enclosed in a chitinous material (Mason, 1926). He considered the main criteria for coloration due to Tyndall blue as (1) minute structure of dimension less than $0.5\,\mu$m, having a different refractive index from that of the surroundings, (2) blue to bluish white scattered light with brown to red, orange or yellow by transmission, and (3) scattered light polarized in a plane perpendicular to the direction of the incident light beam. He exemplified the body of a dragonfly, *Mesothemis simplicicollis*, and the wing of a dragonfly, *Libellula pulchella*. These findings had made a strong impression on scientists long afterward, and the dragonflies had been described (including my recent review (Kinoshita and Yoshioka, 2005b)) as an excellent example for Tyndall blue without performing any further investigation on the quantitative evaluation of the spectrum or polarization characteristics at the reflection.

Veron *et al.* (1974) investigated the physiological color change known for several Odonata species and performed the electron microscopic studies. They employed a male damselfly inhabiting Australia, *Austrolestes annulosus*, as an example and compared the result with those of two different kinds of Odonata, *Diphlebia lestoides* and *Ischnura heterosticta*. *A. annulosus* shows a remarkable color change from dark purple-brown to bright blue, dependent on the environmental temperature. Below 12°C, it shows a dark phase, while above 20°C, it does a blue phase. The color change is found to be more rapid from dark to blue. They performed the electron microscopic observation and found that chromatophores of 11 μm thickness were fixed firmly below the cuticle layer of 6 μm thickness. In the chromatophore, round or nodular, electron-dense pigment bodies with the diameter of 0.15–0.8 μm were distributed at one third or one half of the proximal part of the cell, while spherical or slightly oval light-scattering nanospheres of regular size of 0.25–0.38 μm diameter were packed rather firmly in the rest of the cell (Fig. 4.9(a)).

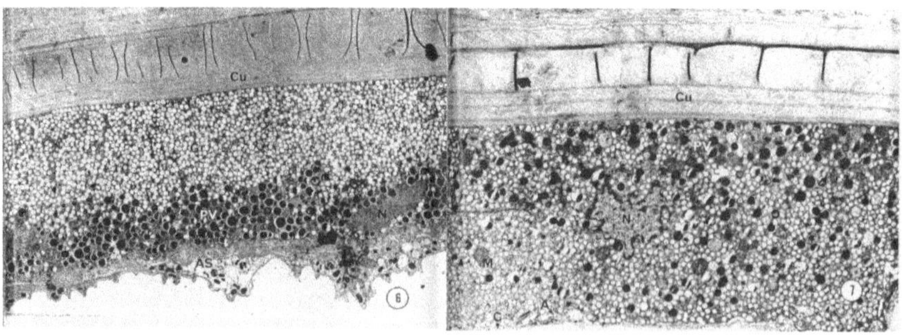

(a) (b)

Fig. 4.9 TEM image of the dorsal abdominal integument of Australian damselfly (*Austrolestes annulosus*, blue ringtail) in (a) blue and (b) dark color phase. Cu: cuticle, Gr: refractive granules, PV: pigment vesicles, N: nucleus, AS: air sac, M: mitochondria, A: nerve axons, and C: collagen. (Reprinted from Veron *et al.* (1974), with permission from Elsevier.)

The distribution of these bodies changes drastically with the color change from blue to dark: The pigment bodies migrate, and are distributed throughout the cytoplasm (Fig. 4.9(b)). The migration is not strictly synchronized in all the chromophores. At the intermediate state of the color change, the pigment bodies are found irregular in shape and the color shows patchy appearance, having a trilaminar membrane of 8 nm thickness. Since they were labile in shape and size, sometimes budding was even found; it was suggested that fluid-filled, membrane-bound vesicles were most probable. On the other hand, a light-scattering nanosphere has a smooth membrane of 8 nm thickness with a central spherical core of 60–65 nm diameter. At the center of the core, a moderately electron-dense region is present, which is surrounded by a very electron-dense wall of 8.5 nm thickness. The radiating electron-dense lines are found in the electron-lucent space between the central wall and peripheral membrane. This suggests that the light-scattering bodies are composed of solid crystalline materials.

They considered that the blue phase was originated from Tyndall scattering due to the densely packed light-scattering bodies, while the migration of pigment bodies changed the color to dark. In a young male, the chromatophores were more irregular with the lack of light-scattering bodies. Thus, neither color change nor blue appearance was seen. The other two species showed essentially the same tendency, though the size and color of the pigment bodies differed from species to species. They also extracted the pigment from several species and confirmed that the mixture of xanthommatin and dihydroxanthommatin was the major component. On the other hand, they deduced that the nanospheres were composed of biocrystalline of purine or of proteinaceous materials.

Recently, Prum *et al.* (2004) performed an experiment on two Odonata species, a familiar bluet damselfly, *Enallagma civile*, and a common green darner, *Anax junius*, as samples. He measured the reflection spectra of the bodies of these two species and found that the reflection spectra had a clear maximum at 475 nm for *E. civile* and at 460 nm for *A. junius*. Since Tyndall blue due to Rayleigh scattering should show the scattering efficiency proportional to the fourth power of the frequency of light, and since the reflection spectrum should show a monotonous increase toward shorter wavelengths, it is clear that the mechanism other than Tyndall scattering should be sought out. He found that within a pigmentary cell located below chitinous cuticle, nanospheres with 200–300 nm in diameter and ommochrome spheres with a much larger size were densely distributed, which was essentially the similar observation reported by Veron *et al.* (1974). The nanospheres were packed densely with the average center-to-center distance of 322 nm and were enclosed in a pocket of endoplasmic reticulum. The spheres were found to be exclusively spherical and highly consistent in diameter, which strongly implied the cause of the structural colors. Further, a sphere was clearly heterogeneous, having a central spot staining darkly and a clear peripheral boundary. They analyzed using a 2D Fourier transform of a transmission microscopic image and found the ring structure of radius $0.00307\,\mathrm{nm}^{-1}$, which was to show the coherent nature of the scattering. He considered the origin of the blue reflection in terms of the quasi-ordered nanospheres or the heterogeneity of the sphere. However, both models failed to explain quantitatively the reflection maximum.

It is clear from the experiment reported by Prum *et al.* (2004) that a simple model based on Tyndall scattering is no longer applicable to explaining the blue coloring on the body, since a definite peak in the reflection spectrum is observed. As they described, recently reported calculations of the reflectivity on quasi-ordered arrays having a short-range order without a long-range order may help us understand the origin of the blue coloring (Jin *et al.*, 2001; Prum *et al.*, 2004).

Another stream of the Odonata researches began from the investigation by Onslow (1923). He investigated the emerald green wings of a damselfly, *Neurobasis chinensis*, and found no particular structure to display the metallic reflection except for the black pigment. His finding was later confirmed by Vukusic *et al.* (2004b). This damselfly shows a transparent forewing and an opaque hindwing. The hindwing is strongly asymmetric: the dorsal side shows conspicuously green reflection with the color change from green to blue by changing the viewing angle, while the ventral side is dull brown without iridescence. Electron microscopic observation shows totally 12 periodic layers of dark- and light-contrasted layer with the gray layer near the ventral side. The surface of the wing is covered with fine wax microsculpture. They analyzed by dividing the layers into three parts and quantitatively explained the reflection spectra for two incident angles: light layers with the refractive index of 1.68, dark layers with that of $1.47+0.06i$, and a gray layer with that of $1.74 + 0.13i$. The inclusion of the imaginary part into the refractive index and the fact that a gray layer lies in the ventral side explain the strongly asymmetric reflection, namely, the hindwing shows the multilayered reflection from the dorsal side with the absorption within and below the layers, while from the ventral side, it turns to be a dark absorptive layer. They explained that the conspicuous green color of the wing was useful for the courtship display, because when a male was alighting briefly on, or flying close to, a water surface beneath overflying females, the hindwings were held motionless to show its striking green color.

Hariyama *et al.* (2005) investigated a damselfly, *Calopteryx japonica*, whose thorax and abdomen were structurally colored with green, golden, and reddish shining patches. The electron microscopic observation showed a multilayered structure with totally 7×2 layers. They converted the measured electron density profiles into the refractive index and calculated the reflection spectrum peaked around 580 nm using the refractive indices between 1.55 and 1.75. Next, they paid attention to its wing coloration. The color polymorphism of this damselfly shows three types: immature and mature females have grayish dull wings, and an immature male shows a gynochromatic color, while a mature male does a bright blue-green color. The reflection spectrum of the mature male solely showed a clear peak around 480 nm. The electron microscopic observation showed a clear multilayer of 3×2 layers for a mature male, while incomplete multilayer was observed for an immature male, and the layered structure was lacking for females. They confirmed in a field experiment that the strikingly glittered wing color was useful for territorial purpose to discriminate an invaded mature male from a female or an immature male.

4.4 Shield Bugs and Cicadas

Shield bugs, cicadas, and leafhoppers belong to the order Hemiptera, within which 55 000 species have been recorded at present. Although shield bugs belonging to the family Scutelleridae show actually metallic brilliancy, it is rare that these species are raised as examples of structural colors.

Onslow (1923) exemplified two types of structurally colored shield bugs and reported his observation. The first one is *Scutellera nobilis* belonging to the family Scutelleridae, which shows metallic color varying from green to yellow. He tried to polish it, and found that the color disappeared quickly, suggesting that the color-producing structure lay within the surface cuticle. A similar feature was also observed in *Chrysocoris stockerus* (Scutelleridae). Another observation was made on *Pycanum rubens* (Pentatomidae), which showed a color change from vivid apple-green to dull purple-brown, when it was dried. He examined the color change for various liquids, and found that only water or water-containing liquid made such a change. He deduced that two or more films of chitin separated by a thin film of water caused a brilliant color, or that otherwise irregular structure made a film

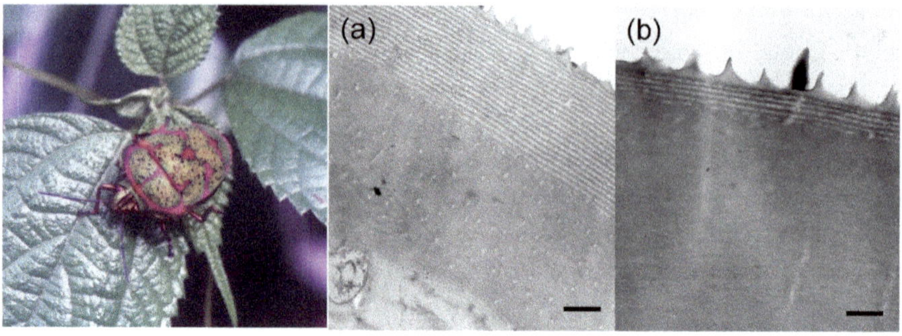

Fig. 4.10 Shield bug, *Poecilocoris lewisi*, and the TEM image of the cross sections of (a) yellowish-green and (b) reddish-red parts of the wing. Scale bar 1 μm. At the cuticle surface, conical protuberances with the height of 900 nm are seen. (Reproduced from Miyamoto and Kosaku (2005) with permission.) See Plate 11 in Appendix B.

opaque without water, but once it was swollen, a homogeneous thin film was restored, causing a color.

Miyamoto and Kosaku examined a beautiful shield bug, *Poecilocoris lewisi* (Scutelleridae, see Fig. 4.10) (Kosaku and Miyamoto, 1994; Miyamoto and Kosaku, 2005). The coloration of the dorsal surface of this bug shows yellowish-green with gold metallic brilliancy, which is sectioned by reddish-brown stripes. It is also noticed that both areas have a slightly light-blue color. The yellowish-green color changes from green to yellow with changing viewing angle, while the reddish-brown stripes and light-blue colors do not change essentially. Further, the yellowish-green color is destroyed when the surface is dried, while the color restores soon when moistened.

From these observations, it is easy to deduce that the reddish-brown color originated from the pigmentation. The reflection spectrum shows two maxima at 550 and 310 nm, and the reflectivity of the latter decreases from 29% to 13% with drying. Electron microscopic study showed a definite multilayered structure in the cuticle for yellowish-green and reddish-brown areas: 20 $\times$ 2 layers for the former, while 5 $\times$ 2 layers for the latter (Figs. 4.10(a) and 4.10(b)). The thickness of electron-dense layer was 110 nm with the gap of 80 nm for the former, while it was 75 nm with the gap of 75 nm for the latter. Below the multilayer, the pigmentary layer was present for the former, while it was absent for the latter. It was further noticed that conical protuberances with the height of 900 nm and the average diameter of 360 nm were distributed in the epicuticle region with 1 μm interval over both the colored surfaces.

They assigned that the brilliant yellowish-green color was due to the multilayer interference with the absorption layer placed beneath, while the multilayer in the reddish-brown area was not effective to produce the structural color, and the pigmentary color was considered as a primary cause. In addition to these colors, they considered that the conical protuberances produced the blue coloring owing to Mie scattering.

In addition to the structural coloration in shield bugs, a few works were reported on cicada's transparent wing. These reported the presence of small regular protuberances functioning as antireflecting coating like a moth-eye-type antireflector. Yoshida *et al.* (1996) investigated the transparent wing of a cicada, *Oncotympana maculaticollis*, and actually found the nipple arrays, which they did not find in the colored wing of a cicada, *Graptopsaltria nigrofuscata*. Watson and Watson (2004) employed atomic force microscopy to observe the nipple array in cicada wings, *Psaltoda claripennis*, *Tomasa tristigma*, and *Cicadetta oldfieldi*, which has already been described in Sec. 3.3.7.

4.5 Other Insects

Diptera and Hymenoptera are the third and fourth largest classes in the order Insecta. About 150 000 species belong to the former that includes crane flies, mosquitoes, flies, horseflies, gnats, and so on, while 130 000 species belong to the latter, including bees and ants. Nevertheless, the researches on the structural colors concerning these two groups are extremely few. In 1927, Mason reported that *Dolichopus canaliculatus* (long-legged fly), *Lucilia* (green-bottle flies), and other blue-bottle flies showed typical metallic iridescence. Since then, no report was found on the structural colors in these species. Only recently, Fung (2005) reported on the iridescence of a blue-banded bee (*Amegilla* sp.) easily seen in Hong Kong: The scales on the back of this bee with a size of $20 \times 100\,\mu\text{m}^2$ showed green structural color with narrow blue patches when the light was polarized along the scale axis, while the color was red-shifted to yellowish green when they were illuminated with a perpendicularly polarized light. The mechanism of this coloration was soon clarified after the electron microscopic observation of the cross-section of the scales. The scale was filled with a hexagonal array of tubules running perpendicular to the scale axis. About 10 tubule layers were seen with the spacing of about 200 nm. The array seemed not so regular and sometimes rotated about their horizontal axis by about 30° within a region of 2 μm or less. Thus, he assigned the change of color as due to the rotation and tilting of the transverse tubules. Similar principle that a photonic crystal in a scale accompanies patch-like domain structures, has been often reported in a butterfly scale such as *Callophrys rubi* (Morris, 1975). Thus, the domain structure in photonic crystal will be a typical expression for the scale-bearing insect to introduce the irregularity into the regular structure.

The surface color patterns of the compound eye of Diptera species have been attracting special attention (Bernard and Miller, 1968; Hinton, 1976; Stavenga, 2002). The Dipteran eyes often show peculiar color patterns such as golden-greenish color with red patches, alternatively arranged red and blue stripes, and irregular greenish patterns. In general, the female eye is more pronounced than that of the male. Bernard and Miller (1968) investigated the cause of this peculiar pattern under optical and electron microscopes. The compound eye of female horse fly, *Hybomitra lasiophthalma* (Diptera, Tabanidae), shows a horizontally striped pattern with five dark stripes, which are bluish in reflected light, and four bright stripes, which are orangish (Fig. 4.11). As the angle of incidence is changed,

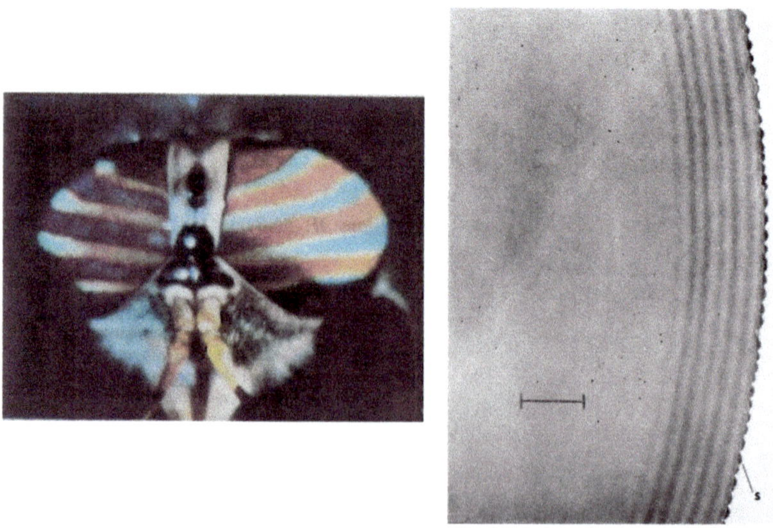

Fig. 4.11 Corneal pattern of the female horsefly, *Hybomitra lasiophthalma*, and the transmission electron micrograph of a bright facet of its compound eye. Corneal surface nipples are indicated as S. Scale bar 1 μm. (Reproduced from Bernard and Miller (1968) with permission.)

the stripe boundaries remain fixed while the stripe colors change. The striped pattern consists of specular reflection from a small flat part (facet) on each ommatidium and their intersections. The electron microscopic observations showed that near the air–corneal interface, bright facet showed six electron-dense layers of about 89 nm thickness with six electron-lucent layers of 112 nm thickness, in addition to the small-scale corneal nipples distributed on the interface (Fig. 4.11). On the dark facets of three central dark stripes, a total of 12, somewhat narrower, layers with 69-nm-thick dense and 86-nm-thick lucent layers were shown. Thus, multilayer interference is anyway the origin of these colors. The colored corneal pattern disappears in death, while it is restored in a moisture condition, implying that the electron-lucent layer is composed largely of water. Using a Mach-Zhender interference microscopy, they determined the average refractive index of the interference layers as 1.548. Assuming the refractive index of electron-lucent layer as 1.40, they obtained that of electron-dense layer as 1.75. The simple calculations showed the maximum reflectivity for the bright stripe as 0.84 at 625 nm, while the dark stripe showed a peak at 480 nm. Quite interestingly, both multilayers almost satisfied the condition of an ideal multilayer. On the contrary, the half of the male corneal surface was covered by the dorsal dark stripe, leaving six stripes densely at the center. These facets involved only six layers of interference multilayer. In the female horse fly, *Tabanus americanus*, the whole region was covered with the bright region with 12 layers of interference multilayer, while in a deer fly, *Chrysops fuligiosus*, 20-layer multilayer gave a noticeably brighter appearance except for the kidney-shaped dark region, where 8-layer multilayer was located.

They deduced that the function of the corneal multilayers was something analogous to the oil drops known in birds and they behaved like a type of contrast filter, changing the intensity spectrum of the incident light. In addition, they denied the possibility that these patterns might work as the display or mutual recognition, because they were too small to be recognized. Later, Stavenga (2002) investigated the interference filter effect in a deer fly, *Chrysops relictus* (Diptera, Tabanidae), and found 60% decrease in transmittance at 585 nm with a rather narrow bandwidth of 60 nm. Owing to this dip, he demonstrated the reduction of the longer wavelength side of the effective absorption spectrum in the photoreceptor, which originally showed a peak at 530 nm. Such surface patterns are known to be very common among Diptera species. In fact, it was described that 72 species belonging to 23 of the 60 Dipteran families show similar patterns on the corneal surface (Bernard and Miller, 1968).

Small and primitive insect of body length ~10 mm, called silverfish, gives a historical topic to the research of the structural colors. This insect is thought to give a vermin damage to make holes in books and is sometimes called "bookworm". However, in reality, the silverfish does not make a hole, but it gives damage to a book as if it licks. The insect to make a hole in book is a species belonging to a beetle, Anobiidae, of body length less than several millimeters. Anyway, the silverfish has been misunderstood as a ringleader owing to its rather large body and to its silvery appearance.

Hooke (1665) was intrigued by this insect and investigated using his microscope. He wrote in his book "Micrographia" as "the appearance of so many several shells or shields that cover the whole body, every one of these shells are covered or tiled over with a multitude of transparent scales, which, from the multiplicity of their reflecting surfaces, make the whole animal a perfect pearl colour." Thus, he thought that the origin of the silvery appearance came from the multiple-piled scales covering the entire body.

About 350 years later, a paper dealing with the silvery appearance of this insect appeared. Large *et al.* (2001) investigated the common silverfish (*Ctenolepisma* sp.) through a reflectance measurement with an integrating sphere and through an electron microscopic observation. They found that the reflectivity amounted to 12–30% in a visible region, grew up to 60% at 1.3 μm and then decreased to 30% at 2.4 μm. Thus, this insect continuously maintains relatively high reflectivity within an entire wavelength region above 400 nm, which gives a silvery appearance to our eyes. In contrast to Hooke's speculation, the reflectivity does not change much, when the scales are removed from the surface, implying that the main cause of the high reflectivity comes from the surface of the insect body. Then, they utilized a transmission electron microscope to observe the cross-section of the surface. They found the regular lamellar structure in an exocuticle region, whose periodicity increased first and then decreased again. They called it a doubly chirped multilayer. The surface of the body was covered with overlapping scales, which were thin with the thickness of 500 nm and had parallel running ridges separated by 1–3 μm on both sides. The ridges clearly contained the pigment possibly to contribute to the absorbing effect in the visible region. They constructed a model for this system by considering the contribution of the surface lamella and scales separately: The surface lamella was considered as a doubly chirped multilayer with high (chitin) and low (cytoplasma and chitin fibers) refractive indices of 1.56 and 1.4, respectively, while a scale was modeled by a set of parallel cylinders with the refractive index of 1.55 with a diameter of 0.25 μm. They found that the contribution of the scales to the total reflection was small enough, which confirmed the experimental result that the reflectivity came mainly from the surface lamella. However, the overall spectral distribution thus obtained did not reproduce well the experimental result, particularly in a near-infrared region. Thus, silverfish will continue to offer an old and yet a new problem to modern science.

Chapter 5

Birds

5.1 Overview

Recently, Sibley *et al.* (1988) and Sibley and Ahlquist (1990) presented a new classification of extant birds of the world, based on DNA–DNA hybridization techniques. This method utilizes the genetic similarity between the DNA strands derived from two species by making hybridization, and then measuring the amount of heat required to break the hydrogen bonds between the base pairs. When two DNA strands have more similarities, firm binding between the homologous base pairs will be formed owing to the corresponding hydrogen bonding. However, when they have genetically lesser similarities, the presence of many unbound sites weakens the binding such that the application of heat easily breaks the binding of the hybrids. Thus, after shearing long-chain DNA strands into fragments of *ca.* 400–600 bases in length, the thermal stability of the hybrid fragments against the temperature rise from 55°C to 95°C gives a melting curve, which offers the information on genetic distance.

If we see the structural colors in view of this new classification, it is easily found that they are widely distributed in various orders as listed in Fig. 5.1, where more than 170 species are recorded, whose microstructures are already known through electron microscopic investigations. It is noticed that the numbers of the species are comparatively localized within the orders Galliformes (peacocks and pheasants), Anseriformes (ducks), Trogoniformes (trogons), Trochiliformes (hummingbirds), Columbiformes (pigeons), and Passeriformes (a large group of small birds). Except for the last one, birds belonging to these orders are famous for their beautiful feathers. The order Passeriformes involves 5300–5500 species, which includes more than half of the avian species. This group normally includes

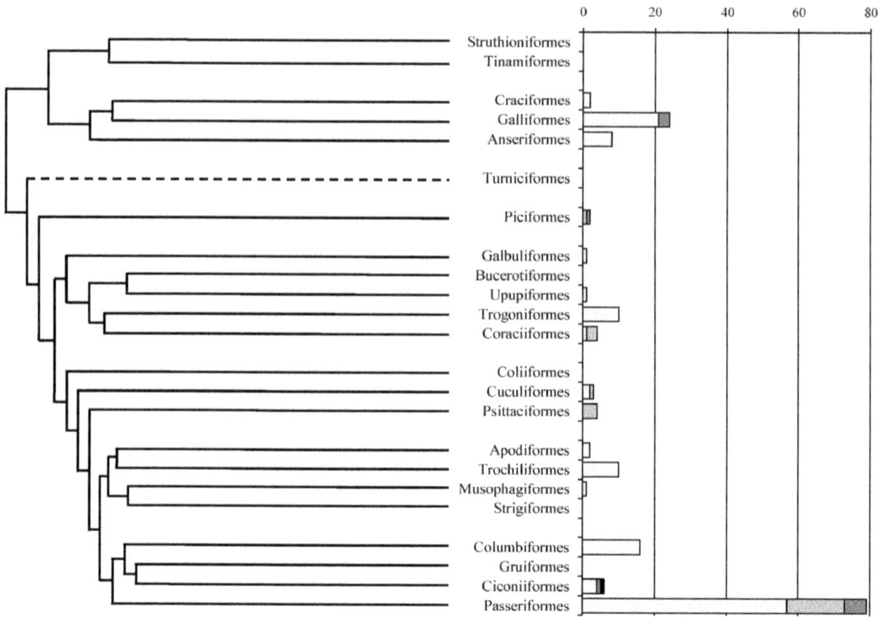

Fig. 5.1 Major divisions of the class Aves (Reproduced from Sibley *et al.* (1988) with permission of Yale University Press.), and the number of species whose structural colorations have been already investigated by electron microscopy. Light gray, gray, dark gray, and black portions indicate the number of species whose structural colorations are found in barbules, barbs, skins, and bills, respectively.

familiar species such as sparrows, crows, starlings, and finches, but also includes beautiful species of cotingas, jays, birds-of-paradise, sunbirds, and tanagers. Anyway, the microstructures variously decorate the feather barbs and barbules, skins and bills of birds. It is also noticed from this figure that extensive investigations have been performed solely on the feather barbules and barbs, and those on the skins and bills are extremely limited, though not a few species show brilliancy in these parts.

Structurally colored barbules in avian species are normally composed of well-ordered melanin granules. These granules originally cause black, gray, and brown colors in feathers, but play an important role in displaying structural colors by arranging themselves into regular structures and also through the enhancement of the colors as a background absorber. In 1960, sophisticated structures in feather barbules were observed under an electron microscope for hummingbirds (Greenewalt *et al.*, 1960a,b; Greenewalt, 1960). Greenewalt *et al.* found the presence of platelet mosaics on the barbules in nearly 50 species of hummingbirds and showed that these platelets were composed of air voids surrounded by melanin layers.

Later, Durrer (1977, 1986) reported systematic investigations on the iridescent barbules and classified the shapes and arrangements of melanin granules into several types. At first, he classified the shapes of melanin granules into five types as shown in Fig. 5.2: (1) S-type (rod-shaped granule), (2) St-type (thin rod-shaped granule), (3) P-type (flattened stick), (4) R-type (hollow tube), and (5) K-type (platelet like a humming bird). Next, the arrangements of the melanin granules were categorized into five types: (1) E-type (monolayer), (2) K-type (close packing), (3) S-type (multilayer), (4) G-type (lattice type), and (5) O-type (surface layer). Until 1977, his elaborate work on totally 106 iridescent avian species revealed that the microstructures of these species were classified into 19 categories (see Fig. 5.3). For example, peacock is categorized into StG-type. Most of the pheasants and ducks belong to StK-type, while trogons to KS or RS-type,

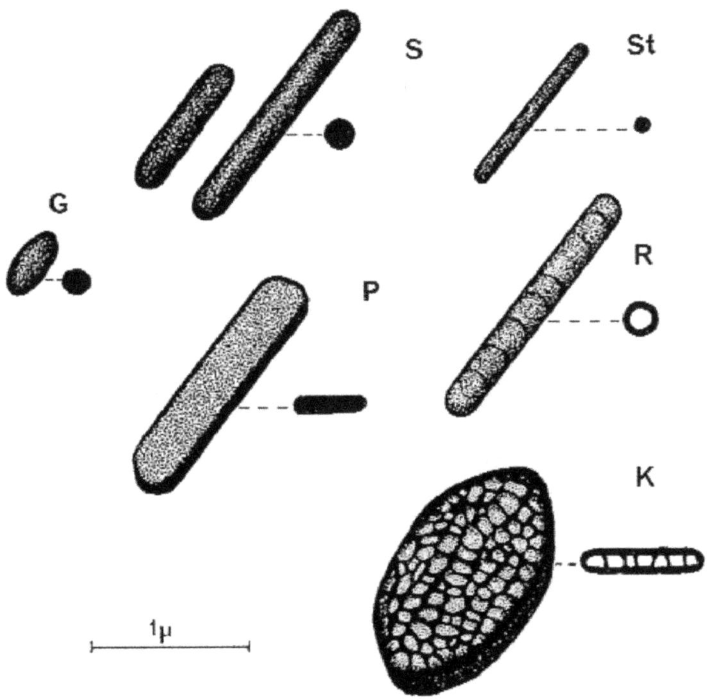

Fig. 5.2 Types of melanin granules used in the classification by Durrer (Reproduced from Durrer (1977) with permission.). S: rod-shaped granule, St: thin rod-shaped granule, P: flattened stick, R: hollow tube, K: hollow platelet, G: basic form.

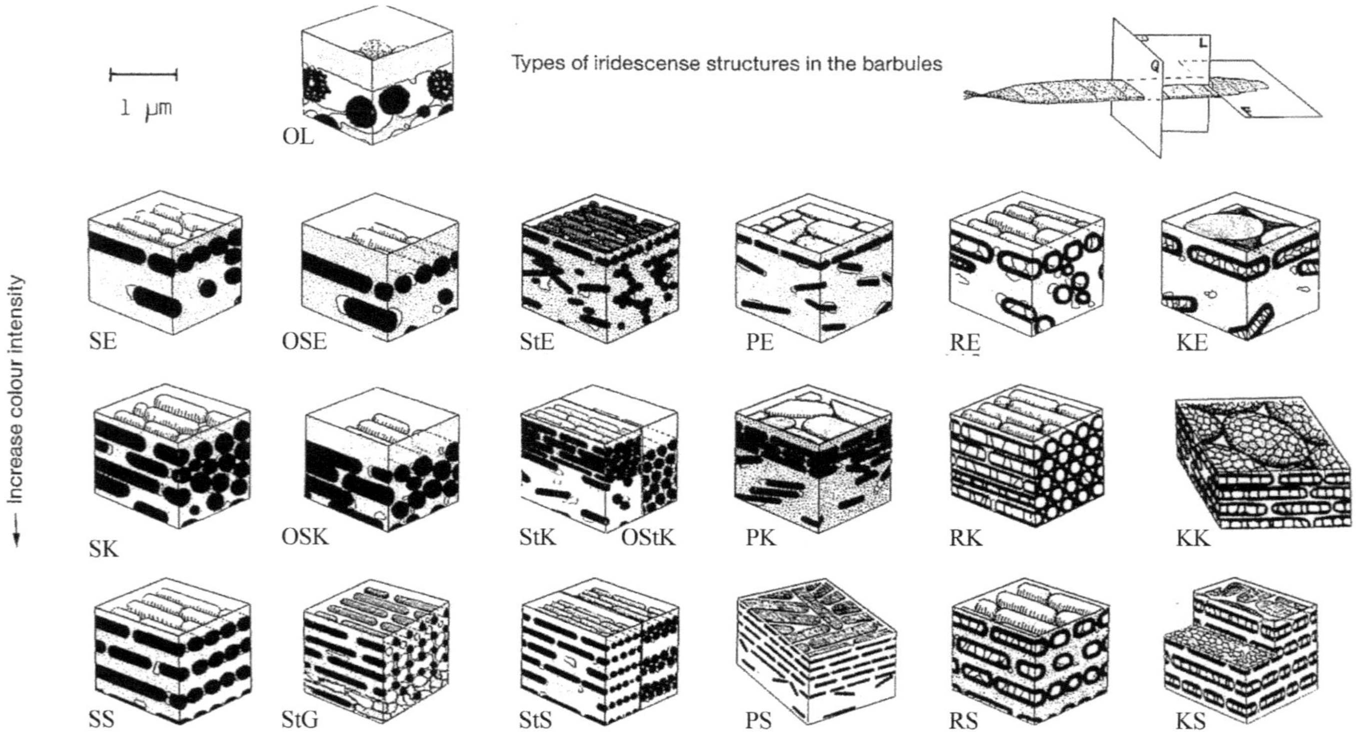

Fig. 5.3 Types and arrangements of melanin granules in feather barbules (Reproduced from Durrer (1986), with kind permission of Springer Science and Business Media.), where the names of the melanin types follow those employed in Durrer (1977).

hummingbirds to KK-type, pigeons to StS-type except for the rock dove, birds-of-paradise to StS-type, and sunbirds to PS-type. Thus, at the present stage, the arrangements of melanin granules in the barbules have been well systematized. We summarize the avian species whose microstructures are already reported in Appendix A. However, physical interpretation have not eventually proceeded since 1960s, when the optical properties were mainly calculated through a thin-layer interference theory.

On the other hand, the structural colors in the barbs are somewhat mysterious, because the barbs are filled with a random network of sticks or random distributions of air bubbles, which do not contribute to the light interference at a glance. In fact, these types of structures show almost no or only weak dependence of viewing-angle-dependent color change and are called noniridescent structural colors. However, recent scientific approach is rapidly progressing even into these wonderful structures, which will be described in Sec. 5.6.

5.2 Peacocks, Pheasants, and Ducks

We will first show the barbules of one of the most famous birds bearing structural colors. The color of the peacock feather has attracted a great deal of scientific attention for more than 300 years (see Plate 12 in Appendix B), and is now known as one of the typical expressions of biological photonic crystals. Since the early observations by Hooke (1665) and Newton (1704), it was in the 20th century that a detailed observation was made by Mason (1923b).

A feather of peacock consists of many barbs sticking out from a main shaft and each barb has a lot of barbules (Fig. 5.4). The barbules are curved along their long axis and slightly twisted from the roots. Each barbule has the shape of connected segments of a typical size of 20–30 μm, the surfaces of which are smoothly curved like a saddle (Fig. 5.5).

Mason (1923b) investigated a tail feather of peacock in conjunction with the conflicting theories of surface and structural colors. He attempted to seek the structure responsible for the iridescence in order to make a decision for the two theories. He observed the cross-section of the barbule microscopically and found that it consisted of three layers, which surrounded a fibrous core of 2 μm thickness, each layer having the thickness of 0.4–0.5 μm. He found that the inner and outer layers were dark, while the middle layer was transparent, and reported that no difference in structure was observed between the barbules of different colors. He observed that the colors of

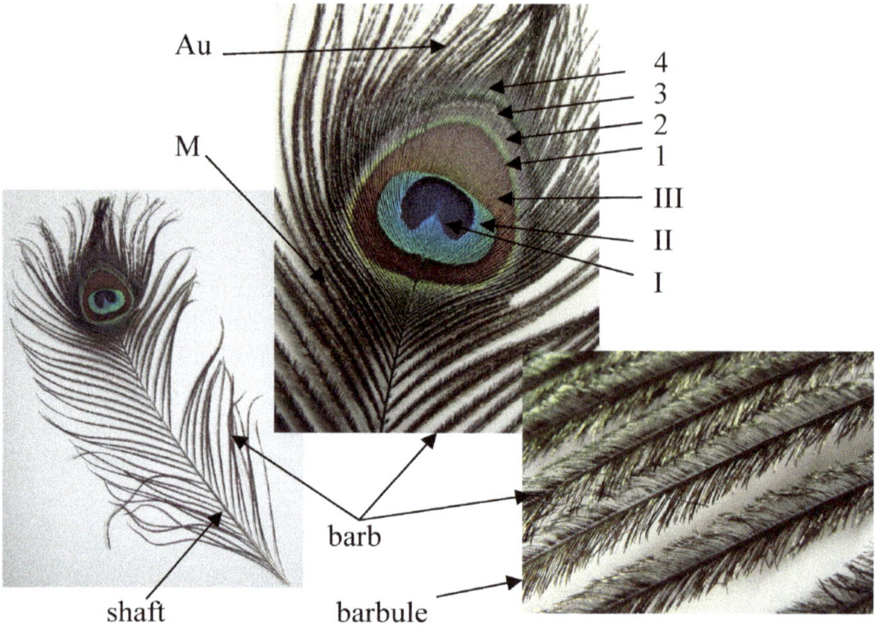

Fig. 5.4 Feather of a peacock.

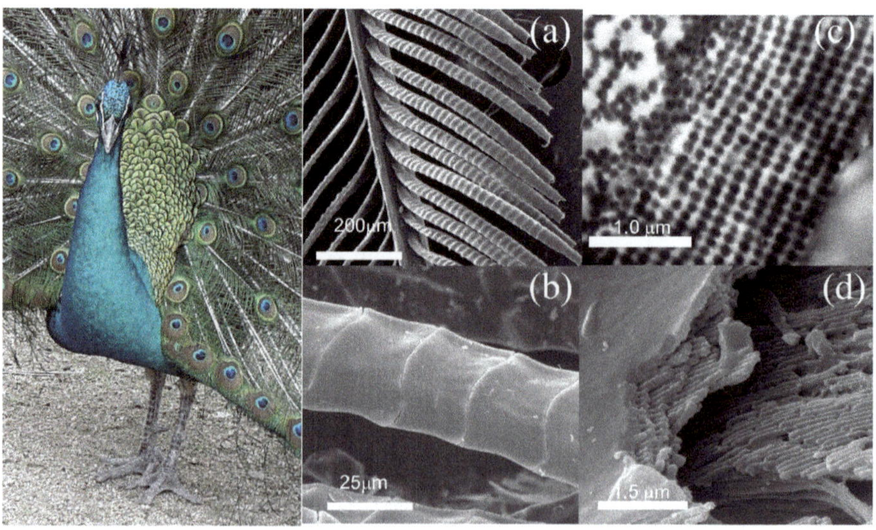

Fig. 5.5 (a, b) Feather barbule of a peacock, and (c, d) its cross section and inside (Yoshioka and Kinoshita, 2002). See Plate 12 in Appendix B.

segmented barbules varied from segment to segment, and even within a segment, and found that the color of barbule changed noticeably to blue when applying the pressure, while the swelling using alkali or ammonia made the color red. The liquid immersion test showed that a liquid of the refractive index between 1.55 and 1.60 destroyed the iridescent color effectively, while alcohol ($n = 1.37$) made the color only less brilliant. He bleached the feather using 3% hydrogen peroxide and found that the color of the feather changed to pale straw color, which was almost colorless under the transmitted light, and yet *their iridescent colors resisted the action of the bleach.* He insisted that the origin of the structural coloration of peacock's feather was due to thin color-producing films of a laminated or plate-like structure, and opposed to the surface-color theory on the basis of the angular dependence of the color change, the effect of bleaching, the presence of the brown or black pigments to enhance the iridescence, and also the swelling and pressure effects.

The sophisticated color-producing structure in peacock's feather was later clarified by Durrer (1962), using an electron microscope. He reported that the cortex surface layer of the barbule consisted of keratin, within which melanin rods of 1 μm length were arranged in a plane parallel to the surface along the longitudinal direction. Below this plane, totally 3–11 layers containing the melanin rods were arranged to form a 2D quasi-square lattice (Figs. 5.5(c) and 5.6(a)). The melanin rods were bound to each other by the keratin band. At the center of each square lattice, an air hole was present with increasing size toward the lower layer (see StG-type in Fig. 5.3). The medullary region of the barbule was filled with keratin filament with a small number of melanin rods and air holes. He found that the rod separation perpendicular to the surface correlated with the apparent color of the feather, while that parallel to the surface did not show such a correlation. The number of layers and the lattice constant were distributed according to the position within a tail feather of Indian peafowl *P. cristatus* (see Fig. 5.4 for notations): in I (dark blue), 9–11 layers were arranged with 0.16 μm separation; in II (turkish green), 9–10 layers with 0.17 μm; in III (red brown), 5–7 layers with 0.21 μm; in 1 (golden yellow), 4–6 layers with 0.208 μm; in 2 (violet), 4–7 layers with 0.21 μm, and in Au/M (green/red), 3–6 layers with 0.21 μm. Later, Durrer (1962) classified this type of melanin arrangement into StG-type (thin rod-shaped granules arranged in lattice), which was characteristic of the genera *Pavo* and *Afropavo*, exemplifying *P. cristatus*, green peafowl *P. muticus* (Zi *et al.*, 2003; Li *et al.*, 2005) and Congo peafowl *A. congensis* (Durrer and Villiger, 1975).

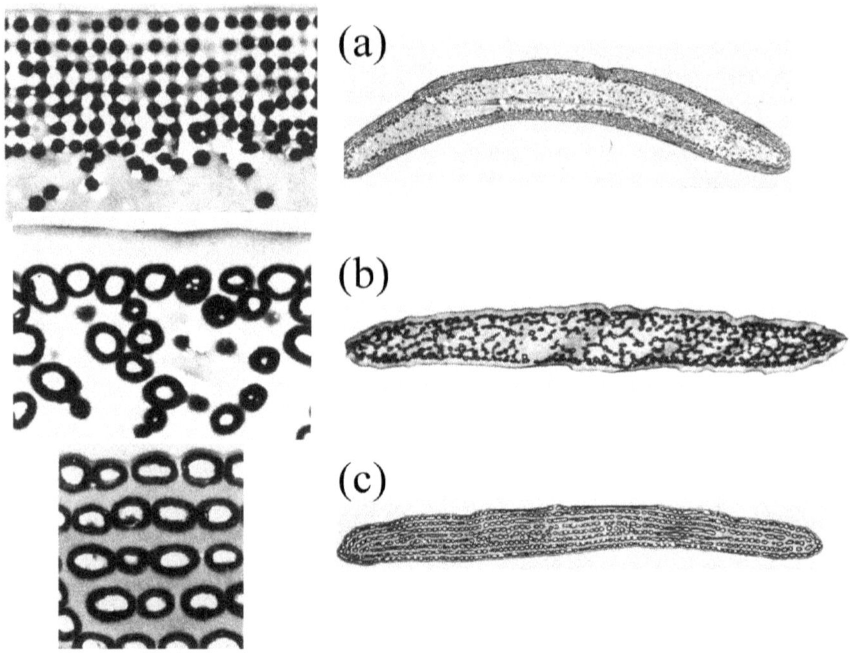

Fig. 5.6 Cross sections of the feather barbules of (a) *Pavo cristatus* (StG-type), (b) *Gallus gallus* (ORE-type), and (c) *Lophophorus impejanus* (RS-type). (Reproduced from Durrer (1977) with permission.)

In Fig. 5.5, we show the barbules observed using SEM (Yoshioka and Kinoshita, 2002). The cross-section of a barbule is crescent-shaped (Figs. 5.5(b) and 5.6(a)). Under high magnification, 8–12 layers consisting of periodically arrayed particles are observed beneath the surface layer (Fig. 5.5(c)), whose diameters range from 110 to 130 nm. The layer intervals are actually dependent on the color of the feather: for example, 140–150, 150, and 165–190 nm for blue, green, and yellow feathers, respectively (Yoshioka and Kinoshita, 2002; Zi *et al.*, 2003; Yoshioka and Kinoshita, 2005b). It is also noticed that the crescent-shaped barbule makes the lattice structure curved along the surface. In contrast to the transverse section, the particles in the longitudinal direction have a long shape with a length of 0.7 μm and are rather randomly distributed (Fig. 5.5(d)). These slender particles are melanin granules, and cause the barbule to bear dark brown when one sees through it.

Zi *et al.* (2003) reported the detailed calculations on this complicated structure. They employed a barbule of green peafowl, *P. muticus*, and observed its microstructure using SEM. They calculated the photonic band structure using the refractive indices of keratin and melanin of 1.54 and 2.0, respectively, and found a partial band gap along the Γ–X direction for two polarization directions. They also reported the reflection spectra for blue, green, yellow, and brown barbules under a microspectrophotometer and found sharp peaks located at 440 and 530 nm for the former two, respectively, while rather broad peaks with side peaks at a blue region for the latter two, indicating that the yellow and brown colors were non-spectral colors. They calculated the reflectance spectra for a finite number of lattice planes with the complex refractive index for melanin rods of $2.0 + 0.01i$ and then found to be fairly in good agreement with the experiment. Further, they paid special attention to a brown barbule, and found that even brown color was of a structural origin (Li *et al.*, 2005).

However, it is clear that even if these sophisticated calculations have been employed, they do not essentially reproduce the actual appearance of the peacock's feather. For example, the angular dependence of the reflectivity cannot be explained if not the crescent-shaped barbule structure superposing the lattice structure is taken into account. We have measured the angular dependence of the reflection spectrum with changing size of the illuminated area (Yoshioka and Kinoshita, 2002, 2005b). When several barbs are illuminated, the reflection spectra are found to be quite smooth and tend to shift toward blue when the angle of the observation is changed, while the reflection spectra are rather irregular and the angle-dependent spectral shift is less conspicuous when a single barbule is illuminated. We have performed the calculation to reproduce the angular dependence taking the variation in a macroscopic scale into account. For this purpose, we introduce the variation of the lattice orientation by assuming a Gaussian distribution for the tilt angle of the lattice. The calculated results seem to be in fairly good agreement with the experiment that was performed on several barbs. The tilt-angle distribution may partly reflect the crescent-shaped cross section of a barbule, and partly does the curved and twisted forms of the barbule. Thus, the macroscopic appearance of the feather should be explained through statistical average of microscopic structures which inevitably suffer from some sort of irregularity. It is quite important to consider both the regularity due to the photonic crystal and the macroscopic arrangement in order to reproduce the actual appearance of the peacock.

As described above, the arrangement of melanin granules like peacock (StG-type) is rather peculiar, and most of the avian species belonging to the family Phasianidae show either OSK- (rod-shaped granules in closed packing with surface layer) or StK/OStK-type melanin arrangement (thin rod-shaped granules in close-packing with/without a distinct surface layer) (Durrer, 1977). For example, ferruginous partridge (*Caloperdix oculea*), crested partridge (*Rollulus roulroul*), and satyr tragopan (*Tragopan satyra*) belong to the former, and black grouse (*Tetra tetrix*), golden pheasant (*Chrysolophus pictus*), crestless fireback pheasant (*Lophura erythrophthalmus*), and green pheasant (*Phasianus versicolor*) belong to the latter. In OSK-type, the rod-shaped melanin granules of 0.19–0.28 μm diameter are rather randomly arranged roughly in 2–3 layers parallel to the surface, over which a distinct keratin layer of 0.3–0.5 μm thickness is present. Thus, it is considered that the main part to produce the weak structural color in these species is the surface layer, and the melanin granules will act as a reflector in addition to the background absorber. Durrer (1977) considered this type as one of the primitive forms of color-producing structures in avian species.

StK or OStK-type shows the close-packed arrangement of thin rod-shaped melanin granules of diameter 0.12–0.14 μm, which produces strong iridescence. In this type, just under the surface keratin layer, thick layers of melanin granules with a triangular lattice are present, while in the inner part, only a small number of granules are randomly distributed. The melanin granules are bound with the neighboring granules by keratin bands and are surrounded by a thin air layer. Thus, the color-producing structures are mainly due to the photonic crystal with an additional contribution from the surface layer.

The exceptional cases for this family are found in ocellated turkey, *Agriocharis ocellata*, and wild turkey, *Meleagris gallopavo* (ORK-type: hollow tubes in close packing with surface layer), red junglefowl, *Gallus gallus*, (ORE-type: hollow tubes in a monolayer with surface layer; see Fig. 5.6(b)), and Himalayan monal, *Lophophorus impejanus* (RS-type: hollow tubes in multilayer; see Fig. 5.6(c)). All these types have air void inside the melanin granules, the diameters of which are distributed from 0.12 to 0.25 μm. These structures tend to show strongly iridescent colors, because the refractive index change between the melanin layer ($n \sim 2.0$) and air void ($n = 1.0$) is extremely large. Schmidt and Ruska (1962b) reported the microscopic structure of strongly brilliant feather of *L. imperiyaus*, using optical and electron microscopy. In this study, they reported that a barbule of *L. imperiyaus* has 6–7 layers of melanin rods having air voids inside, whose diameter was 0.28 μm. The thickness of the keratin layer between melanin rods was around 0.07 μm. The melanin rods were placed along a longer side of the barbule, but disorders were sometimes observed in the orientation. The number of the melanin layers decreased at the side edge of the barbule, which made a transparent appearance under transmitted illumination and under cross-polarized observation. Later, Durrer (1977) reported the thicknesses of the melanin layer and air void within a granule, and also that of the surface keratin layer for three typical feather colors as 0.064, 0.143, and 0.104 μm for turkish, 0.061, 0.129, and 0.098 μm for green, and 0.064, 0.137, and 0.168 μm for red feathers.

Rutschke (1966a,b) investigated brilliant feathers, showing blue, green and copper red, of 11 duck species using optical and electron microscopies. He showed an electron microscopic image of a feather of spectacled duck, *Anas specularis*, in which the rim of barbule observed in the cross section was covered completely with a regular arrangement of melanin rods. These rods were hexagonally packed like a honeycomb, and totally 7–8 and 5–6 layers were observed in the dorsal and ventral sides of the barbule. Each rod had a shape of a hexagonal prism with a rough diameter of 0.12–0.17 μm, and was surrounded by a thin keratin layer. At the surface of the barbule, two types of keratin layers with the total thickness of 0.09–0.34 μm were found to be present. This type of melanin arrangement was later classified into StK- or OStK-type (Durrer, 1977). Rutschke noticed that a melanin rod was not homogeneous, but consisted of many pigment-carrying microgranules. He also investigated the correlation between the diameter of melanin rods and the apparent color; however, he found no particular correlation except for the large diameter found in the case of coppery red feather.

If we compare these structures with that of a peacock's feather, one easily notices that the two-dimensional photonic crystal in ducks is hexagonally close-packed with a smaller number of layers without air holes, while that of peacock has a square lattice of 8–12 layers with an air hole at the center. The comparison clearly offers the speculation that the structure of a duck's feather may reduce the reflectance in a wide wavelength and may give an impression of a relatively dark color, because the melanin rods largely occupy the space. However, up to now, no experiment has appeared to clarify the optical properties of a duck feather in conjunction with the arrangements of the melanin rods.

5.3 Hummingbirds

In early days of the 20th century, it was believed that brilliant colors of bird feather were generated through thin-film keratin structure. On the contrary, Schmidt (1952) clarified that using dark field optical microscope, the melanin granules were largely responsible for the color-producing structures, and demonstrated that even a single granule generated the iridescent color. He found that barbules of several birds, including hummingbirds, were composed of melanin granules of various shapes. In a hummingbird (amethyst-throated sunangel, *Heliangelus clarissae*, now called *H. amethysticolis*), the shape of a melanin granule was described as a thin platelet, which covered almost the whole region of the barbule.

Greenewalt *et al.* investigated totally 50 species of hummingbirds and confirmed using an optical microscope that the colored surface of barbules consisted of a beautiful mosaic of colored platelets *resembling a tiled floor* (see Fig. 5.7(a)) (Greenewalt *et al.*, 1960a,b; Greenewalt, 1960). The platelets were of an elliptic shape with 2–3 μm length and 1–1.5 μm width (Fig. 5.7(b)). They further measured the reflection spectrum from the feather in a wide wavelength range between 0.3 and 2.2 μm, and found that a very sharp peak appeared only in a visible region with the width of 60–90 nm, the maximum reflectivity of which amounted to 50%. They investigated many hummingbirds' feathers and classified their reflection spectra into three cases: the spectrum showed (1) a single peak, (2) an additional blue peak, which disappeared when immersion oil was added, and (3) diverse anomalies. They exemplified fiery topaz, *Topaza pyra*, for case (1), Anna's hummingbird, *Calypte anna*, for case (2), and three species (not shown) for case (3). They noticed that the reflectance spectrum showed an extraordinarily narrow width in spite of high peak reflectance, which was not explained in terms of thin-film or multilayer interference theory.

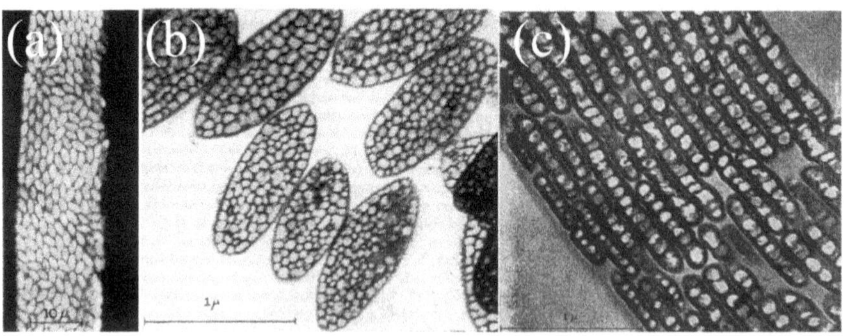

Fig. 5.7 Platelets covering a feather barb of hummingbird. (a) *Heliangelus viola*, and (b) the section containing a platelet plane and (c) the cross section of a barbule of *Chrysolampis mosquitus*. (Reproduced from Greenewalt *et al.* (1960a) with permission of The Optical Society of America.)

Greenewalt *et al.* measured electron microscopic image of the barbule and found that a platelet was not homogeneous, but had the appearance of *an elliptical pancake made from dough to which baking soda had been liberally added* (Fig. 5.7(b)). They found *a matrix containing air bubbles which in cross section resembled a very thin foam mattress.* Actually, 7–15 layers of discrete platelets lay in parallel with the surface in *Thalurania furcata.* The platelet consisted of air voids sandwiched by melanin layers (Fig. 5.7(c)). The air voids were partitioned into small chambers by melanin structural reinforcements that were thick near the upper and lower ends, while they were thin in the middle (Schmidt and Ruska, 1962b). Viewed from the feather surface, the air voids in a platelet formed irregular foam-like pattern. The thickness of the platelets ranged from 100 to 220 nm with the average of 150 nm.

They considered that the platelet constituted a half-wave interference plate by changing the thickness of air voids to accommodate the specific color. In order to reproduce the narrow spectral width of this system, they analyzed the structural feature as follows: A thin uniform keratin layer is located in the outer surface with a discontinuous rise of the refractive index at the interface between the keratin and the melanin layer; then the refractive index smoothly changed, decreasing toward the center of the platelet and then increasing again. Thus, periodic modulation in refractive index occurred. The reason for this continuous change in the refractive index owed mostly to the shape of the melanin structural reinforcements described above. Riccati differential equation was adopted for this system and the reflection spectrum was calculated (Greenewalt *et al.*, 1960a). They considered three layers of platelets with the refractive index of melanin as 2.0 and attained good agreement with the experiment. Particularly, the types of reflectance designated as cases (1) and (2) were assigned as the difference in the thickness of the surface layer. Thus, the blue peak observed in case (2) was concluded to arise from a very thin surface layer.

Schmidt and Ruska (1962b) also reported the sophisticated structure in a gorgeted sunangel (*Heliangelus strophianus*) in detail using optical and electron microscopy. Durrer (1977) classified this type of melanin structure as KK-type (platelets in close-packing), which was characteristic only in hummingbirds. He investigated a blue-tailed hummingbird (*Amazilia cyanura*) and reported the thicknesses of the melanin layer and air void as 0.032 and 0.123 μm. He also implied that the limitation of reflectance angle found in a hummingbird was due to the structure of a platelet divided into small chambers. Recently, Osorio and Ham (2002) investigated the strong directional reflection from a magnificent hummingbird *Eugenes fulgens* using a measuring system capable of moving the position of the light source and three-axis rotation of the sample. The maximum of the reflection spectrum from a crown feather was located at 440 nm within a rotation angle of the sample of $\sim 15°$; otherwise, the feather showed mat black. Only a small peak shift linear to the variation of the incidence angle was noticed.

However, in spite of the first success in explaining a narrow reflection spectrum using a model of continuous refractive index, no effort has been made to further explain the optical properties quantitatively. It is clear that the angular dependence characterized by the relatively small color change with varying viewing angle, has not been clarified yet, although the shape of the platelet clearly contributes to the anisotropic diffraction effect, and the melanin layer definitely plays a role of absorption contributing both to the spectral narrowing and background reduction as described in Sec. 2.3. It is strongly expected to construct a new model to explain these properties in a unified manner.

5.4 Trogons

Trogons are avian species belonging to the order Trogoniformes, which contains only one family Tronoidae. The trogons inhabit widely in Africa, Asia, and Central and South America. The feathers of male individual are known to glint with metallic green, violet, and red. The family Tronoidae contains six genera, *Apaloderma*, *Pharomachrus*, *Euptilotis*, *Priotelus*, *Trogon*, and *Harpactes*, among which species belonging to *Pharomachrus* are called quetzal.

Durrer and Villiger (1966) first investigated the feathers of five genera belonging to this family using the electron microscope and discussed the origins of the structural colors. They selected seven species as samples and compared the microstructures. The structural colors in these species are located in barbules, within which melanin granules are arranged.

For yellow-green dorsal feather of resplendent quetzal, *Pharomachrus mocinno*, and in cuprous to golden-green crown feather and blue-green tail feather of pavonine quetzal, *P. pavoninus*, they found that 5–7 layers of melanin platelets were arranged in the dorsal side of the barbule (Figs. 5.8(a), 5.8(d), and 5.8(e)). A platelet had an oval shape with 0.8–1.15 μm in length, 0.27 μm in width, and 0.18–0.27 μm in thickness. Within

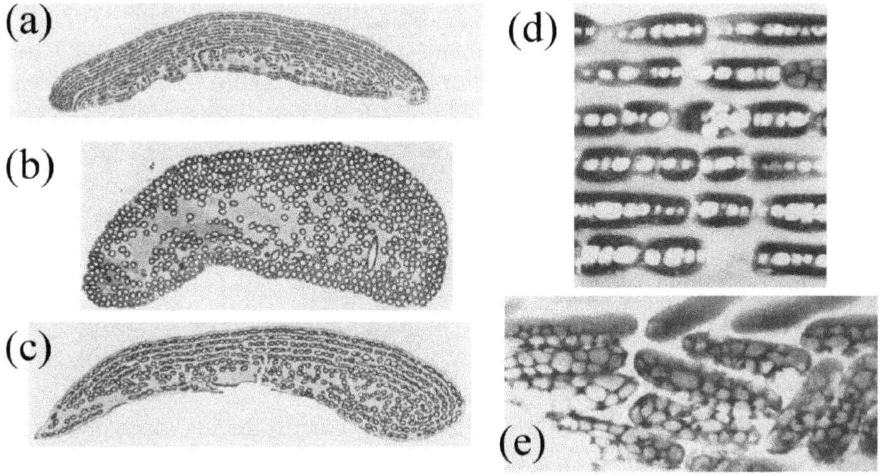

Fig. 5.8 Cross-sections of barbules in (a) *Pharomachrus pavoninus*, (b) *Trogon rufus*, and (c) *Apaloderma narina*. (d) A longitudinal section and (e) the section containing a platelet plane of a barbule of *Pharomachrus pavoninus* are shown. (With kind permission from Springer Science + Business Media: Durrer and Villiger (1966).)

the platelet, air void was present, which was surrounded with melanin layers. The air void was chambered into small rooms by thin keratin walls. The thicknesses of melanin layers and air void were found to be 0.07–0.08 μm and 0.05–0.1 μm, respectively. The overall appearances are reminiscent of those found in hummingbird's feather. However, the arrangement of the platelets is quite different. In quetzal's feather, the 2D layers of platelets are separated by 0.1–0.14 μm thick keratin layer, while in that of hummingbird, the platelets are almost in contact with each other. Thus, the arrangement of platelets in quetzal looks like a multilayer system. Later, Durrer (1977) defined this type of arrangement as KS-type (platelets in multilayer).

On the other hand, the feathers of black-throated trogon, *Trogon rufus*, and biolaceous trogon, *Trogon violaceus*, showed very different microstructures. The melanin granules had a shape of hollow tube closed at both ends, which were close-packed to become a pile of 3–8 layers (Fig. 5.8(b)). The diameter of the closed tube was found to be 0.19–0.23 μm, with the thickness of the melanin layer of 0.036–0.053 μm and the diameter of air cavity of 0.096–0.127 μm. The melanin tube had a length of 1.25–1.5 μm, within which thin melanin bands divided the cavity into small chambers. Between the adjacent tubes, air bubbles were sometimes observed. This type was then called RK-type (hollow tube in close-packing) (Durrer, 1977). In narina trogon, *Apaloderma narina*, the melanin tubes were found to be arranged in layers to constitute the multilayer of totally 4–5 layers (Fig. 5.8(c)). Between the layers, keratin layer of thickness of 0.11 μm was present (RS-type: hollow tube in multilayer) (Durrer, 1977). In contrast to the species belonging to the above three genera, which display strong metallic colors, the species in the remaining two genera show rather weak reflectance. In cuban trogon, *Priotelus temnurus*, only 1–3 layers of hollow tubes are present, while in blue-tailed trogon, *Harpactes reinwardtii*, only one layer of hollow tube is separated from the other seemingly random distribution of underlying hollow tubes.

Durrer and Villiger (1966) analyzed the color-producing mechanisms of these trogons by considering the conditions for constructive interference using a thin-layer interference model. Using the refractive indices of 1.5 and 2.0 for keratin and melanin layers, respectively, they compared the results with the observed color of the feather. From this analysis, they concluded that the color-producing structure in *Trogon* was the repetitive structure of the hollow tube, while those in *Pharomachus* and *Apaloderma* were due to the multilayer structure of the platelets or hollow tubes. On the other hand, in case of *Harpactes*, they implied that the sum of the hollow tube and wide keratin layer caused the higher-order interference. However, since no quantitative study has been made after then, the coloring mechanisms of trogons seem to be still hidden behind the veil.

5.5 Pigeons

Next, we will show that even the simplest structure of a thin film yet produces the ingenious color effect to human and possibly animal's eyes (Yin *et al.*, 2006; Yoshioka *et al.*, 2007). The neck feather of the commonly spotted rock dove, *Columba livia*, shows unusual optical characteristics: the green (purple) feather located at the neck suddenly changes its color into purple (green) only by slightly shifting the viewing angle, which is quite in contrast to the gradual color change observed in usual iridescence. The color of the rock dove's feather is mainly due to numerously sticking barbules, whose length, width, and thickness are typically 350, 40, and 3 μm, respectively. The cross section of a barbule is crescent-shaped, which produces a very broad reflection pattern in one plane under light illumination (Fig. 5.9(a)). It is remarkable that the reflection pattern from a single

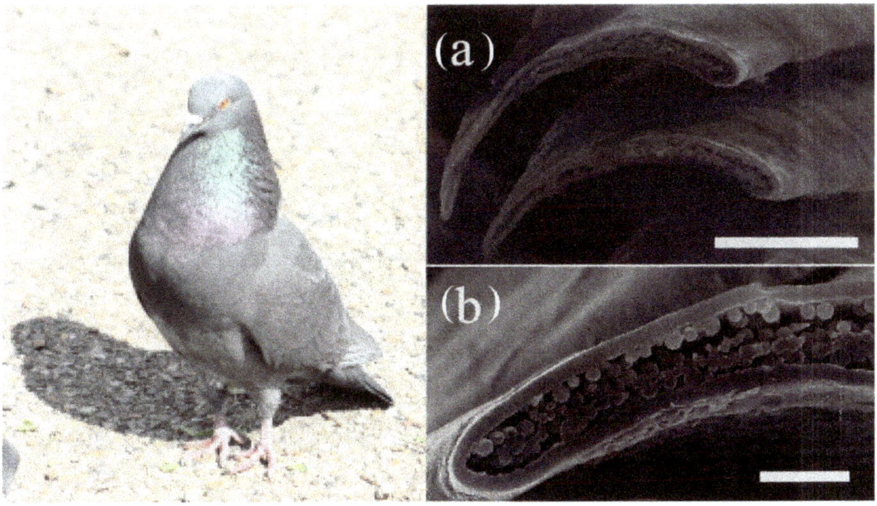

Fig. 5.9 Cross sections of a neck feather of rock dove, *Columba livia* (Yoshioka *et al.*, 2007). See Plate 13 in Appendix B.

barbule of a green (purple) feather shows the sudden change from green (purple) to purple (green) with a rather distinctive gray boundary.

We have investigated the electron microscopic images of the cross section of a barbule and have found that a lot of melanin granules are enclosed below the outer cortex layer (Fig. 5.9(b)). The diameters of the melanin granules are randomly distributed from 500 to 750 nm, and their arrangement is also irregular. Thus, they will not contribute directly to the coloration. Thus, the cortex layer with a rather uniform thickness becomes the one and only candidate for the coloration, whose thickness ranges are 600–700 nm for a green feather and 480–580 nm for a purple feather. In fact, the reflection spectra from a single barbule displaying green, gray, and purple colors show multiple sinusoidal reflection spectra (Fig. 5.10(a)), which are gradually shifted according to the perceptible color. These spectra are found to be in very good agreement with an expectation of thin-film interference, assuming, for example, a layer thickness and refractive index of 650 nm and 1.5 for a green feather.

Thus, the neck feather of rock dove is actually caused by the simplest mechanism of thin-film interference. However, the optical interference occurs on higher orders, which produce multiple peaks in the visible wavelength range: a green feather shows $m = 3$, 4, and 5 interference peaks at 700, 520, and 420 nm under the incidence angle of 30°. The most important

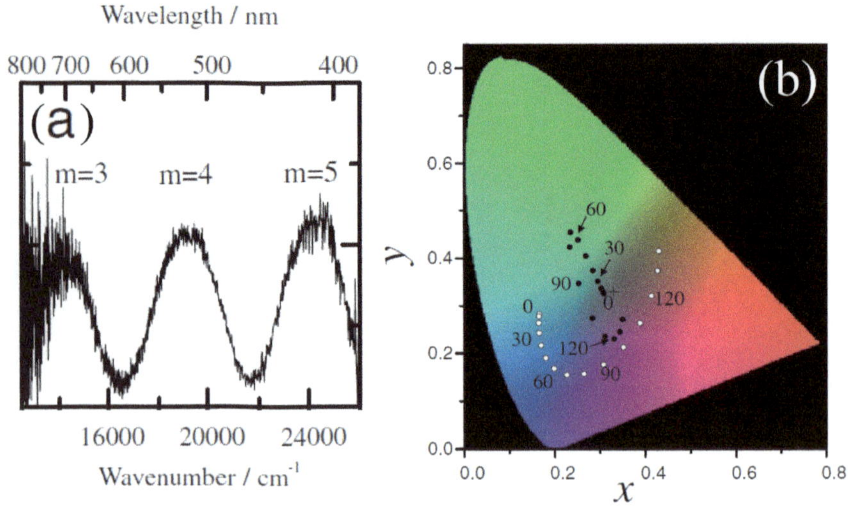

Fig. 5.10 (a) Reflection spectrum from a green neck feather and (b) chromaticity diagram showing loci of the reflection from the thin-layer interference at various angles. The thicknesses of the layers are 650 nm (closed circle) and 400 nm (open circle) with a refractive index of 1.5 (Yoshioka *et al.*, 2007).

point to produce this *two-color iridescence* is based on the fact that the separation of the two adjacent peaks in wave number unit coincides with that of the color-matching functions of human vision for the blue and red ones. Thus, when this condition is satisfied, one may perceive purple color, whereas when the viewing angle is changed and the reflection peak shifts from red to green and also from blue to ultraviolet, one may perceive only green color. Among these colors, one feels the gray color because all the colors are perceived almost equally.

This situation is visually understood in terms of chromaticity diagram, where two-color iridescence is expressed by a linear movement of the loci crossing a white (gray) zone ($x = 1/3$ and $y = 1/3$), as shown in Fig. 5.10(b). It is clear that the multiple peaks due to thin-layer interference are suitable to enhance the two-color iridescence, because the largest peak shift is obtained when the locus crosses a gray zone. However, to satisfy these conditions, it is necessary to strictly control the thickness of the cortex layer, because if the layer is thicker, the locus only moves near the gray zone, and then no particular color sense is stimulated, while if it is thinner, the locus draws a circle like an ordinary iridescent thin film. The two-color iridescence will be more effective to avian tetrachromatic vision, because the color-matching functions in avian vision are much more well-defined owing to the presence of the color filter (oil drop).

5.6 Non-iridescent Colorations — Kingfishers, Parakeets, Cotingas, and Jays

Scattering of light is one of the important sources for the structural colorations, which is completely different from thin-layer and photonic-crystal types, because it has a root on the irregularity of the structure. Light scattering has been thought of as a cause of blue coloring in a wide variety of random media. The origin of blue sky was explained first by Lord Rayleigh as due to light scattering by atmospheric molecules. The color of the suspension involving small colloidal particles is called Tyndall blue. The blue colors in these cases are based on the fact that the scattering cross section is dependent on the fourth power of the light frequency. These phenomena are generally categorized into Rayleigh scattering, where light scattering due to particles of a sufficiently small size as compared with the wavelength of light occurs. With increasing size of the particle, Mie scattering occurs, whose wavelength dependence differs considerably from Rayleigh scattering as described in Sec. 2.5. Thus, the different coloration should appear through the latter process.

The light scattering or Tyndall blue has often appeared in literature as a typical example of non-iridescent structural colors. Mason (1923a, 1926) reported that the colors of some birds and insects were due to Tyndall blue, and exemplified the feather of a blue jay, and the body/wing of dragonflies. He considered six necessary conditions for a medium to give Tyndall blue: (1) inhomogeneities in the refractive index; (2) dimensions comparable to the wavelength of light; (3) blue scattering and red transmission; (4) size-dependent depth of blue; (5) polarized scattering, and (6) inversely proportional to the fourth power of the wavelength.

At that time, it was a central problem to seek out the microstructures responsible for Tyndall blue. In 1939, the first attempt was made to clarify the mechanism of blue coloring in the feathers of ivory-breasted pitta, *Pitta maxima*, and turquoise tanager, *Tangara mexicana*. Frank and Ruska (Hermut Ruska) applied "Siemens Super Microscope", the first marketable electron microscope developed in this year by Ruska (Ernst Ruska) and Borries, to the problem of Tyndall scattering (Frank and Ruska, 1939; Frank, 1939). Within a barb of blue feather, they found a spongy structure consisting of keratin and air at the inner wall of the medullary cell. The spongy structure was made from air-filled space of 0.1–0.25 μm, which could not be observed under an optical microscope. The space dividing the adjacent air spaces was found to be very thin with the thickness of 0.01–0.015 μm.

Later, more detailed observation on the spongy structure was made by Schmidt and Ruska (Helmut Ruska) (1962a) on violet-red feather of Eurasian jay, *Garrulus glandarius*, deep-blue feather of purple-breasted cotinga, *Cotinga cotinga*, and light-brown feather of blue-and-yellow macaw, *Ara ararauna*. They observed that spongy structures sometimes consisted of numerous spherical or oval vacuoles (foam) with the mean diameter of 0.2 μm (see Fig. 5.12(a)). The walls in this case were so thin that it could not become a framework of the spongy structure. On the other hand, the foam structures sometimes deformed considerably to attain a network structure with hard sticks and air holes.

The spongy structure (known as "cloudy medium" under optical microscope (Auber, 1957, 1971/1972)) is now known to be widely distributed in avian species. Kingfishers (see Fig. 5.11), parakeets, cotingas, jays, manakin, and bluebird are famous examples of this type. Their structural colors are apparently characterized by (1) weak or no change in color with varying view angles, and (2) rather low reflectivity with the lack of directionality. Most of the spongy structures are found in the medulla of the feather barb, which consists of several medullary cells running along the longer axis of the barb, and is covered by the cortex. The spongy structure mostly lies in the inner part of the medullary cells, but sometimes covers almost the whole barb. At the center of the barb, some vacuoles are distributed, which correspond to the nuclei of medullary cells, around which small melanin granules are distributed.

Owing to the above detailed observations using electron microscopy, it was generally believed that Tyndall blue was the only cause for the blue coloration in the spongy structures. Yet, there had been no report to measure accurately the wavelength and angular dependence of the scattering cross section and its polarization characteristics, which were the necessary conditions for Tyndall blue, proposed by Mason in 1920s (Mason, 1923a). Thus, the problem whether pure Rayleigh/Tyndall scattering determined the colors of animals or not was actually unknown at that time.

However, it was gradually recognized that light scattering was somehow modified and contributed effectively to the animal colors. The first implication on the presence of interference phenomenon on this problem was made by Raman (1934a), who threw doubt on Tyndall scattering in the feather of

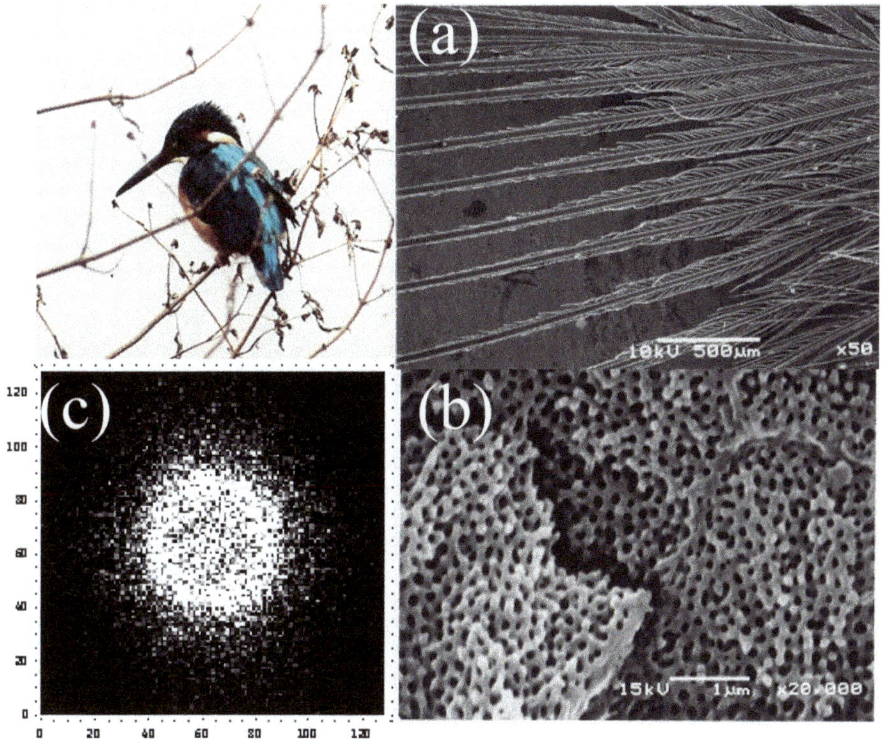

Fig. 5.11 (a) Barbs and (b) the cross section of a barb of a common kingfisher, *Alcedo atthis*, and (c) 2D Fourier image of the TEM image of its spongy structure (Kinoshita and Yoshioka, 2005b). See Plate 13 in Appendix B.

Indian roller, *Coracias indica*. He observed under Leitz "ultra-opak" microscope and found that the barb involved a series of cells, polygonal in shape, and the color varied individually from cell to cell. Further, he noticed that when the barb was fully damp, the dark blue color changed to bright green, while light blue to light red. He compared these results with the light-scattering phenomena known for particles of various sizes and concluded the inapplicability of the light scattering theory. He suggested, without a final decision, the possibility of diffraction due to cavities not small compared with the wavelength or the interference of light due to a film or extended cavity.

Dyck (1971a,b) performed detailed anatomical and optical investigations on the feathers of rosy-faced love bird, *Agapornis roseicollis*. He found that the inner part of the barb was completely filled up with spongy

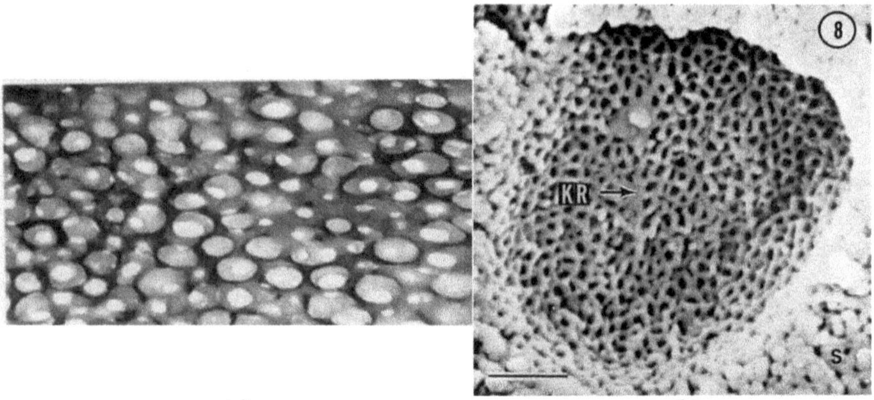

(a) (b)

Fig. 5.12 (a) Foam-like spongy structure in *Cotinga cotinga* (With kind permission from Springer Science + Business Media: Schmidt and Ruska (1962a).), and (b) rod- and air-channel type spongy structure in *Cotinga maynana*. (With kind permission from Springer Science + Business Media: Dyck (1971b).) KR indicates keratin rod, with the scale bar of 1 μm for the latter.

structure and the nuclear vacuoles surrounded by melanin granules were distributed in the central part of the barb. The spongy structure consisted of three-dimensional network of connecting keratin rods of 0.1 μm wide, which supplied empty spaces of 0.1 μm in diameter among them (see Fig. 5.12(b)). He investigated the spongy structures in various avian species and found that the widths of the rods and air spaces were strikingly uniform in size, while their orientations were definitely random. The type of the spongy structures varied among species: for example, in blue feather of *Cotinga maynana*, in contrast to the case in *A. roseicollis*, the structure consisted of spherical cavity whose openings are mostly polygonal with the width of 0.05–0.25 μm. Dyck (1976) measured the reflection spectra of violet, blue, and green barbs of three structurally colored avian species, and found a clear peak from ultraviolet to visible regions, which were in good accordance with the apparent colors of the feathers. Further, he found that the sizes of keratin rods and air spaces varied strongly in correlation with the reflection color. He considered the interference of light due to a small unit as the cause for the structural colors and proposed a "hollow cylinder model" in order to explain the interference phenomena in the spongy structure. Namely, the spongy structure was considered to consist of randomly oriented hollow cylinders with the diameters of 200–400 nm. He compared the wavelengths giving the maximum reflectivity with those of the constructive interference conditions for normally incident light, and found a relatively good correlation between them.

Along this stream, Prum *et al.* (1998) performed a spatial Fourier transformation of the digitized TEM image of the cross sections of the medullary spongy structure from the blue feather barbs of *C. maynana* and found a clear ring structure of 0.0059 nm^{-1} in diameter around the origin in momentum space. If spongy structures with a wide variety of sizes were distributed, a Gaussian-like distribution would be obtained around the origin. The presence of a ring structure clearly indicated that the uniform characteristic size of the matrix exists actually, which should correspond to the reflection maximum observed at 500–520 nm. They estimated the reflection spectrum from the 2D Fourier power spectrum by averaging the 2D power spectrum radially and by converting it into real space using the mean refractive index of 1.283. The result is fairly in good agreement with the experiment. The ring structure in momentum space is easily reproduced by placing randomly oriented broken multilayers, whose layer thicknesses coincide with the widths of the air channel and keratin rod, respectively. It is intuitively understandable that the medullary keratin matrix functions to produce brilliant hues similar to that of broken multilayer reflectors as in *Morpho* butterfly, although the reflection is nearly omnidirectional owing to the random orientation.

They applied 2D Fourier method to the feathers of various avian species such as rose-faced lovebird (*Agapornis roseicollis*), budgerigar (*Melopsittacus undulatus*), Gouldian finch (*Poephila guttata*) (Prum *et al.*, 1999b), and blue whistling thrush (*Myiophonus caeruleus*) (Prum *et al.*, 2003), and found a similar ring structure in the 2D Fourier image. Thus, it is quite clear that the spongy structures generally possess a characteristic length that contributes to the generation of a well-defined reflection spectrum through constructive interference. They also investigated avian skins of various species, which appeared as quasi-ordered arrays of parallel collagen fibers, and found similar spatial order (Prum *et al.*, 1994, 1999a; Prum and Torres, 2003a,b).

It is now clear that a simple model based on Rayleigh/Tyndall scattering is no longer applicable to explaining the non-iridescent blue colorations, since a definite peak is observed in the reflection spectrum (Dyck, 1971a, 1976; Finger and Burkhardt, 1992; Finger, 1995; Andersson, 1996, 1999; Andersson and Amundsen, 1997; Shawkey *et al.*, 2003; Osorio and Ham, 2002; Prum *et al.*, 1998, 1999b), and also clear spatial correlation exists in seemingly random structures. It appears that these structures are widely distributed in the animal world, particularly in birds (Prum, 2006) and insects (Tilley and Eliot, 2002; Kinoshita and Yoshioka, 2005b; Prum *et al.*, 2004, 2006). Their structural units appear sometimes in the form of network, of stacks having random holes and sometimes of randomly distributed spheres. These findings are quite important because the structures actually possess uniform characteristic length, which changes the incoherent scattering into coherent scattering, with enforcing reflection to a large extent.

Although the 2D Fourier transform method has been widely used to find out the spatial correlation within seemingly random structures (Shawkey *et al.*, 2003; Doucet *et al.*, 2004; Shawkey and Hill, 2006), as for the analysis of their optical properties, many problems still remain unsolved. Prum *et al.* attempted to combine the 2D Fourier images with the reflection spectrum using an average refractive index calculated from the areas of keratin rods and air channels. However, this treatment essentially neglects the multiple scattering, which is essential to cause the interference of light within regularly arranged structures and hence to produce highly reflective band in the reflection spectrum. Thus, in spite of the usefulness and easiness of the 2D Fourier method, it should always be kept in mind that the 2D Fourier image does not give essentially an answer to the actual scattering problem.

Chapter 6

Fish

6.1 General Description

Fish is one of the well-known animals abundant in structural colors. Although almost all the fish species anyway show the structure-based optical phenomena, the color-producing structures and their mechanisms have not been well understood as compared with those in insects and birds. The most characteristic point in fish is that those structures are embedded in living cells. Thus, the color-producing structures are dynamically controlled under nervous systems, and also under hormonal and physiological conditions. This makes the researches in this field quite complicated.

The structural colors in fish appears as brilliant colorings and splendid reflectors in skins, scales, corneas, and tapeta. In the present chapter, we will mainly deal with those in the first three, which are characteristic of fish. The colors of fish skins are generally controlled by specialized cells called chromatophores. The beautiful colors and patterns, silvery white franks, and spectacular color changes are all achieved by the combined action of these pigmentary cells. Generally, the chromatophores are classified into several types according to the pigmentary granules involved and the microstructures inside the cells (Fujii, 1993a,b). Above all, the following five chromatophores are widely known and are universally distributed in fish: melanophore, xanthophore, erythrophore, leucophore, and iridophore. Among these, the last two are belonging to structural colorations, while the first three contribute to pigmentary colorations. The first four are known to be dendritic in shape, while the iridophores are usually round.

Melanophores are widely distributed in poikilothermic vertebrates and contain small granules of $\sim$0.5 μm in diameter called melanosomes, which include melanin pigments and make the skin dark brown or black. The

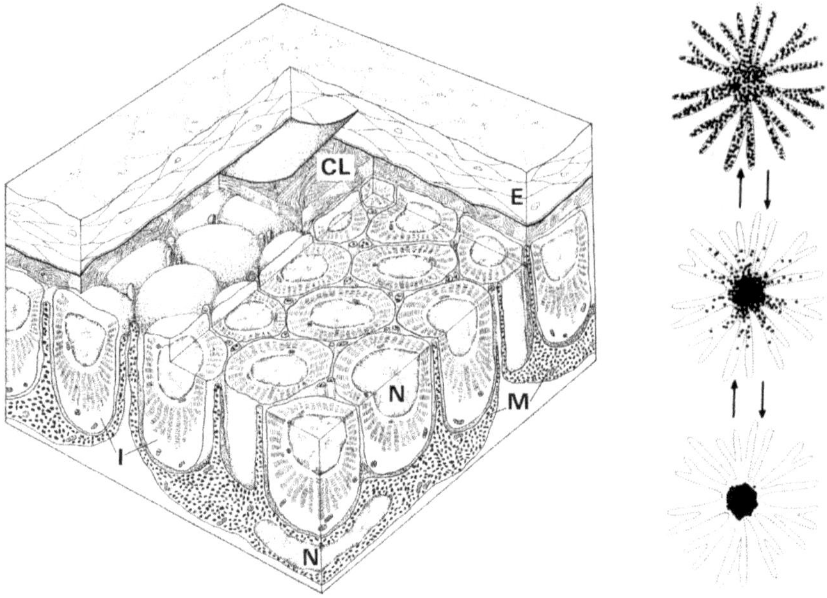

Fig. 6.1 Schematic illustration of the dermal chromatophore units of blue damselfish (left) (Reproduced from Oshima (2005) with permission.) and motile responses of a chromatophore of the dendritic type (right) (Reprinted from Fujii (1993a), with permission from Elsevier.). CL: collagenous layer, E: epidermis, I: iridophore, M: melanophore and N: nucleus.

mature melanosomes are usually round, while they are more slender in immature granules. These granules are quickly translocated to the dendritic processes when the dark coloration in skin occurs, whereas they are concentrated around a center when the other colorations are restored (Fig. 6.1). The chromatophores other than melanophore are sometimes called "bright-colored chromatophores." Xanthophores and erythrophores are belonging to pigmentary chromatophores that contribute to yellowish and reddish colorations, respectively. The shapes of these cells are dendritic, which are almost coincident with that of a melanophore. These chromatophores contain pigmentary granules called xanthosomes and erythrosomes, which normally include the pigments of water-insoluble carotenoids and water-soluble pteridines.

Leucophores are sometimes included in iridophores and are known to bear the whitish appearance owing to the light scattering. These cells include the light-reflecting organelles called leucosomes of 0.5–0.8 μm in diameter, which are translocated to dendritic processes similarly to the other chromatophores. A leucophore is usually located concentrically just beneath a melanophore to behave quite oppositely in response and the combination of these chromatophores is sometimes called melaniridosomes. This means that when the leucosomes are contracted, the melanosomes are expanded and make the skin in dark color. On the other hand, the skin color changes to whitish, when the inverse process occurs. The leucophores are known to be distributed among limited species such as killifish (Menter *et al.*, 1979), medaka (Fujii, 1993a), and guppy (Takeuchi, 1976).

Iridophores play a central role in the structural colorations within fish skin. They usually involve definite stacks of purine crystals, i.e. guanine, hypoxanthine or uric acid, which constitute a multilayer to reflect the light with specific colors and/or broad silvery whitish colors. Usually the iridophores can be further divided into static and motile ones according to the response to the physiological change. The most of iridophores distributed in the skin are belonging to static ones, which appear as silvery whitish flank, brilliant blue skins and so on. This type of iridophore was extensively studied since 1900s. On the contrary, the presence of motile iridophores was noticed in 1980s and up to now, not so many species are known to exhibit the motilities against the neurotransmitters, hormones and light stimulations. Neon tetra, damselfish, and goby are such well-known examples.

6.2 Static Iridophores

Guanine crystals found in iridophore are generally so large that they are easily observed under an optical microscope as a form of hexagonal platelet. This made it possible to investigate the platelets in the first half of 1900s and even the thicknesses of the crystals, though they are less than the wavelength of light, were measured using developing interference microscopy (Fox, 1976). However, since the stacking of the platelets was less than the resolution of optical microscopy, the exact observation had to wait for the electron microscopic observation after 1960's. Kawaguti *et al.* reported a series of electron microscopic studies on various living objects (Fig. 6.2). They reported on blue parts of the Japanese wrasse, *Halichoeres poecilopterus* (Kawaguti, 1965), blue spots on the Japanese porgy, *Diadema setosum* (Kawaguti and Kamishima, 1964; Kawaguti, 1966), blue back of blue-scaled herring, *Harengula zunasi* (Kawaguti and Kamishima, 1966), and silvery layer of the Japanese sardinella, *Sardinella zunasi* (Kawaguti, 1967), and observed that guanine crystals of 20–100 nm thickness were stacked with the spacing of ~170 nm.

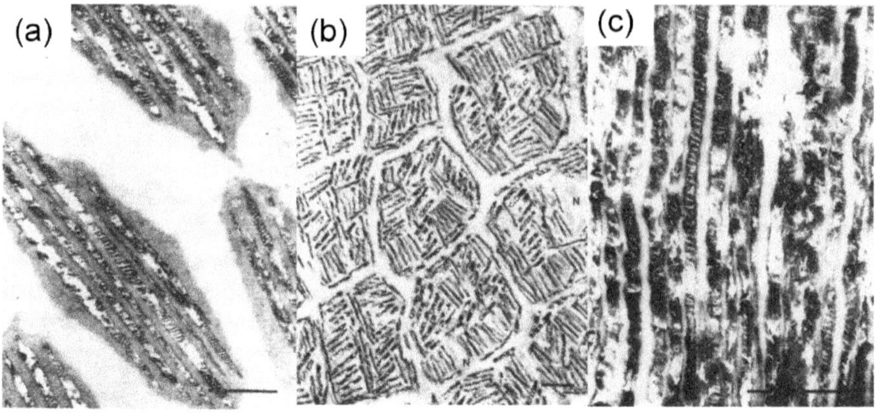

Fig. 6.2 Electron microscopic images of guanine platelets in iridophores at (a) ventral silvery part and (b) ventral blue part of clupeid fish, *Sardinella zunasi*, and (c) blue spot of Japanese poggy, *Chrysophrys major*. (Reproduced from Kawaguti (1967) with permission. Copyright (1967) The Zoological Society of Japan.)

Denton and Land observed ventral scales of juvenile sprat, *Clupea sprattus*, and found that on the surface of coppery-colored scale, 4–5 layered thin platelets of $25 \times 5\,\mu m^2$ were arranged parallel to the scale. The platelets were found to be arranged so as to make an angle of 15° to the plane of the scale (Denton and Land, 1967, 1971; Denton, 1970, 1971). Since under the transmitted light, they were seen as the complimentary color, the structural colors were thought to be a main cause of the coppery colors. They detached a scale and immersed it into water, which caused the drastic color change into violet, while in the ordinary Ringer solution, the color did not change. The reflectivity calculated from the measurement of the transmittance showed a broad peak around 650–800 nm with the maximum reflectivity of 50%. Using the thickness of the crystal estimated from an interference microscope was 100 nm with the refractive index known for guanine crystal of 1.83, they confirmed that the peak position and reflectivity were reproduced fairly well assuming an ideal multilayer reflector. They also confirmed that the change of salt concentrations caused the change in the cytoplasmic thickness between the multilayered platelets.

Denton and Land also investigated its silvery whitish flank and found that the reflection colors changed with the positions on a scale (Denton and Land, 1971; Land, 1972). Namely, in the anterior part, it looks orange, while in the middle and posterior parts, it looks purple and green, respectively. In a Ringer solution with doubling concentration of salt (double-strength

Ringer solution), the colors are changed into blue + yellow, orange, and purple, respectively. They found that using the thicknesses of the platelets of 98, 109, and 104 nm from the anterior to posterior, and assuming an ideal multilayer, the thicknesses of the cytoplasm between the platelets were estimated at 83, 149, and 278 nm, which shrunk into 45, 75, and 166 nm after adding the double-strength Ringer solution. These considerations led to the conclusions that the purple color seen in Ringer and double-strength Ringer solutions are due to the nonspectral color resulted from the first- and second-order interference peaks, while green color seen in Ringer solution was due to the second-order interference. In an ideal multilayer of $(\lambda/4 + \lambda/4)$-system, where the optical path lengths for the two layers were both equal to $\lambda/4$, the second-order reflection did not occur in general. He considered that $(\lambda/4+3\lambda/4)$-system for either the higher or lower refractive index layer were most suitable to give the maximum constructive interference under nonideal condition. In the epidermis, these scales are geometrically piled in threefolds with slightly shifting their positions. Thus, the silvery color in sprat was caused by skillfully distributed platelets to generate the optical mixing of the reflecting colors through the geometrical overlapping of the different parts of the scales.

Similar considerations were made for horse mackerel, *Trachurus trachurus*, in which the reflecting platelets were found to be located in the dermis (Denton and Land, 1971). In the dorsal blue part, the platelets of 66–86 nm thickness were considered to generate blue color under the assumption of ideal multilayer, while in the flank, two iridophore layers with the long axes of the crystals crossed each others produced the reflection of red and yellow-green colors, which causes the silvery color by optical mixing.

On the other hand, in a scale of the roach, *Rutilus rutilus*, the reflected color appeared to be nearly uniform silvery white (Denton and Land, 1971). Since the thickness of the crystal lay around 100 nm similar to the ventral scale in sprat, which corresponded to the reflection maximum of 740 nm, he considered a random distribution of the thicknesses of the cytoplasms might function as a broadband reflector. Thus, he concluded that the silvery color was produced in the following three ways: (1) three $\lambda/4$ stacks, each reflecting one-third of the visible spectrum, (2) a single stack with systematic variation of optical thickness, and (3) a single stack with random distribution of the intervals. At present, similar static iridophores are widely known in fish. Among these, the random stacking proposed by Denton and Land is now known as "chaotic" broadband reflector, which is widely distributed in silvery fish (Parker, 2005a). Denton and Nicol investigated various teleosts bearing silvery surface and determined the inclination angles of the platelets against the surface of the fish skin through optical and histological investigations (Denton and Nicol, 1965a,b,c, 1966). They found that on the broad flank of the fish, most of the reflecting platelets located beneath the scales were situated nearly parallel to the fish skin and hence almost perpendicular to the sea surface, while on the curved dorsum, they were inclined considerably from the skin surface, but maintained their directions nearly perpendicular to the sea surface. They considered the reflecting platelets played a role of camouflage by effectively reflecting the light from upward in the environments such as in clear oceanic waters, or in shallow or turbid rivers, where light was distributed almost symmetrically about the vertical to the surface.

6.3 Motile Iridophores

The color changes under the change in physiological conditions are well known in poikilothermic vertebrates, in which the contractive and extensive actions of the chromatosomes in various pigmentary chromatophores play a major role. Thus, at the beginning, people believed that the light-reflecting iridophore played only a passive role in these color changes. Foster reported the color change in the reflecting patch of killifish, *Fundulus heteroclitus* (Foster, 1933). When the fish was stimulated by direct sunlight, the reflecting patch changed the color very rapidly from blue/green to orange/wine-red within 5 s, while the reverse change took 40 s to 2 min. In addition to the mechanical and electrical stimulations, the change of the osmotic pressure was also found to change the color. He confirmed that light stimulated the iridophores directly, which was independent of nervous system. He considered that the mechanism of the color change was that the laminae of the crystals, if existed, were in some way thickened or thinned progressively during the reaction phase.

The stacking structures of motile iridophores in neon tetra, *Paracheirodon innesi*, inhabiting in the Amazon River, were first observed by Kawaguti using electron microscope (Kawaguti, 1967). He reported that iridophores in brilliant blue lateral stripe of neon tetra containing large guanine platelets of 70–100 nm thickness, which were stacked parallel with the interval 30–70 nm. The fact that thin platelets were densely packed was considered as an origin of brilliant blue coloring through light interference and with the help of melanophores. Denton and Land reported that the iris of neon tetra had two reflective systems (Denton and Land, 1971): that in the upper side reflected blue light and contained platelets of $25 \times 7\,\mu m^2$ with the thickness of 62–66 nm, while that in the lower part reflected red color and contained platelets of $20 \times 3\,\mu m^2$. They considered an ideal multilayer system for both cases. Rohrlich investigated cardinal tetra, *Paracheirodon axelrodi*, the iridophores of which showed green color in daytime, while it changed into dark bluish-violet without iridescence at night (Rohrlich, 1974). Under an electron microscope, she observed the regular parallel stacking of broad, thin crystal platelets in such a physiologically active iridophore. She speculated that thin 6.5 nm filaments linking successive crystal platelets played a role in the movement of the iridophore components.

Lythgoe and Shand (1982, 1989) noticed the pronounced color change between day and night in neon tetra. In daytime, it shows a brilliant blue-green stripe, while it does dull violet-blue color at night times. They performed various experiments and found that the color change occurred under light stimulation and the blue color changed into violet after placing it for 25–35 min in the dark. Later, using an immunofluorescence technique, Lythgoe *et al.* clarified the responses to light was related to an opsin-based photopigment, e.g. rhodopsin or porphyropsin, within the iridophore (Lythgoe *et al.*, 1984).

The color changes was also observed under various physiological conditions such that osmotic changes to hypotonic conditions changed the color into orange, while the addition of adrenaline changed it toward the longer wavelength. They described that two types of light-reflecting crystal were distinguished in the dermis: one is a wide crystal distributed solely in the stripe zone with the thickness of 5–10 nm, and the other is a slender crystal of 43–108 nm thickness, which is distributed in silvery dermis, ventral region, and even in lateral stripe and iris. The former is clearly active platelet, while the latter is inactive. They found a clear correlation between the spacing of the former platelets and the apparent colors. They measured the change in the reflection spectrum due to osmotic change and compared it with the calculation assuming the multilayer interference (Clothier and Lythgoe, 1987). The thickness of guanine platelet is as thin as 5–10 nm and thus the multilayer system is strongly nonideal. Thus, the high reflectivity is attained only by piling up a large number of layers. They confirmed that owing to the high refractive index of guanine crystal of 1.83, 20 crystal platelets of 10 nm thickness were enough to attain 80% reflectivity, which inevitably caused narrow width of the reflection band and differed considerably from the experiment. They considered the stimulating light caused sodium channels in the plasma membrane to open, leading osmotically to an increase in thickness of the cytoplasmic layers. In addition, they described that the filaments reported by Rohrlich (Rohrlich, 1974) was not visible in their preparation of the sample.

Nagaishi and Oshima reported the detailed observation on the iridophores in the blue stripe of neon tetra (Nagaishi and Oshima, 1989, 1992; Nagaishi *et al.*, 1990; Oshima and Nagaishi, 1992). According to their descriptions, the iridophores have a rectangular flat shape with the dimension of 60–70 μm length, 15–20 μm width, and 1–2 μm height, and lie in two layers between collagenous lamellae and melanophores below epidermis. Many guanine platelets in a hexagonal shape are enclosed by membranous structure and arranged regularly inclining at a constant angle against the surface (Figs. 6.3(a) and (b)). A pair of the stacks are positioned above a thin triangle-shaped nucleus located at the bottom of the cell. The inclination of the platelets may benefit to reflect the light to a horizontal direction when illuminated from the dorsal direction.

They performed an experiment to measure the change of the inclination angle of the crystal platelet in order to investigate the mechanism of the color changing. For this purpose, they measured the intensity of

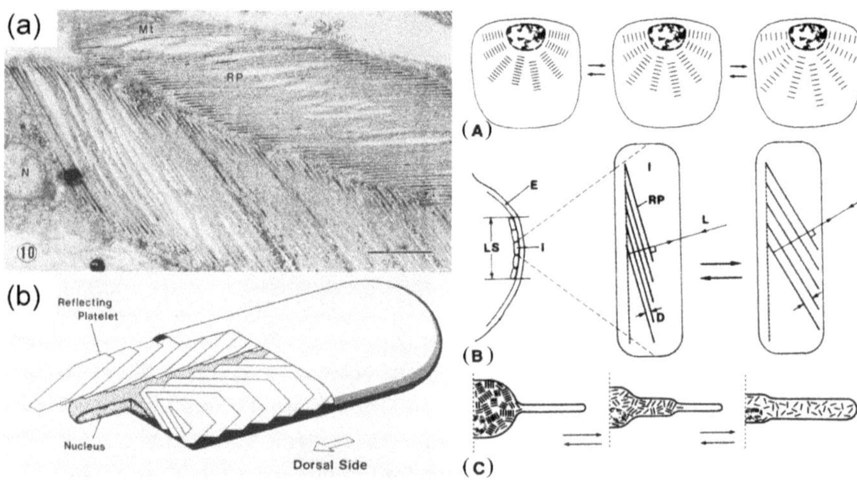

Fig. 6.3 Left: (a) Electron microscopic image of guanine platelets in an iridophore at the blue stripe of neon tetra and (b) illustration of the iridophore. (Reproduced from Nagaishi and Oshima (1992) with permission. Copyright (1992) The Zoological Society of Japan.) Right: Three types of motile iridophore found in (A) blue damselfish, (B) neon tetra, and (C) goby. (Reprinted from Fujii (1993b), with permission from Elsevier.) See Plate 14 in Appendix B.

backscattered light while varying the incident angle to the skin surface by employing a fiber assembly combining the fibers for the illumination and detection, and observed a reflection spectrum at each incident angle. They measured an angle giving the maximum reflectance for various colors caused by adding K^+-rich saline and found the good correlation between the peak wavelength and the incident angle giving the maximum reflectance. From this result, they concluded that the major cause for the color change was due to the change in the inclination angle of the crystal platelets. They named this mechanism "Venetian blind" mechanism and deduced that the motility of the crystal inclination was controlled by tublin–dynein system.

Similar thin platelets forming a multilayer were found by Oshima *et al.* in coral-reef fish with brilliant cobalt-blue coloration named blue damselfish, *Chrysiptera cyanea* (Oshima *et al.*, 1985; Oshima and Fujii, 1987; Kasukawa *et al.*, 1987; Oshima, 2005). The reflection peak from the integument is normally 465 nm, while under stress, it shifts to 380 nm, and when the fish is feeding, it shifts to 500–530 nm with greenish tone. Under electron microscopy, the iridophores with an ellipsoidal shape of 11 μm diameter are located in a monolayer under collagenous layer below epidermis. In this species, only iridophores and melanophores are present, with the melanophores underlying the iridophore layer and inserting their dendritic processes into the spaces among the iridophores (Fig. 6.1). Within an iridophore, a number of piles of thin guanine platelets are arranged radially around a nucleus that is located at the top of the cell. The separations between the platelets are strikingly uniform around 120–130 nm with the thickness of the platelet of 5 nm. The platelet was found to be composed exclusively of guanine. They confirmed that the nervous stimulation or norepinephrine on the excised skin shows the progressive shift of the reflection peak from 372 to 510 nm, while the reverse process occurred by the addition of adenosine. The color change was believed to occur synchronously in every pile by changing the separations between the platelets. The similar stacking and succeeding color changes were found in blue-green damselfish, *Chromis viridis* (Fig. 6.4(a)), where four types of iridophores were recorded (Oshima, 2005; Fujii, 1993b); a common sergeonfish, *Paracanthurus hepatus*, which showed a double iridophore layer (Goda and Fujii, 1998; Oshima, 2005); and a domino damselfish, *Dascyllus trimaculatus* (Goda and Fujii, 2001). Recently, rapid color changes have been reported on the reflecting stripes in paradise whiptail, *Pentapodus paradiseus*, but the arrangement of the stacks of 10 platelets, on an average, within an iridophore are circular and

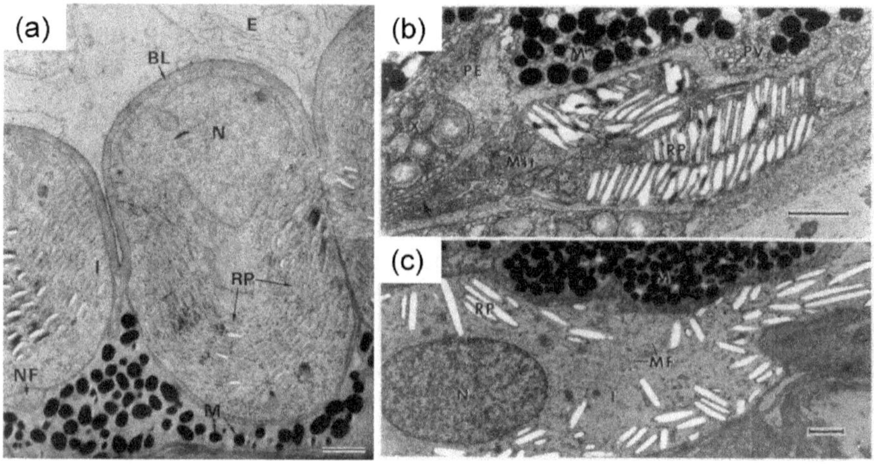

Fig. 6.4 (a) Electron micrograph showing iridophores in blue-green damselfish, *Chromis viridis* (Reproduced from Oshima (2005) with permission.); and those in goby, *Odontobutis obscura*, (b) in a punctate state and (c) in a dendritic state (With kind permission from Springer Science + Business Media: Iga and Matsuno (1986).).

the platelets are rather thick with the mean thickness of 51 nm (Mäthger *et al.*, 2003).

Completely different type of motile iridophore was found in a freshwater goby, *Odontobutis obscura*, by Iga and Matsuo (Iga and Matsuno, 1986; Iga *et al.*, 1987; Maeno and Iga, 1992). This type of fish contained three kinds of chromatophore in the dermis: melanophores, xanthophores, and iridophores, in addition to the epidermal melanophores. The iridophore was exceptionally of a dendritic type and contained reflecting platelets of 1–2.5 μm in length with the thickness of 60 nm (Figs. 6.4(b) and (c)). They observed that the reflecting platelets in the iridophores could translocate under the responses to certain stimuli. The platelets were regularly arranged in stacks of 10–20 layers in the perinuclear region of the cell. However, once K^+ ions, or prospective transmitters and hormones were added, the platelets were translocated into the processes in a rather random way. This response is quite slow compared with the melanophore and the other chromatophores, and takes about 1 h to complete. The apparent color of the skin is bluish when the platelets are aggregated, while it is yellowish when they are dispersed (Fujii, 1993a). Similar type of iridophore to produce white appearance of the skin has been reported for paddlefish, *Polyodon spathula* (Zarnescu, 2007). However, the detailed information on the motility is not clear.

6.4 Corneal Iridescence

Marine teleost fishes living on the bottom or among rocks in shallow water sometimes display marked iridescence in corneas under particular condition of illumination. This is known as corneal iridescence. Lythgoe and the staff of the Royal Society Research Station (Aldabra) performed an *in situ* observation of the corneal iridescence at 15–25 m depth in the Indian Ocean immediately after the explosion (Lythgoe, 1971). They recorded the direction from which corneal iridescence was visible using a glass sphere filled with sea water and marked with lines of latitude and longitude. They examined several species and found that the iridescence was visible only from the center to the upper direction. Locket also investigated the cornea of serranid fish in coral reefs, *Nemanthias carberryi*, and noticed that the iridescence was remarkable when the illumination was made from the dorsal direction, and the change of the color from orange to green-blue was observed when the viewing angle was changed from the normal to the cornea to the dorsal side (Locket, 1972). Electron microscopic observation showed that the endothelial cells in the corneal endothelium had lamellar structures which were oblique by 40–60° to the surface of the cornea. The lamellae consisted of electron dense parts with the uniform thickness of 71 nm and gap between them with the thickness of 125 nm. He speculated that the gaps filled with water and the electron-dense parts consisting of proteins gave a function of multilayer and this structure might be derived from modified endoplasmic reticulum. Lythgoe further investigated 13 species in various families and confirmed the above observations (Lythgoe, 1975).

In general, the structure of teleost cornea is composed of three layered structures: epithelium, stroma composed of very uniform collagen fibrils, and endothelium of a single cell thick. Between the stroma and endothelium, amorphous layer known as Descemet's membrane is present, while between stroma and epithelium, Bowen's membrane is known to exist. The structures giving corneal iridescence are distributed in various parts of the cornea such as epithelium, stroma, Descemet's membrane, and endothelium. Lythgoe and Shand classified the types of corneal iridescence into three according to the origin of the structure (Lythgoe, 1975; Shand, 1988): (1) connective tissue, (2) endoplasmic reticulum, and (3) whole cell. Each category is further divided according to the location. Type 1 is composed of collagen fibrils in stroma or amorphous material in Descemet's membrane, while type 2 contains rough endoplasmic reticulum in endothelial cells or in the cells within stroma (see Fig. 6.5). Shand included epithelial

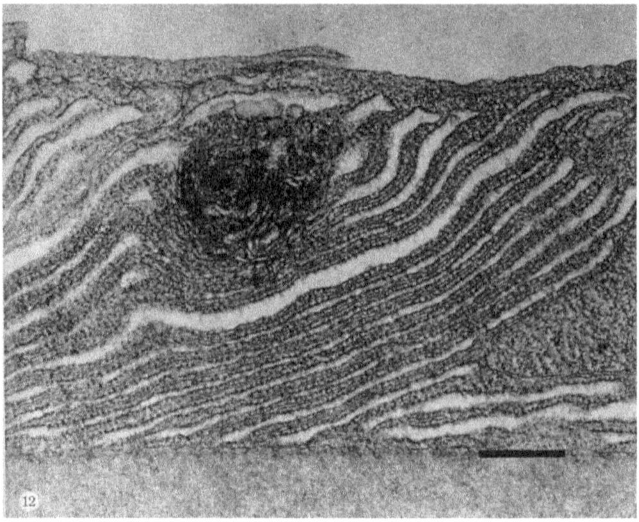

Fig. 6.5 Regularly arranged modified endoplasmic reticulum producing the corneal iridescence situated in the endothelium of *Anthias* sp., which belongs to type 2. Scale bar is 500 nm. (Reproduced from Lythgoe (1975) with permission of Royal Society Publishing.)

cells accompanying the infloldings of plasma membrane into this category (Shand, 1988). Type 3 is composed of flat plate of the cytoplasm and the matrix around the cell, which is situated between Descemet's membrane and stroma. Actually, type 3 is the most popular among the species investigated. Because of the varieties in orders, morphologies and locations, Lythgoe considered the corneal interference had some biological importance (Lythgoe, 1975). In any case, these lamellae are not so regular and not so hard as the other multilayer systems and are running obliquely to the plane of cornea, which will be of benefit to reflect the light illuminated from the dorsal side and to reduce the intra-ocular flare. This reduction of the flare was partly evidenced by the fact that the downwelling light was stronger by 10^4 than upwelling one in the shallow condition.

Shand and Lythgoe reported the light-induced color changes in corneal iridescence on sand goby, *Pomatoschistus minutus* (Shand and Lythgoe, 1987). Generally, the corneal coloration in teleost was determined by corneal chromatophore including carotenoid pigment and by reflective structure. The former was known to reduce the chromatic aberration and to enhance the contrast, while the latter was to reduce the intra-ocular flare and partly to camouflage the cornea. They found the marked red shift of the corneal color after the eye was excised, while under dark adaptation for 2 h, no trace of corneal iridescence was seen. However, after 2 min in the light, dark blue iridescence became visible and after about 15 min it became green or blue-green in color.

Shand examined totally 31 teleosts species, within which 13 species showed light-induced color change (Shand, 1988). Ashcroft and Lythgoe investigated the color change in sand goby, *P. minutus*, more in detail (Ashcroft and Lythgoe, 1991). The corneal iridescence in this species was classified into type 3 (whole cell), which composed of stratified layers of flat plates of cytoplasm and corneal matrix, and situated between Descemet's membrane and stroma. The cornea of undisturbed sand goby showed blue-green color when illuminated from the dorsal side. Once the fish was aroused, the color changed into red-bronze color with the reflection peak of 668 nm, while under 2 h dark adaptation, the reflectivity was greatly reduced with the peak of 519 nm. Then under the illumination, it changed gradually to 580 nm within 20 min. They considered a net inflow of water into the cornea would result in a shift to longer wavelength of the reflected light and the redistribution of water between cellular elements and surrounding matrix would result in a change of the reflectivity. Shand considered that in the dark environment where the reduction of the flare was unnecessary, it would be better to abolish the structure, since the lamellar structure anyway reduced the transmission through the cornea (Shand, 1988).

Before closing this chapter, I will briefly comment on the structural colorations in fish. Compared with those in insects and birds, the extremely different point in this species is that the color-producing structures are dynamically maintained in living systems. This situation causes various interesting physiological change in structure and hence in color. At a glance, the structures viewed from electron microscopic images are rather irregular as compared with those in the other species. However, we have to consider that these structures are dynamically stabilized under proper physiological conditions. If the average structure are regular, but the instantaneous structure shows the displacement around the mean structure, the former plays a role of a rigid reflector, while the latter does a role of a scatterer, the ratio of which are dependent on the degree of the displacement. Thus, the strict optical measurement including the dynamical aspect of the structure should be extremely important.

Finally, it is now clear that extremely thin guanine crystal platelets of 5–10 nm thickness, which is far beyond our imagination, play an important role in motile activity in fish coloring. The physiological condition of the crystal growth, which is strictly controlled by the membrane system, may allow the extremely directional growth of this crystal. Although the developmental studies concerning the growth of guanine crystal within iridophore was limited (Gundersen and Rivera, 1982), the research from this aspect is strongly encouraged to clarify one of the typical cases of self-organization processes in living systems.

Chapter 7

Plants

Before entering this chapter, I should be honest to confess that I have never seen the structural colors at any part of plant as described below. Hence, this chapter owes mostly to the pioneering work done by Lee and the other researchers, who have extensively studied in this unexplored field.

When he was in the University of Malaya in Kuala Lumpur, Lee was led to a walk of exploration in the forested mountain outside the city, where he was astonished by the leaf color of understory fern, *Selaginella*, because their leaves bore the shimmering electric blue (see Figs. 7.1 and 7.2) (Lee, 1997). He had thought that the green color of leaf was inevitable for photosynthesis to capture the appropriate light energy by chlorophylls. He investigated the plants in the deep shade of Malaysian rain forests and found more than 10 species displaying blue or green iridescence. He considered that the green iridescence was originated from the convex nature of the outer wall of the epidermal cell, which played a role of microlens array to focus the incident light to the chloroplasts, from which green light was selectively reflected (see Fig. 7.5(left)). On the other hand, the blue iridescence was considered to come from thin-film interference at the leaf surface (Lee, 1977).

Later he published a paper with his colleague to clarify the origin of blue coloring in the leaves and discussed the physiological and ecological meanings of the blue colorations (Lee and Lowry, 1975). He selected one of the blue iridescent spikemosses, *Seleginella willdenowii*, which was blue only under extremely shady condition in humid tropical forests (Fig. 7.2). The blue coloration was found to distinguish when immersed into water and also disappeared when the plants were aged or cultivated under sunlight. They measured the reflection spectra from the blue and green leaves, and calculated a difference spectrum, which showed a positive peak at 405 nm

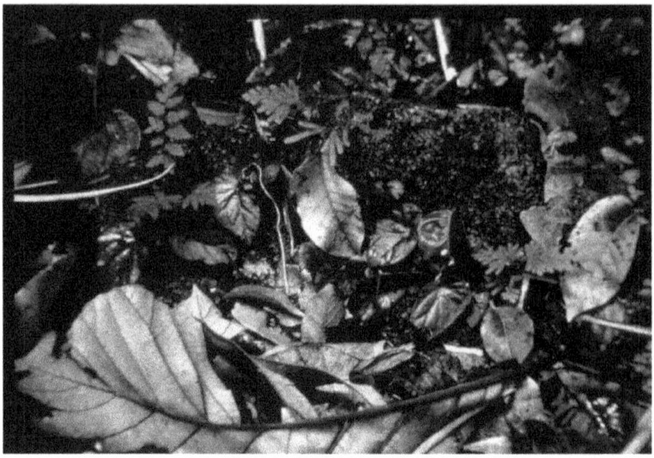

Fig. 7.1 Shady understory plants in Malaysian rain forest. (Reproduced from Lee (1997) with permission.)

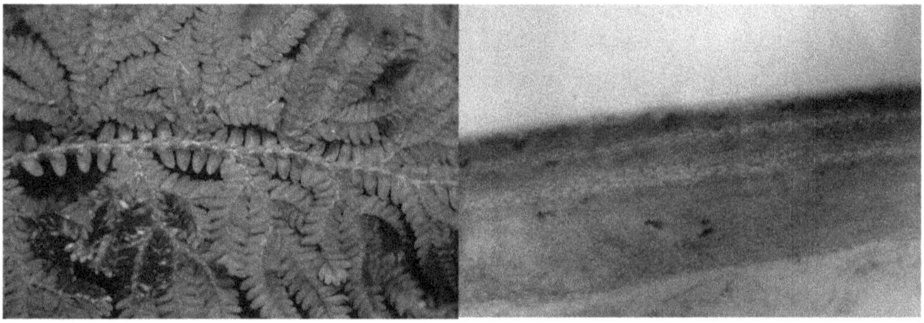

Fig. 7.2 Left: iridescent Malaysian fern, *Seleginella willdenowii*. Right: the cross section of the epidermis showing the lamellar structure in the outer portion of the epidermal cell wall. (Reproduced from Lee (1997) with permission.)

and became negative in the wavelength range of 500–700 nm. Hébant and Lee performed the optical and electron microscopic studies on two blue iridescent ferns, *S. willdenowii* and *S. uncinata* (peacock fern), and found the presence of the lamellar structures in the outer portion of the epidermal cell wall of blue leaves, while they were absent in green leaves. (Hébant and Lee, 1984). The lamellar structure consisted of two electron dense layers and a translucent layer between them with the thicknesses of the former of 87–94 and 74–98 nm, respectively, in the two species (Fig. 7.2). They speculated that the role of the iridescent leaf was to reflect the blue light unnecessary for photosynthesis, and to reduce the surface reflection in the photosynthetically active range of 600–680 nm. They also analyzed the egg-shaped epidermal cells as an efficient tool to collect the light (Lee and Lowry, 1975; Bone *et al.*, 1985).

On the other hand, Graham *et al.* investigated the iridescent leaves of a juvenile fern, *Danaea nodosa*, occurring in Central American and Amazonian rain forests (Graham *et al.*, 1993). They noticed that the iridescent color did not change when immersed in water. From the optical measurement, the iridescent leaf was found to show the blue reflection at 440 nm in addition to the normal green reflection at 550 nm. The blue color was found to originate from the epidermal cell wall, and the electron microscopic investigation showed 18–30 alternate translucent and opaque layers with the thicknesses of 84 and 49 nm, respectively. Moreover, the translucent layers showed a characteristic parabolic pattern (Fig. 7.3). This pattern will be one of the strong evidences for the presence of helicoidal structure as in cholesteric liquid crystal, which is reported generally in living creatures and plays an essential role in coloring the elytra of scarabaeid beetles (Sec. 4.2.1). This was the first report that the similar helicoidal structure contributing to the coloring in plants. Gould and Lee reported two other fern species, glade fern (*Diplazium tomentosum*) and necklace

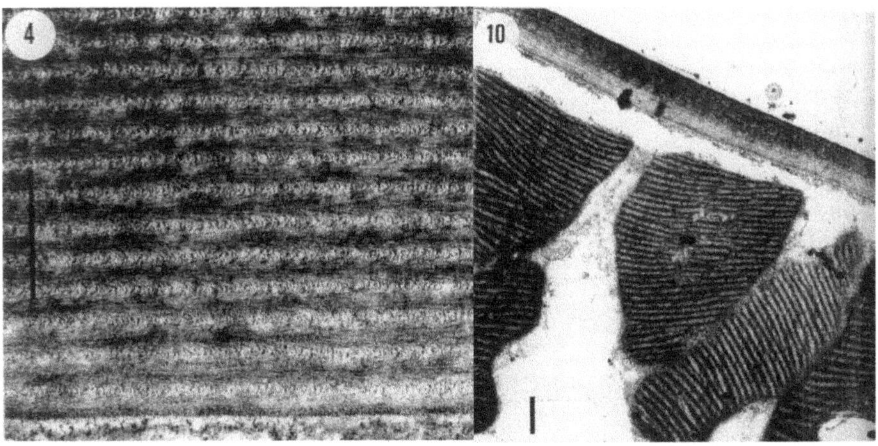

Fig. 7.3 Left: ultrastructure of an iridescent Malaysian fern, *Danaea nodosa*, showing parabolic pattern in the epidermal cell wall. (Reproduced from Graham *et al.* (1993) with permission of Botanical Society of America.) Right: structurally modified chloroplasts in the epidermal cells of an iridescent Malaysian fern, *Trichomanes elegans*. (Reproduced from Graham *et al.* (1993) with permission of Botanical Society of America.)

fern (*Lindsaea lucida*) found in Malaysian rain forest, and found the similar helicoidal structures in the cell wall (Gould and Lee, 1996). The leaf reflection of *D. tomentosum* showed a peak at 446 nm, and the epidermal cell wall showed 20–23 helicoidal layers with the spacing of 141 nm, while that of a green leaf showed only 6–7 layers with much larger spacing of 268 nm. Simple estimation showed that the former contributed to the blue coloring, while the latter to the reflection in near infrared region. Green iridescent *L. lucida* showed a reflection peak at 536 nm with 17 helicoidal layers with the spacing of 192 nm.

The other coloring type in plant was found in bristle fern, *Trichomanes elegans*, of Central American and Amazonian rain forests (Graham *et al.*, 1993), and in peacock begonia, *Begonia pavonina*, and Topak Sulaiman (Solomon's sole), *Phyllagathis rotundifolia*, of Malaysian rain forests (Gould and Lee, 1996). These species possessed structurally modified chloroplasts in the epidermal cells. The epidermal cell layer of *T. elegans* consisted of 2–3 cell layers. Adjacent to the lower cell wall in the epidermal cell layer, slender-shaped chroloplasts were distributed, whose inner part consisted of regular alternate layers of grana and electron-translucent stroma with the thicknesses of 91 and 86 nm, which selectively reflected blue–green light around 550 nm (Fig. 7.3). A granum consisted further of five closely appressed thylakoids. Similar structures were found in the other two species. The authors described that the selective advantage for the color production seemed to be a by-product of ultrastructure, but there was a possibility that the iridescence altered the internal light environments in these shade-adapted leaves to be advantageous to these plants.

Lee also investigated brilliant blue fruit of silver quandong, *Elaeocarpus angustifolius*, occurring in Asian tropics and Australesia (Lee, 1991). The color of this fruit was found to distinguish when contacted with water. The optical measurement showed a reflection peak at 439 nm and electron microscopic investigations showed a remarkable structure beneath the outer cell walls of the epidermis (Fig. 7.4). The structure consists of a roughly parallel network of strands of 78 nm in thickness, which are separated by lacunae of 39 nm in thickness. In addition, the paradermal section revealed the presence of straight rows of 62 nm wide and 62 nm apart. Thus, the structure in the epidermal cell of this fruit is a kind of three-dimensional lattice similar to the wing scale of *Morpho* butterfly. He called this structure *iridosome* based on analogous structure in animals. Histochemical staining revealed that the layers consisted of polysaccarides, at least partly of cellulose. The selective advantage of the iridescently blue coloring is partly

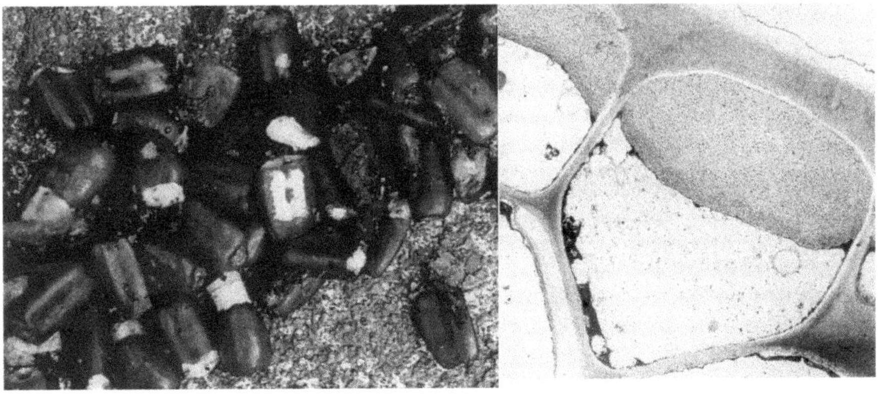

Fig. 7.4 Left: iridescent blue fruit of *Heliconia psittacorum*. Right: the "iridosome" located at the upper part of the epidermal cell in *Elaecocarpus angustifolius* fruit. (Reproduced from Lee (1997) with permission.)

important for the diets of cassowary, pigeons, and small mammals, because the conspicuous color persists even when the mesocarp is almost completely senescent. The photosynthetic carbon-fixation in ripe fruit may be another cause of blue iridescence. As described in spikemoss, the blue iridescence coloring has an advantage to reduce the reflection loss at the photosynthetically active wavelength region, while the pigmentarily blue coloring absorbs the light in the same wavelength region. The similar structure was found in the brilliant iridescent blue fruit of blue nun (*Delarbrea michieana*) found in North Queenland understory rain forest (Lee *et al.*, 2000).

The different type of iridescent color was reported in marine algae and spores of fern. Gerwick and Lang reported the structural, chemical, and ecological studies on iridescent algae, *Iridaea floccida* and *I. cordata*, grown profusely in the intertidal zone along the Northern California coast (Gerwick and Lang, 1977). These species become especially conspicuous when submerged in tide pools due to their iridescent blue, green, and red colors. The iridescence greatly diminishes when they are dried, and restores when they become wet. The microscopic observation showed that the iridescence was seen in teardrop-shaped spots of 1–2 μm long and beneath its surface, a multilayer of 17 alternate electron opaque and translucent layers of 35–67 and 75–250 nm thickness, respectively, was observed. The multilayer was located within the surface cuticle layer of 0.5–1.6 μm thickness and below it the layer of polysaccharide of 3–8 μm was located. Below the multilayer, the densely packed layer with granules was located, in which the electron opaque layer was formed by the aggregation of granules. On the other hand, the outer surface layer was rather rough so that some type of wearing-away of the outer cuticle layer took place. Thus, the forming and wearing-away process maintained the surface layer dynamically. They considered that the striking iridescence seemed to be a by-product to form a cuticle comprised of multiple layers deposited at particular rates, since the cuticle layer originally functioned as a protective covering in *Iridaea*. On the other hand, Hemsley *et al.* reported the iridescent spores in the fern *Selaginella* species (Hemsley *et al.*, 1994). The electron microscopic observation showed a striking ordered structure made of particles of 0.24 μm diameter, which was very similar to colloidal crystal, was found in the exine of the spores. They discussed the self-assembly process in terms of colloidal crystal formation in biological systems.

Finally, we briefly comment on the structural modifications found in the plant world, which are by far larger than the wavelength of light, but anyway have a special effect on the optical function (Vogelmann, 1993). One is a lens effect found in the leaves of tropical rain-forest shady plants. Bone *et al.* investigated the convexly curved epidermal cell walls in a number of shade-adapted plants and considered the lens effect to focus the photon flux into the lower part of the epidermis (Fig. 7.5), where chloroplasts were located (Bone *et al.*, 1985). They assumed an appropriate function to reproduce the surface curvature of the epidermal cell and employed a ray tracing method to estimate the degree of the concentration of light beam within a leaf. When the leaves were irradiated with collimated light, the concentration factor at the focal region ranged up to 8–20, while under the diffuse radiation it amounted to 1.5–2.5. Thus, the enhancement of the light collection efficiency is anyway expected from the shape of the epidermal cell, which will strongly depend on the condition of irradiation.

Bernhard *et al.* described, referred to the earlier work, that the cells of the surface epithelium of the petals of pansies and roses were cone-shaped, and the surface of the petal was completely covered with such cones (Fig. 7.5) (Bernhard *et al.*, 1968). The height of the cone was 35 μm, and below this cone, so-called tapetum, a lamellar system of flat air-filled spaces, was formed. Both the basal region and the cone were pigmented. A ray tracing analysis showed that the light hitting on the surface of the petal was complicatedly refracted and reflected in these systems to reduce the surface reflection and as a result to elongate the optical length. Thus, these structure may function as a spectral enhancement of the flower color.

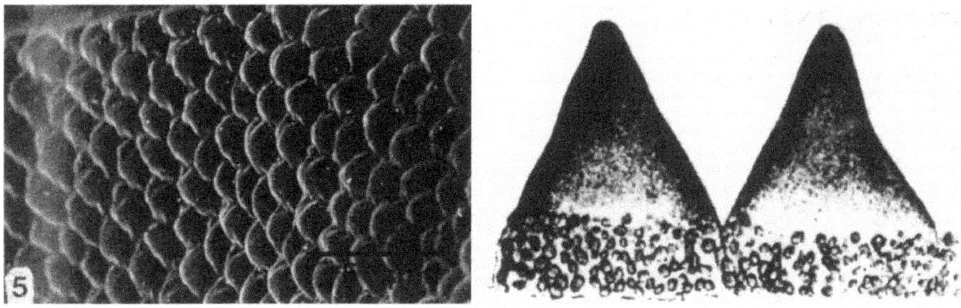

Fig. 7.5 Left: convexly curved surface of the upper epidermal cells of a green leaf of *Selaginella willdenowii.* (Reproduced from Hébant and Lee (1984) with permission of Botanical Society of America.) Right: cone-shaped modification of the outer surface of epithelial cells in a pansy petal. (Reproduced from Bernhard *et al.* (1968) with permission from Cambridge University Press.)

Thus, the researches in the structural colorations and structurally enhanced optical functions in plant world are just beginning, and various types of micro- and macro-structure will come out from now on, although to date, its biological meaning is somewhat suspicious and is sometimes assigned to a by-product to form the structures having the other functions.

Chapter 8

Miscellaneous

8.1 Shells

Pearl and mother-of-pearl have offered interesting topics on structural colorations. In the middle of 19th century, Brewster first observed the optical phenomena concerning mother-of-pearl (Brewster, 1854). Under the illumination of a candle, he observed that the images of the candle reflected from the surface of shell showed several images on the screen, whose intensity distribution clearly changed after polishing the shell. Further, if the surface of the shell was covered with bee's wax, the surface of the wax was shining with prismatic colors after removing from the shell. He considered that the colors of mother-of-pearl was "communicated" to the wax surface and called "communicable" colors. He noticed that the surface of the shell had a grooved structure of 3000 lines per inch (118 lines/mm), which originated from the cross-section of numerous strata within the shell. Later, Lord Rayleigh, FRS, stressed the presence of a series of equally spaced parallel plates, which reflected distinguishable colors due to multilayer interference (Rayleigh, 1923). He measured the transmission spectrum through a thin plate of shell and found the presence of two dark bands located at 590 and 430 nm. He analyzed it in terms of the reflection from the stratified layer of the shell corresponded to the second- and third-order reflection bands, respectively.

The stratified plates in pearl and mother-of-pearl are known as thin single crystal of aragonite ($CaCO_3$) with a thickness of $\sim$0.5 μm and a diameter of $\sim$10 μm, which are cemented by organic materials called conchiolin (complex proteins) with a thickness of $\sim$25 nm (Tan *et al.*, 2004), which are secreted from epithelium. These layers are stratified one another to strengthen the body like a brick wall.

The origins of the structural colors in mother-of-pearl were, thus, conflicting among diffraction grating, multilayer interference, and the combination of both. Raman played a decisive role on this problem (Raman, 1934b,c,d; Raman and Krishnamurti, 1954a,b). First, he characterized three special features of shell iridescence: (1) *the internal laminations are inclined to the external surface*, (2) *the laminae are granular and not homogeneous in structure*, and (3) *considerable thicknesses of the material have to be considered.* He stressed the importance of a relationship between the diffraction grating at the surface and that of multilayer interference under the surface, because the outcrops of the laminar structure were the origin of the diffraction grating.

He proved that reflected light due to these two processes is directed into the same directions (Raman, 1934c). The proof is easy through geometrical inspection. If the layered structure with the periodicity of d is inclined by an angle α to the surface, a diffraction grating with a pitch of $d/\sin\alpha$ is naturally formed. When incident light is inclined by an angle θ from the normal to the surface, the interference condition will be given as

$$\frac{d(\sin\chi - \sin\theta)}{\sin\alpha} = \pm m\lambda, \tag{8.1}$$

where χ is the angle of the diffracted light and m is a positive integer (see Fig. 8.1(a)). On the other hand, for the inclined multilayer within a material (refractive index n), the incident angle to this multilayer will be

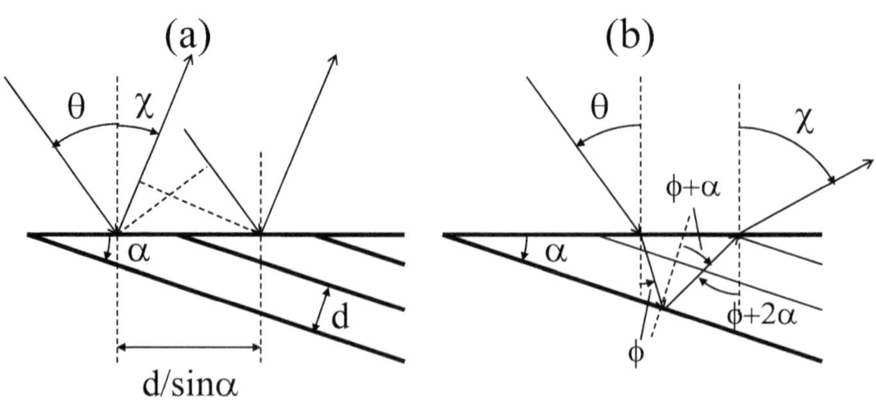

Fig. 8.1 Geometries for (a) diffraction grating at the surface of shell and (b) multilayer interference due to obliquely running layers.

changed into $\phi + \alpha$ due to the refraction at the surface using the relation of $\sin\theta = n\sin\phi$. Thus, the multilayer interference occurs with the condition

$$2nd\cos(\phi + \alpha) = m\lambda, \tag{8.2}$$

and then the angle of the outgoing light at the surface should satisfy the relation $\sin\chi = n\sin(\phi + 2\alpha)$ due to the refraction (see Fig. 8.1(b)). If Eq. (8.1) is transformed as

$$\begin{aligned}\frac{d(\sin\chi - \sin\theta)}{\sin\alpha} &= \frac{nd(\sin(\phi + 2\alpha) - \sin\phi)}{\sin\alpha} \\ &= 2nd\cos(\phi + \alpha) = m\lambda,\end{aligned} \tag{8.3}$$

we can attain the same result as Eq. (8.2). Thus, the diffraction at the surface and the multilayer reflection of the same order are essentially indiscernible in direction, though their spectra are basically different except for the center of reflection band. Using the observed data for *Turbo* shell, which emitted greenish-blue iridescence, such as $\theta = 0°$, $\chi = 24.1°$, $m = 3$, and $\lambda = 500\,\text{nm}$, Raman deduced $d = 456\,\text{nm}$, $\alpha = 7.14°$ and $d/\sin\alpha = 3.66\,\mu\text{m}$, which was in good agreement with the observed grating pitch. Raman and Krishnamurti investigated various shells and found that in *Margaratifera*, the fourth-order diffraction spots superposed on the reflection from the layered structure, while in *Turbo* the third-order, and in *Nautilus* and *Haliotis*, the second-order (Raman and Krishnamurti, 1954a). They also analyzed the strong diffusion halo known for the reflection from the layered structure and the transmission through a thin plate of shell. They assigned its origin partly as the diffraction effect due to the narrow width of crystal plate and partly as due to the difference of the optical path length between the crystal and conchiolin when a light wave propagated to the oblique direction.

It is also noticed that the form of the outcrop surface will strongly affect the intensity distribution of the diffraction spots and also will affect the incidence angle to the multilayer. Thus, the intensity profile of the reflected light is strongly affected by the surface condition, whether it is as grown, polished, or cracked. Furthermore, relatively large lamellar thickness to give the higher-order interference, will supply two reflection peaks in a visible region. In addition, thin layer of conchiolin makes a strongly nonideal multilayer which results in relatively sharp reflection peaks. As a result, the shell will display green/purple two-color iridescence more vividly than that which was found in rock dove (Sec. 5.5).

Recently, several interesting papers have appeared concerning the iridescence of shell. Liu *et al.* investigated black-lipped pearl oyster *Pinctada margaritifera* and found a grooved structure of 3.38 μm pitch in the outer surface of the shell, while in the inner surface, somewhat rough structure with a pitch of 11.5 μm was seen (Liu *et al.*, 1999; Nassau, 2001). They observed the diffraction pattern from the outer surface and found it consisted of many spots from −2nd to +8th orders, with the diffuse bright area superposed between 3 and 8 orders. Tan *et al.* investigated *Haliotis glabra* and found 2–8 μm grooved structure on the inner surface. They found, in addition to that, rough grooved structure with a wider spacing of 30–50 μm running perpendicular to the fine grooves (Tan *et al.*, 2004). They observed the diffraction spots for both grooves and found that bright diffuse spots overlapped on the higher-order diffraction spots. They separately analyzed the effects of diffraction and multilayer reflection using the values of the thickness and refractive index of aragonite layer as 500 nm and 1.6, while those of conchiolin layer as 25 nm and 1.3. They showed that greenish color was due to the third-order reflection peak. Brink *et al.* investigated small iridescent spots dispersed on the outer surface of *Helcion pruinosus* (Brink *et al.*, 2002). They found tilted multilayer structure by 24° to the surface was located at 50 μm below the surface, which was only observed below the iridescent spot. They performed optical measurements and found the polarization-dependent reflection peaks from the multilayer. They confirmed that the multilayer was made of birefringent aragonite crystal, whose anisotropy changed the apparent optical thicknesses of the multilayer for *s*- and *p*-polarizations.

8.2 Spiders

The structural colors in spiders have been rarely reported in literatures. Oxford and Gillespie presented a review on spider colors, in which two origins of structure-based colors were reported (Oxford and Gillespie, 1998). One is based on guanocytes lying beneath the hypodermis, in which guanine crystals are distributed to make the body whitish. Particularly, among the species in the genera *Tetragnatha* and *Leucauge* (Tetragnathidae), *Argyrodes* (Theridiidae), and *Theridiosoma* (Theridiosomatidae), extremely thin guanine crystals make the spider a silvery appearance. They reported that the guanocytes are also responsible for the physiological darkening as a result of guanine retraction or contraction. The other type of the structural colors appears as those located at the surface sculpturing such as modified setae or scales, and the genera *Argiope* (Araneidae), *Lycosa* (Lycosidae), and *Josa* (Anyphaenidae) are exemplified. Townsend and Felgenhauer investigated cuticular scales of lynx spiders (Oxyopidae) and jumping spiders (Salticidae) using electron microscope, and found that the iridescent scales in the male spiders definitely lacked interior structures with a narrow lumen formed by the upper and lower laminae, which were quite different from noniridescent scales (Townsend and Felgenhauer, 1999). However, these authors did not discuss the origins of their iridescence.

On the other hand, Parker and Hegedus investigated two species of spiders, *Castaneira* sp. (Corinnidae) and *Cosmophasis thalassina* (Salticidae) (Parker and Hegedus, 2003). The males and females of the former showed conspicuous metallic appearance on the dorsal surfaces of cephalothorax (head and thorax) and abdomen. On the other hand, those of the latter showed metallic color from the dorsal surfaces of the bodies and legs. Particularly, strong color was observed from two broad stripes on the cephalothorax. They observed the body surface using SEM and found the cuticle surface in metallic region was modulated by a series of linear grooves and ridges of 150 nm periodicity with 80 nm in depth, while for the latter, the structures were of 460 nm in periodicity and 150 nm in depth. The cross-section showed the multilayer of totally 4 μm thickness for the former, while in the latter, 50 layers was observed below the homogeneous outer layer of 4 μm thickness. They analyzed using a model that the multilayer whose outer surface was modulated by a diffraction grating. In *Cataneira* sp., the periodicity of the grating was so small that it behaved like a zeroth-order grating to function as an antireflector particularly for ultraviolet light. Thus, the metallic green reflection was mainly originated by the underlying multilayer. On the other hand, in *Cosmophasis thalassina*, the surface modulation worked as a first-order grating to diffract away the blue light nearly parallel to the surface, while the chirped multilayer reflected the green to red light toward specific direction, which resulted in a gold appearance.

8.3 Marine Animals

One of the conspicuous findings in structural colors have revealed from marine animals. Parker *et al.* reported that brilliantly arranged two-dimensional photonic crystal was found in a kind of sea mouse *Aphrodita* sp. (Polychaeta: Aphroditidae) (Parker *et al.*, 2001; McPhedran *et al.*, 2001, 2003; Welch, 2005), whose lower part of the body was decorated by many iridescent hair and spines called chaetae. Sea mice with 15–20 cm in length and 5 cm in width are known to live along continental shelves at depths of 1 m–2 km, whose natural history is hardly known (McPhedran *et al.*, 2001). The first electron microscopic observation of the *Aphrodite aculeata* was performed by Lippert and Gentil, who observed numerous regularly-arranged capillaries within a hair of this animal (Lippert and Gentil, 1963). However, the electron micrographs obtained by them contained considerable disorder and hence complete understanding of the structure was not attained at that time. Parker *et al.* investigated thicker iridescent spines, which showed a hollow cylinder shape with 0.21 and 0.1 mm in external and inner diameters, respectively, and found that within the cylinder, a tremendous number of small hollow cylinders are hexagonally close-packed like a photonic crystal fiber (Parker *et al.*, 2001). In a wider view, these cylinders are arranged circularly along the inner surface of the large cylinder and the number of the circular layer amounts to 88. The diameters of the cylinder are somewhat dependent on the depth of the layers and tend to increase gradually from 0.109 to 0.365 μm from the surface, while the nearest-neighbor distance is rather constant at 0.51 μm (McPhedran *et al.*, 2001). The supported material is chitin, which has a refractive index of 1.54 and the cylinder is filled with water of 1.33. Thus sophisticated structure produces brilliant colors even under small difference in refractive index of 0.21. They calculated the reflectivity by modeling this system as a stack of 88 gratings, each of which consisted of a row of parallel cylinders. The obtained spectrum had large reflection band around 650 nm, with slightly higher reflectivity for *s*-polarization (electric field vector parallel to the axis of cylinder) (Parker *et al.*, 2001; McPhedran *et al.*, 2001). The photonic band calculation showed the partial photonic band gap of the first band appearing in a infrared region and the second band, which are complicated by mixing of several modes, appeared at a visible region, corresponding to the red reflection in the reflection spectrum (McPhedran *et al.*, 2001). They also attempted to measure the reflection spectrum from thin green-gold hair using a light source of supercontinuum generation from a femtosecond Ti:sapphire laser and found the presence of high-reflectivity region above 590 nm (McPhedran *et al.*, 2003).

Comb-jellyfishes are known to have eight rows of comb-plates running radially on the body. These comb-rows are composed of numerous locomotory cilia fused each other to form a comb-shaped swimming plate. These combs fantastically reflect iridescent light and when they are swimming, moving combs make rows of wavy light like neon signs. Welch *et al.* attracted the iridescence of comb-jellyfishes and investigated three species, *Bolinopsis infundibulum*, *Boroë cucumis*, and *Hormiphora cucumis* (Welch, 2005).

In these species, photonic crystals are found to be due to periodic arrangement of cilia. In *B. infundibulum*, the periodically arranged microtubules with the radius of 0.31 μm forms a parallelogram with one side 0.97 μm and the other 0.66 μm in length, and the included angle of 72°. In *B. cucumis*, a unit cell is a similar parallelogram with 0.215 and 0.195 μm with the including angle of 77° (Welch, 2005). Furthermore, each lattice point is found to be composed of tightly packed microtubules form so-called $9 + 2$ arrangement, nine microtubules are arranged concyclically with two combined microtubules placed at the center. Thus, two-dimensional photonic crystal with quite a low symmetry. They considerably simplified the model system and calculated the reflection spectrum using a transfer matrix method. The resulting spectrum shows a sharp peak at 615 nm under normal incidence, and the peak position changes drastically over a whole visible region when the angle of incidence is changed. Furthermore, the photonic band calculation showed the reflection in a visible region is originated from the first order (McPhedran *et al.*, 2003). They also discuss the relation with the bioluminescence, which emits light at 489 nm, and the comb will contribute to increasing the transmittance at this wavelength (Welch *et al.*, 2006).

Probably, marine animals are the most fascinating ones from a viewpoint of structural colorations, although only a few species have been investigated up to now. These include swimming crabs (Parker *et al.*, 1998c), squids (Arnold *et al.*, 1974; Cooper and Hanlon, 1986; Hanlon *et al.*, 1990; Cooper *et al.*, 1990; Shashar *et al.*, 2001), sea urchins (Millott and Manly, 1961), and marine planktons (Parker, 1995; Chae and Nishida, 1994; Chae *et al.*, 1996; Chae and Nishida, 1999). Because of the limitation of time and page, I cannot describe even these interesting species here and I hope these will be presented elsewhere in future.

Chapter 9

Mathematical Background

9.1 Calculations of Multilayer Reflection

9.1.1 *Transfer Matrix Method*

Several methods have been introduced to calculate the optical properties of multilayer since the first derivation by Lord Rayleigh (Rayleigh, 1917). Here, we will show three typical methods, which are often employed to calculate the reflection properties of multilayer. First, we show a transfer matrix method.

Put an electric field of y-polarized light propagating toward z direction as $E_y(z)$ as shown in Fig. 9.1. Then, the light is expressed as the sum of waves traveling toward the positive and negative directions along z-axis as

$$E_y(z)\mathrm{e}^{-\mathrm{i}\omega t} = (C_+\mathrm{e}^{\mathrm{i}nk_0z} + C_-\mathrm{e}^{-\mathrm{i}nk_0z})\mathrm{e}^{-\mathrm{i}\omega t}, \tag{9.1}$$

where $k_0 = \omega/c_0$ is the wave number in vacuum, and ω, c_0, and n are the angular frequency and the velocity of light in vacuum, and the refractive index of the medium, respectively. From the Maxwell equations, the magnetic field has only x-component and is expressed as

$$H_x(z)\mathrm{e}^{-\mathrm{i}\omega t} = -\frac{n}{\eta_0}(C_+\mathrm{e}^{\mathrm{i}nk_0z} - C_-\mathrm{e}^{-\mathrm{i}nk_0z})\mathrm{e}^{-\mathrm{i}\omega t}, \tag{9.2}$$

with the radiation impedance of vacuum $\eta_0 = \sqrt{\mu_0/\epsilon_0}$, where ϵ_0 and μ_0 are permittivity and permeability of vacuum, respectively. At $z = 0$, Eqs. (9.1) and (9.2) become

$$\begin{aligned} E_y(0) &= C_+ + C_-, \\ H_x(0) &= -(n/\eta_0)\cdot(C_+ - C_-). \end{aligned} \tag{9.3}$$

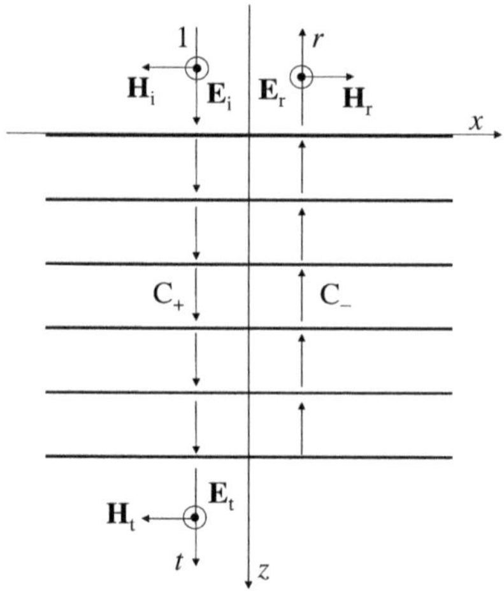

Fig. 9.1 Coordinates employed in transfer matrix method.

Combining Eqs. (9.1) and (9.2) with Eq. (9.3), we can express the above relations in a matrix form:

$$\begin{pmatrix} E_y(0) \\ H_x(0) \end{pmatrix} = \begin{pmatrix} \cos nk_0 z & -\mathrm{i}(\eta_0/n)\sin nk_0 z \\ -\mathrm{i}(n/\eta_0)\sin nk_0 z & \cos nk_0 z \end{pmatrix} \begin{pmatrix} E_y(z) \\ H_x(z) \end{pmatrix}$$
$$\equiv M(z) \begin{pmatrix} E_y(z) \\ H_x(z) \end{pmatrix}. \tag{9.4}$$

In a multilayer consisting of N layers, the above relation holds in each layer and at the boundary between two adjacent layers, the tangential components of the electric and magnetic fields in one layer connect with those in the other layer. Then, the electric and magnetic fields just after transmitting the multilayer are connected with the incident fields simply as

$$\begin{pmatrix} E_{y1}(0) \\ H_{x1}(0) \end{pmatrix} = M_1(d_1)M_2(d_2)\cdots M_N(d_N) \begin{pmatrix} E_{yt}(d_N) \\ H_{xt}(d_N) \end{pmatrix}$$
$$\equiv \begin{pmatrix} A & B \\ C & D \end{pmatrix} \begin{pmatrix} E_y(d_N) \\ H_x(d_N) \end{pmatrix}, \tag{9.5}$$

where d_j is the thickness of the jth layer. If we put the amplitude reflectivity and transmittance of the multilayer as r and t, respectively, as shown in

Fig. 9.1, and define the transfer matrix as

$$\begin{pmatrix} 1+r \\ (n_\mathrm{i}/\eta_0)(1-r) \end{pmatrix} = \begin{pmatrix} A & B \\ C & D \end{pmatrix} \begin{pmatrix} t \\ (n_\mathrm{t}/\eta_0)t \end{pmatrix}, \tag{9.6}$$

then the calculation of this matrix gives the following result:

$$r = \frac{n_\mathrm{i}A + (n_\mathrm{i}n_\mathrm{t}/\eta_0)B - \eta_0 C - n_\mathrm{t}D}{n_\mathrm{i}A + (n_\mathrm{i}n_\mathrm{t}/\eta_0)B + \eta_0 C + n_\mathrm{t}D}, \tag{9.7}$$

$$t = \frac{2n_\mathrm{i}}{n_\mathrm{i}A + (n_\mathrm{i}n_\mathrm{t}/\eta_0)B + \eta_0 C + n_\mathrm{t}D}. \tag{9.8}$$

The (power) reflectivity and transmittance are expressed, using these expressions, as $R = |r|^2$ and $T = (n_\mathrm{t}/n_\mathrm{i})|t|^2$, respectively, with the refractive indices of the incident and transmitted media n_i and n_t.

For oblique incidence, the result is dependent on the polarization of the incident light and the above expression is slightly modified. For the polarization perpendicular to the xz incident plane (s-polarization), the continuation of the electric field E_y and the magnetic field H_x should be satisfied at the boundary. The transfer matrix in the jth layer becomes

$$M_j^{(s)}(z) = \begin{pmatrix} \cos(g_j^{(s)}k_0 z) & -\{\omega\mu_0/(g_j^{(s)}k_0)\}\sin(g_j^{(s)}k_0 z) \\ -\{g_j^{(s)}k_0/(\omega\mu_0)\}\sin(g_j^{(s)}k_0 z) & \cos(g_j^{(s)}k_0 z) \end{pmatrix}, \tag{9.9}$$

where $g_j^{(s)} \equiv n_j \cos\theta_j$ with the angle of refraction in the jth layer θ_j. The amplitude reflectivity and transmittance are

$$r_s = \frac{g_i^{(s)}A + (g_i^{(s)}g_t^{(s)}/\eta_0)B - \eta_0 C - g_t^{(s)}D}{g_i^{(s)}A + (g_i^{(s)}g_t^{(s)}/\eta_0)B + \eta_0 C + g_t^{(s)}D}, \tag{9.10}$$

$$t_s = \frac{2g_i^{(s)}}{g_i^{(s)}A + (g_i^{(s)}g_t^{(s)}/\eta_0)B + \eta_0 C + g_t^{(s)}D}, \tag{9.11}$$

respectively. On the other hand, for the polarization parallel to the incident plane (p-polarization), the continuation of the electric field E_x, and the magnetic field H_y should be satisfied at the boundary and then the transfer matrix similar to Eq. (9.9) after replacing $g_j^{(s)}$ with $g_j^{(p)} \equiv n_j/\cos\theta_j$ gives

$$r_p = -r_s(g_j^{(s)} \to g_j^{(p)}), \quad t_p = t_s(g_j^{(s)} \to g_j^{(p)}). \tag{9.12}$$

The features of transfer matrix method are summarized as follows: (1) The reflectivity and transmittance from an arbitrary multilayer are obtained without any periodicity in layer thickness and refractive index, and (2) when the optical thickness of the layer satisfies a simple relation, the reflectivity can be expressed as a simple analytical form as described in Secs. 2.3 and 9.3.1, whereas (3) complicated matrix calculations are sometimes needed.

9.1.2 *Iterative Method*

Next, we show an iterative method using the result obtained in thin-film interference. Consider the jth layer, within which the field amplitudes are denoted as u_j and v_j according to the propagation directions toward $+z$ and $-z$ directions, as shown in Fig. 9.2. We can describe their amplitudes as

$$u_j = t_{j-1,j}u_{j-1}e^{i\phi_j/2}\kappa_j + r_{j,j-1}t_{j+1,j}v_{j+1}e^{i\phi_j}\kappa_j, \tag{9.13}$$

$$v_j = r_{j,j+1}t_{j-1,j}u_{j-1}e^{i\phi_j}\kappa_j + t_{j+1,j}v_{j+1}e^{i\phi_j/2}\kappa_j, \tag{9.14}$$

where $t_{j,k}$ and $r_{j,k}$ are the amplitude transmittance and reflectance, derived from the Fresnel's law, from the jth to kth layer, and $\kappa_j = 1/(1 - r_{j,j+1}r_{j,j-1}\exp[i\phi_j])$ (Eq. (2.42)) and $\phi_j = 4\pi n_j d_j \cos\theta_j/\lambda$ (Eq. (2.43)).

For simplicity, we replace Eqs. (9.13) and (9.14) into the following forms:

$$u_j = a_j u_{j-1} + b_j v_{j+1}, \tag{9.15}$$

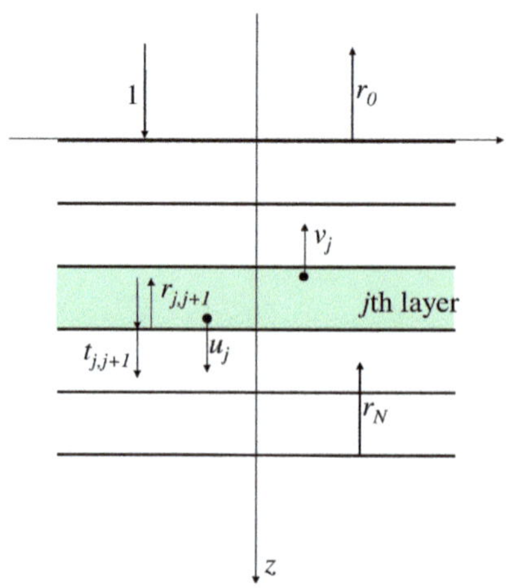

Fig. 9.2 Coordinates employed in iterative method.

$$v_j = c_j u_{j-1} + d_j v_{j+1}, \tag{9.16}$$

and put $v_j/u_j = \rho_j$. Eliminating u_{j-1} and v_{j+1} from Eqs. (9.15) and (9.16), we obtain

$$(1/a_j - \rho_j/c_j)u_j = (b_j/a_j - d_j/c_j)u_{j+1}\rho_{j+1}, \tag{9.17}$$

$$(1/b_j - \rho_j/d_j)u_j = (a_j/b_j - c_j/d_j)u_{j-1}. \tag{9.18}$$

Replacing j with $j+1$ in the second formula yields

$$(1/b_{j+1} - \rho_{j+1}/d_{j+1})u_{j+1} = (a_{j+1}/b_{j+1} - c_{j+1}/d_{j+1})u_j. \tag{9.19}$$

Inserting u_{j+1} obtained from Eq. (9.19) into Eq. (9.17), we obtain

$$(1/a_j - \rho_j/c_j)u_j = \frac{(b_j/a_j - d_j/c_j)(a_{j+1}/b_{j+1} - c_{j+1}/d_{j+1})}{(1/b_{j+1} - \rho_{j+1}/d_{j+1})}u_j\rho_{j+1}, \tag{9.20}$$

and further multiplying $a_j c_j$ to both sides of Eq. (9.20) yields

$$c_j - a_j\rho_j = -|A_j| \cdot \frac{|A_{j+1}|}{d_{j+1} - b_{j+1}\rho_{j+1}}\rho_{j+1},$$

and then

$$\rho_j = \frac{1}{a_j}\left(c_j + \frac{|A_j| \cdot |A_{j+1}|\rho_{j+1}}{d_{j+1} - b_{j+1}\rho_{j+1}}\right), \tag{9.21}$$

where $|A_j| \equiv a_j d_j - b_j c_j$.

Using the relations

$$c_j/a_j = r_{j,j+1}\mathrm{e}^{\mathrm{i}\phi_j/2},$$

$$|A_j|/a_j = d_j - b_j c_j/a_j,$$

$$\frac{|A_{j+1}|}{d_{j+1} - b_{j+1}\rho_{j+1}} = \frac{t_{j,j+1}\mathrm{e}^{\mathrm{i}\phi_{j+1}/2}}{1 - r_{j+1,j}\mathrm{e}^{\mathrm{i}\phi_{j+1}/2}\rho_{j+1}},$$

we finally obtain the following iterative form for the reflection at the jth layer:

$$r_j = \frac{r_{j,j+1} + e^{\mathrm{i}\phi_{j+1}}r_{j+1}}{1 + r_{j,j+1}e^{\mathrm{i}\phi_{j+1}}r_{j+1}}, \tag{9.22}$$

where $r_j \equiv v_j \exp[-i\phi_j/2]/u_j$, and we have used the relations, $r_{j,j+1} = -r_{j+1,j}$ and $r_{j,j+1}^2 + t_{j,j+1}t_{j+1,j} = 1$, derived from the Fresnel's law.

The calculation of the reflectivity is performed as follows: For an N-layer system, the initial value for the iteration starts from the calculation of the reflectivity at an interface between the Nth layer and the outer space as $r_N = v_N \exp[-i\phi_N/2]/u_N = r_{N,N+1}$. Then, by inserting this result into Eq. (9.22), r_{N-1} is obtained and so forth. Finally, r_0 calculated in this way results in the total amplitude reflectivity, where suffices 0 and $N+1$ denote the incident and transmitted media of outer space, respectively. Since the reflectance and transmittance at each interface are simply derived from the Fresnel's law, Eq. (9.22) is applicable for both s- and p-polarizations under oblique incidence.

The features of iterative matrix method are summarized as follows: (1) The reflectivity and transmittance from an arbitrary multilayer are obtained without any periodicity in layer thickness and refractive index, and (2) the numerical calculation is always easy, while (3) it is difficult to derive any physical meaning from the expression.

9.1.3 *Huxley's Method*

Huxley derived a completely different expression for a periodic multilayer (Huxley, 1968). He considered p layers (designated as b) of a higher refractive index embedded in a material (designated as a) of a lower refractive index as shown in Fig. 9.3. Then, the recurrence equations for the field amplitudes evaluated at the middle of a layers become

$$a_{j-1}^- = \rho a_{j-1}^+ + \tau a_j^- \quad \text{and} \quad a_j^+ = \tau a_{j-1}^+ + \rho a_j^-, \tag{9.23}$$

where a_j^+ and a_j^- are the field amplitudes for $+z$ and $-z$ propagating waves with respect to the jth layer. ρ and τ are the amplitude reflectivity and transmittance of a single layer evaluated at the middle of the a layer and are obtained similarly to the thin-film interference of Eqs. (2.40) and (2.41) after replacing the notations as $r_{ba} = r_{bc} \to r$, $r \to \rho$, and $t \to \tau$:

$$\rho = -r\frac{1-e^{-2i\phi_b}}{1-r^2e^{-2i\phi_b}}e^{-i\phi_a}, \tag{9.24}$$

and

$$\tau = \frac{(1-r^2)e^{-i(\phi_a+\phi_b)}}{1-r^2e^{-2i\phi_b}}, \tag{9.25}$$

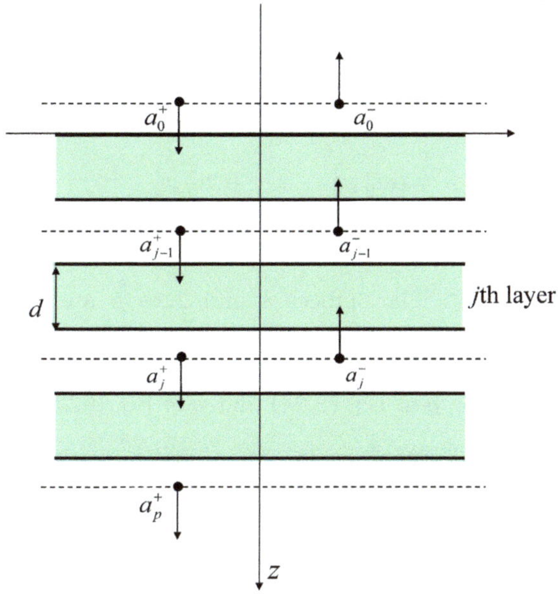

Fig. 9.3 Coordinates employed in Huxley's method.

where we replace ϕ in Eq. (2.43) with $-2\phi_m$ and thus $\phi_m = -2\pi n_m d_m/\lambda$ with $m = a$ and b. n_m and d_m are the refractive index and thickness of the mth layer.

He considered the following trial functions for the solutions of the recurrence equations:

$$a_j^+ = a_0^+ \mu^j \quad \text{and} \quad a_j^- = h a_0^+ \mu^j, \tag{9.26}$$

where μ and h are unknown constants. Inserting these functions into Eq. (9.23), we obtain

$$h = \rho + \tau h \mu, \tag{9.27}$$

$$\mu = \tau + \rho h \mu. \tag{9.28}$$

Further, eliminating either μ or h, we finally obtain the following two quadratic equations:

$$\mu^2 - \mu(1 + \tau^2 - \rho^2)/\tau + 1 = 0, \tag{9.29}$$

$$h^2 - h(1 - \tau^2 + \rho^2)/\rho + 1 = 0. \tag{9.30}$$

By eliminating the product μh from Eqs. (9.27) and (9.28), we confirm that μ and h are connected with each other through the linear relation: $h = (\tau\mu - \tau^2 + \rho^2)/\rho$. Since Eq. (9.23) is linear with respect to a_j^+ or a_j^-, the superposition of the two solutions, $h_{1,2}$ or $\mu_{1,2}$, gives the general solutions:

$$a_j^+ = \alpha a_0^+ \mu_1^j + (1-\alpha) a_0^+ \mu_2^j \tag{9.31}$$

$$a_j^- = \alpha h_1 a_0^+ \mu_1^j + (1-\alpha) h_2 a_0^+ \mu_2^j, \tag{9.32}$$

where α is a constant. The suffices 1 and 2 to μ and h correspond to the negative and positive signs in the solutions of Eqs. (9.29) and (9.30), respectively. The following relations are generally derived from Eqs. (9.29) and (9.30) by inserting ρ of Eq. (9.24) and τ of Eq. (9.25):

$$\mu_1 + \mu_2 = \frac{(1 + \tau^2 - \rho^2)}{\tau} = 2\frac{\cos 2\phi - r^2}{1 - r^2}, \tag{9.33}$$

$$\mu_1 \mu_2 = 1, \tag{9.34}$$

and

$$h_1 + h_2 = \frac{(1 - \tau^2 + \rho^2)}{\rho} = -\frac{\cos\phi}{r}, \tag{9.35}$$

$$h_1 h_2 = 1, \tag{9.36}$$

where we assume an ideal multilayer with $\phi_a = \phi_b \equiv \phi$ for simplicity.

The boundary condition $a_p^- = 0$ in Eq. (9.32) gives $\alpha h_1 a_0^+ + (1-\alpha) h_2 a_0^+ = 0$, which results in

$$\alpha = \frac{h_1}{(h_2 - m^2 h_1)}, \tag{9.37}$$

with $m^2 = (\mu_1/\mu_2)^p$. Thus, the power reflectance and transmittance are obtained as

$$R = \left|\frac{a_0^-}{a_0^+}\right|^2 = |\alpha h_1 + (1-\alpha) h_2|^2 = \left|\frac{1 - m^2}{h_2 - m^2 h_1}\right|^2, \tag{9.38}$$

and

$$T = \left|\frac{a_p^+}{a_0^+}\right|^2 = |\alpha \mu_1^p + (1-\alpha)\mu_2^p|^2 = \left|\frac{(h_2 - h_1)m}{h_2 - m^2 h_1}\right|^2. \tag{9.39}$$

The physical meaning of Eq. (9.38) is explained as follows: First, using the relations Eqs. (9.33)–(9.36), we have converted Eqs. (9.29) and (9.30) into

$$\mu^2 - 2\frac{\cos 2\phi - r^2}{1 - r^2}\mu + 1 = 0, \tag{9.40}$$

$$h^2 + 2\frac{\cos\phi}{r}h + 1 = 0. \tag{9.41}$$

Then, the discriminants of the above two quadratic equations, D_μ and D_h, become

$$D_\mu/4 \equiv \left(\frac{\cos 2\phi - r^2}{1 - r^2}\right)^2 - 1 = \frac{4(1-\cos^2\phi)(r^2 - \cos^2\phi)}{(1-r^2)^2},$$

$$D_h/4 \equiv \left(\frac{\cos^2\phi}{r}\right)^2 - 1 = \frac{\cos^2\phi - r^2}{r^2}.$$

Thus, it is clear that D_μ and D_h are complementary with respect to the term $\cos^2\phi - r^2$. Hence, when $\mu_{1,2}$ are real, $h_{1,2}$ become complex and *vice versa.*

(1) Case for $\cos^2\phi - r^2 < 0$

The solutions $\mu_{1,2}$ are real, while $h_{1,2}$ become pure imaginary through the relation Eq. (9.36). The reflectivity in this case is further calculated as

$$\begin{aligned} R &= \left|\frac{1 - m^2}{(h_1 + h_2)(1 - m^2)/2 - (h_1 - h_2)(1 + m^2)/2}\right|^2 \\ &= \frac{(1 - m^2)^2}{(h_1 + h_2)^2(1 - m^2)^2/4 - (h_1 - h_2)^2(1 + m^2)^2/4} \\ &= \left\{1 - \frac{4m^2}{(1 - m^2)^2}\left(1 - \frac{\cos^2\phi}{r^2}\right)\right\}^{-1}, \end{aligned} \tag{9.42}$$

where $m^2 = (\mu_1/\mu_2)^p < 1$ and we use the relation $(h_1 - h_2)^2/4 = (\cos^2\phi - r^2)/r^2$. With increasing number of layers p, $m^2 \to 0$ and hence the reflectivity becomes unity. Thus, the relation $\cos^2\phi - r^2 < 0$ gives the range of reflection band, which just corresponds to one-dimensional photonic band gap described in Secs. 2.5 and 9.1.4, and also offers another method to estimate the reflection bandwidth as will be described in the following section.

(2) Case for $\cos^2\phi - r^2 > 0$
The solutions $h_{1,2}$ are real, while $\mu_{1,2}$ are pure imaginary. Putting $\mu_{1,2} \equiv \exp[\mp i\theta]$, we calculate the reflectivity as

$$R = \left| \frac{(h_1 + h_2)}{2} - \frac{1+m^2}{1-m^2}\frac{(h_1 - h_2)}{2} \right|^{-2}$$
$$= \left\{ 1 + \frac{\cos^2\phi - r^2}{r^2 \sin^2 p\theta} \right\}^{-1}, \tag{9.43}$$

where we use the relation $(1+m^2)/(1-m^2) = -i\cot p\theta$. The term including $\sin^2 p\theta$ expresses the oscillation appearing in the side lobe of the reflection band. When $\sin^2 p\theta = 0$, $R \to 0$, while in case of $\sin^2 p\theta = 1$, $R \neq 0$, which gives an envelope function for the side lobe:

$$R_{\mathrm{env}} = \frac{1}{1 + (\cos^2\phi - r^2)/r^2}. \tag{9.44}$$

In Fig. 9.4, we show an example of the reflectivity calculated using Eq. (9.38) and the envelope function R_{env} using Eq. (9.44).

Huxley's expression for multilayer reflectivity has the following features: (1) The reflectivity is obtained through arithmetical computation without

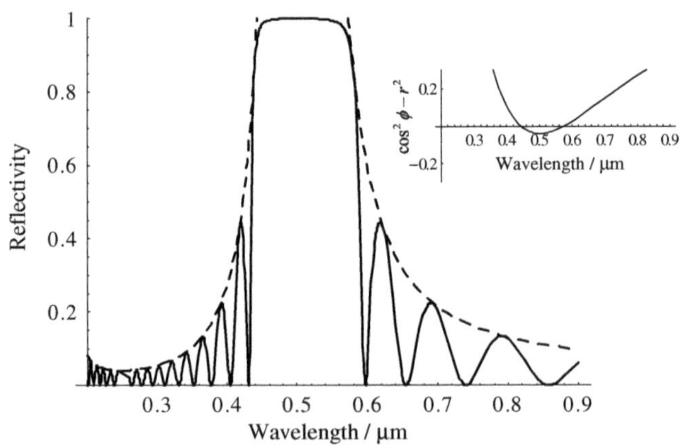

Fig. 9.4 Reflectivity from an ideal multilayer consisting of 10 layers of 0.0833 μm thickness with the refractive index of 1.5, which are embedded in a medium with the refractive index of 1.0. Solid line is calculated using Eq. (9.38), while dashed line is calculated using Eq. (9.44). The parameter, $\cos^2\theta - r^2$, characterizing the reflection properties is also plotted in the inset.

iteration or matrix calculation, and (2) the reflection bandwidth and oscillation appearing in the side lobe are obtained analytically, whereas (3) the applicable multilayers are considerably limited.

9.1.4 *Estimation of Reflection Bandwidth*

In order to estimate the reflection bandwidth from a multilayer, it is useful to consider a simple one-dimensional (1D) photonic crystal depicted in Fig. 9.5, which is essentially equivalent to a multilayer except that the number of layers is infinite (Sakoda, 2001). The governing equations in a one-dimensional case corresponding to Eq. (2.62) become

$$\sum_{G'} b_{G-G'}(k+G')^2 E_{G'} = \omega^2 \mu_0 E_G. \tag{9.45}$$

Further, we expand the reciprocal permittivity as

$$1/\epsilon(x) = b_0 + b_1 e^{iG_1 x} + b_{-1} e^{iG_{-1} x} + \cdots , \tag{9.46}$$

with $G_j = 2\pi j/a$. Here, we will consider only the first reflection band, i.e. $k \approx \pi/a$ and $\omega \approx b_0 k/\sqrt{\mu_0}$, with a the period of the multilayer. Considering only the terms involving E_0 and E_{-1}, we obtain the following relation (see Fig. 9.6):

$$\begin{pmatrix} b_0 k^2 & b_1(k+G_{-1})^2 \\ b_{-1}k^2 & b_0(k+G_{-1})^2 \end{pmatrix} \begin{pmatrix} E_0 \\ E_{-1} \end{pmatrix} = \mu_0\omega^2 \begin{pmatrix} E_0 \\ E_{-1} \end{pmatrix}. \tag{9.47}$$

Solving the following eigen equation with respect to ω^2,

$$\begin{vmatrix} b_0 k^2 - \mu_0\omega^2 & b_1(k+G_{-1})^2 \\ b_{-1}k^2 & b_0(k+G_{-1})^2 - \mu_0\omega^2 \end{vmatrix} = 0, \tag{9.48}$$

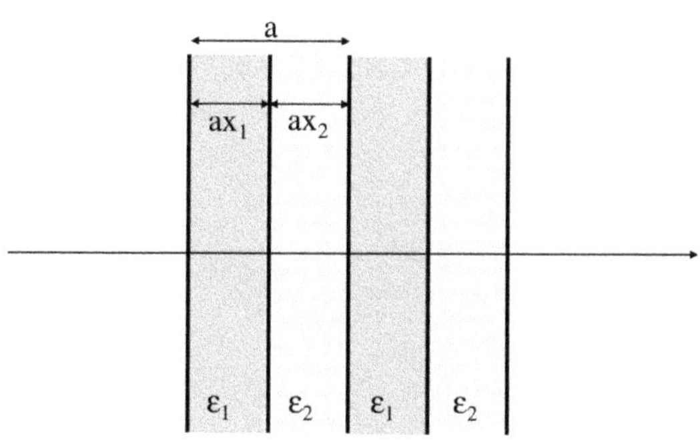

Fig. 9.5 One-dimensional photonic crystal.

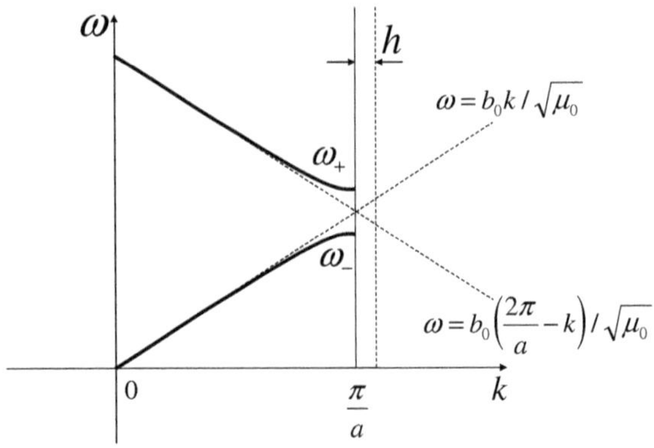

Fig. 9.6 Schematic diagram for the dispersion curve for one-dimensional photonic crystal.

we obtain the expression for the dispersion relation near the first band gap:

$$\mu_0\omega^2 = \frac{1}{2}b_0\{k^2 + (k+G_{-1})^2\}$$
$$\pm\frac{1}{2}\sqrt{b_0^2\{k^2-(k+G_{-1})\}^2 + 4b_1b_{-1}k^2(k+G_{-1})^2}. \quad (9.49)$$

In order to investigate the behavior of the dispersion curve near the first band gap, we put $k = \pi/a + h$ and expand the above expression within the second order of h:

$$\omega_\pm \approx \frac{\pi}{a\sqrt{\mu_0}}\sqrt{b_0 \pm |b_1|} + \frac{a}{2\pi\sqrt{\mu_0}\sqrt{b_0 \pm |b_1|}}\left(b_0 \pm \frac{2b_0^2 - |b_1|^2}{|b_1|}\right)h^2, \quad (9.50)$$

where we use the relation $b_1b_{-1} = |b_1|^2$.

The schematic diagram for the dispersion relation for one-dimensional photonic crystal is shown in Fig. 9.6. The band gap corresponds to the reflection band in multilayer reflection, whose band center and bandwidth are thus obtained as

$$\omega_c = \frac{\pi\sqrt{b_0}}{(a\sqrt{\mu_0})}, \quad (9.51)$$

and

$$\Delta\omega = \omega_+ - \omega_-, \quad (9.52)$$

respectively.

For periodic double layers, whose dielectric constants are expressed as $\epsilon_1 = n_1^2$ and $\epsilon_2 = n_2^2$, the expansion coefficients are obtained as follows:

$$\begin{aligned} b_0 &= \frac{x_1}{n_1^2} + \frac{x_2}{n_2^2} \\ b_1 &= \frac{1}{\pi} \mathrm{e}^{-\mathrm{i}\pi x_1} \sin(\pi x_1) \left(\frac{1}{n_1^2} - \frac{1}{n_2^2} \right), \end{aligned} \tag{9.53}$$

where the thicknesses of the layers, 1 and 2, are put as ax_1 and ax_2, respectively, with $x_1 + x_2 = 1$. Inserting these relations into Eq. (9.51), we obtain the band center in a wavelength unit, λ_c, as

$$\begin{aligned} \lambda_c &= \frac{2\pi c}{\omega_c} \\ &= 2a \Bigg/ \sqrt{\frac{1}{n_2^2} \left\{ 1 + \frac{n_2^2 - n_1^2}{n_1^2} x_1 \right\}}. \end{aligned} \tag{9.54}$$

When $n_1 \approx n_2$, this relation reduces to

$$\lambda_c = 2a(n_1 x_1 + n_2 x_2), \tag{9.55}$$

which just corresponds to that we have already derived in an analogy with the thin-film interference.

The bandwidth is obtained as

$$\begin{aligned} \Delta\omega &= \omega_+ - \omega_- \\ &= \frac{\pi}{a\sqrt{\mu_0}} (\sqrt{b_0 + |b_1|} - \sqrt{b_0 - |b_1|}) \\ &\approx \frac{\pi}{a\sqrt{\mu_0}} \frac{|b_1|}{\sqrt{b_0}}, \end{aligned} \tag{9.56}$$

where the last relation is derived under the assumption of $|b_1| \ll b_0$. The bandwidth in a wavelength unit, $\Delta\lambda$, is obtained as

$$\Delta\lambda = \frac{2\pi c}{\omega_-} - \frac{2\pi c}{\omega_+} = \frac{2\pi c \Delta\omega}{\omega_+ \omega_-} \approx \frac{2a|b_1|}{(b_0)^{3/2}}. \tag{9.57}$$

Under the conditions $x_1 \approx x_2$ and $n_1 \approx n_2$, it reduces to $\Delta\lambda \approx 4a|n_2 - n_1|/\pi$. Thus, the bandwidth is proportional to the difference of the refractive indices of the two. In Fig. 9.7, we show the calculated results for λ_c and $\lambda_\pm (\equiv 2\pi c/\omega_\mp)$, for small and large differences in the refractive indices

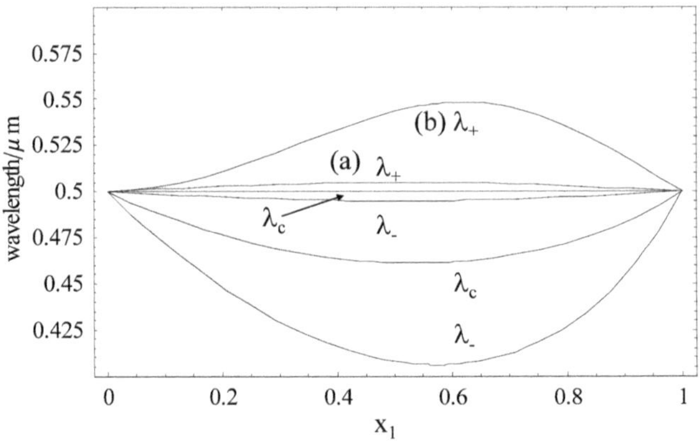

Fig. 9.7 Photonic band gap for alternate multilayer against the thickness of a higher refractive index x_1. Light is assumed to propagate perpendicularly to the layer. The refractive indices employed are the combinations of (a) 1.6 and 1.55, and (b) 1.6 and 1.0, respectively. λ_+, λ_-, and λ_c are longer and shorter wavelength edges of the band gap and its center.

between the two layers. Thus, the reflection bandwidths are clearly dependent on the difference of the refractive indices, which tend to be larger when the multilayer comes close to ideal one. It is noticeable the band center itself deviates from that determined by the volume fraction of one component, when the difference of the refractive indices is large.

We further compared the result with the exact photonic band calculation. The result is shown in Fig. 9.8. The agreement is quite good for the difference of refractive index of 0.5. Thus, the approximate expressions, Eqs. (9.51) and (9.52), will be useful to estimate the reflection center and bandwidth in actual multilayers. We also plot a band center deduced from the average refractive index calculated according to Maxwell–Garnett model, which will be explained in Sec. 9.4.

9.2 Model for *Morpho* Butterfly Scale

9.2.1 *"Shelf Structure" Model*

In the present section, we will explain a simple model to explain the reflection and diffraction properties of light in the scale of the *Morpho* butterflies by taking the geometrical shelf structure into account (Kinoshita *et al.*, 2002a,b). At first, we assume N ridges standing equidistantly as shown in Fig. 9.9(b). Each shelf in a ridge is assumed to be infinitely thin, and to

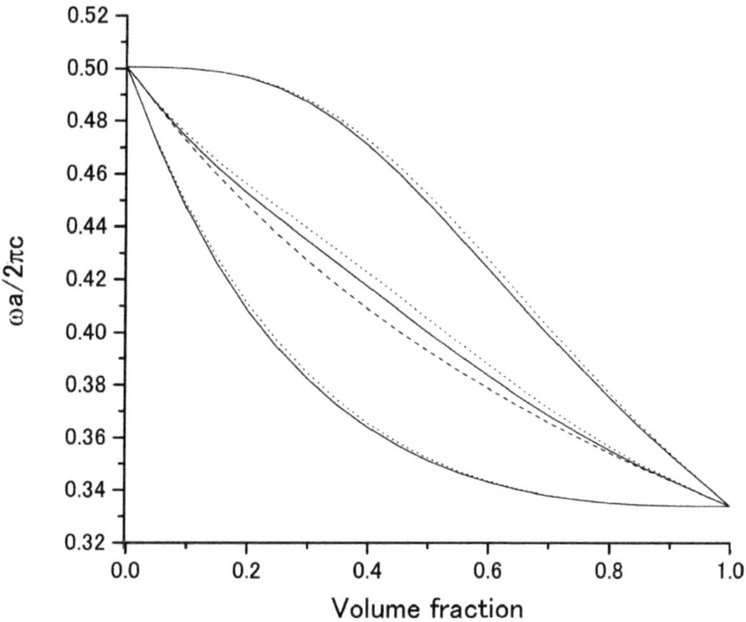

Fig. 9.8 Photonic band center and edges calculated for alternate multilayer against the fraction of high-refractive index portion (Fujimura, 2008). Light is assumed to propagate perpendicularly to the layer. The refractive indices employed are 1.5 and 1.0, respectively. Solid curves are obtained by 1D photonic band calculation with the band center obtained by the average, while dotted curves are obtained through approximate expressions, Eqs. (9.50) and (9.51). Dashed line is an extrapolated value of the average refractive index calculated using the Maxwell–Garnett model, Eq. (9.92).

have a narrow width of a in one direction and a long length of l in the other. A plane wave is incident on a ridge, and is scattered at each layer. We assume the incident and diffracted light is neither subjected to reflection nor refraction while passing through the other. Then, the light diffracted by each layer interferes in far-field region. Although this simplified assumption does not take into account the refraction and reflection at each shelf and also neglects the multiple reflections, it offers an opportunity to analyze the characteristics of structural color without missing the physical essence, and also partly reflects the finite and somewhat irregular structure of an actual shelf structure.

Consider first a case where a single shelf is located at a position, where the center of the shelf is expressed by a 2D Cartesian coordinate (x, y) as shown in Fig. 9.9(a). The shelf is assumed to be placed in parallel with the x-axis. For the diffracted light within the xy plane toward the direction of

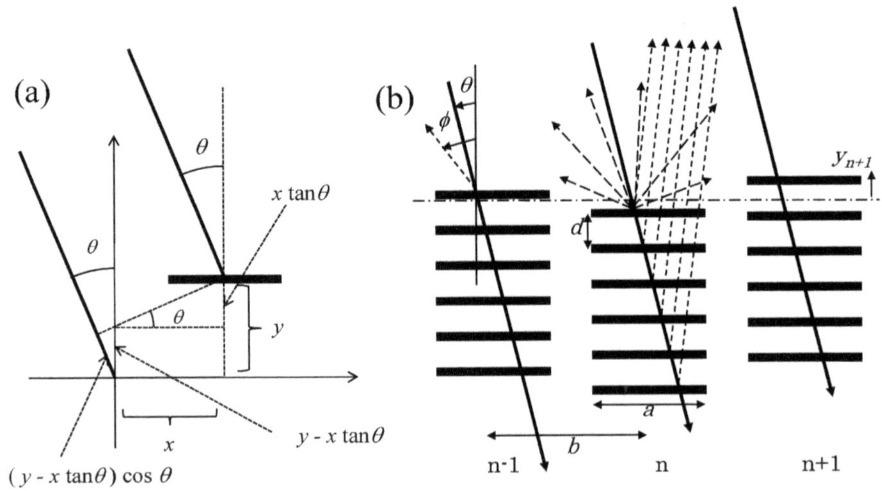

Fig. 9.9 Shelf structure to model an iridescent scale of the *Morpho* butterfly. (a) Single shelf and (b) M shelves standing equidistantly with a pitch of b along x direction.

ϕ under the illumination of incident light from the direction of θ, the path differences between the light incident on the origin and the center of the shelf are expressed as

$$z = -\{(y - x\tan\theta)\cos\theta\} - \{(y - x\tan\phi)\cos\phi\}$$
$$= x(\sin\theta + \sin\phi) - y(\cos\theta + \cos\phi) \equiv xu - yv, \tag{9.58}$$

where

$$u = \sin\theta + \sin\phi, \tag{9.59}$$
$$v = \cos\theta + \cos\phi. \tag{9.60}$$

Then, the electric field in far field is calculated as

$$E \propto \int_{x-a/2}^{x+a/2} \mathrm{e}^{\mathrm{i}k(x'u-yv)}\mathrm{d}x' BE_{\mathrm{in}}\cos\theta = \frac{\mathrm{e}^{\mathrm{i}kua/2} - \mathrm{e}^{-\mathrm{i}ua/2}}{\mathrm{i}ku}\mathrm{e}^{\mathrm{i}k(xu-yv)}BE_{\mathrm{in}}$$
$$= \frac{a\sin(kua/2)}{kua/2}\mathrm{e}^{\mathrm{i}k(xu-yv)}BE_{\mathrm{in}}, \tag{9.61}$$

where E_{in} is the electric field of incident light with the inclination factor $B = (\cos\theta + \cos\varphi)/(2\mathrm{i}\lambda) = v/(2\mathrm{i}\lambda)$. If M shelves are arranged periodically along the y-axis such that $y = y_0 - md$ (Fig. 9.9(b)), where $m = 0, \ldots, M-1$,

$$\begin{aligned} E &\propto \sum_{m=0}^{M-1} \mathrm{e}^{ikmdv} \frac{a\sin(kua/2)}{kua/2} \mathrm{e}^{\mathrm{i}k(xu-y_0v)} BE_{\mathrm{in}} \\ &= \frac{1-\mathrm{e}^{\mathrm{i}kMdv}}{1-\mathrm{e}^{\mathrm{i}kdv}} \cdot \frac{a\sin(kua/2)}{kua/2} \mathrm{e}^{\mathrm{i}k(xu-y_0v)} BE_{\mathrm{in}} \\ &= \frac{\sin(kdvM/2)}{\sin(kdv/2)} \cdot \frac{a\sin(kua/2)}{kua/2} \mathrm{e}^{\mathrm{i}k(xu-y_0v)+\mathrm{i}kdv(M-1)/2} BE_{\mathrm{in}}. \end{aligned} \tag{9.62}$$

Furthermore, if N such shelf structures stand equidistantly with a pitch of b along x direction (see Fig. 9.9(b)):

$$E \propto \frac{\sin(kdvM/2)}{\sin(kdv/2)} \cdot \frac{a\sin(kua/2)}{kua/2} \sum_{n=0}^{N-1} \mathrm{e}^{\mathrm{i}k\{(nbu-y_nv)+dv(M-1)/2\}} BE_{\mathrm{in}}, \tag{9.63}$$

where y_n is the deviation of the nth ridge height from the average. Then, the cycle-averaged intensity of diffracted light becomes

$$\begin{aligned} I_\phi &= \frac{1}{2}|E|^2 \\ &= \frac{a^2}{2} \cdot \frac{\sin^2(kdv^2/M/2)}{\sin^2(kdv/2)} \cdot \frac{\sin^2(kau/2)}{(kau/2)^2} \cdot F_R \cdot I_0 v^2/4\lambda^2, \end{aligned} \tag{9.64}$$

with

$$F_R \equiv \left| \sum_{n=0}^{N-1} \exp[\mathrm{i}(kbun - \psi_n)] \right|^2, \tag{9.65}$$

where $k = 2\pi/\lambda$ with the wavelength of light λ. ψ_n is the phase difference at the nth ridge due to the height distribution of the ridge and is expressed as

$$\psi_n = ky_nv. \tag{9.66}$$

Here, the terms $\sin^2(kdvM/2)/\sin^2(kdv/2)$, $\sin^2(kau/2)/(kau/2)^2$, and F_R express the interference within a ridge, the diffraction due to a single shelf, and the interference between different ridges, respectively. If there is no spatial correlation among ridge heights, then the factor F_R reduces to

$$F_R = N\langle|\exp[\mathrm{i}(kbun - \psi_n)|^2\rangle = N, \tag{9.67}$$

and any interference term between different ridges disappears, where $\langle\cdots\rangle$ indicates the ensemble average. This means that the diffracted light

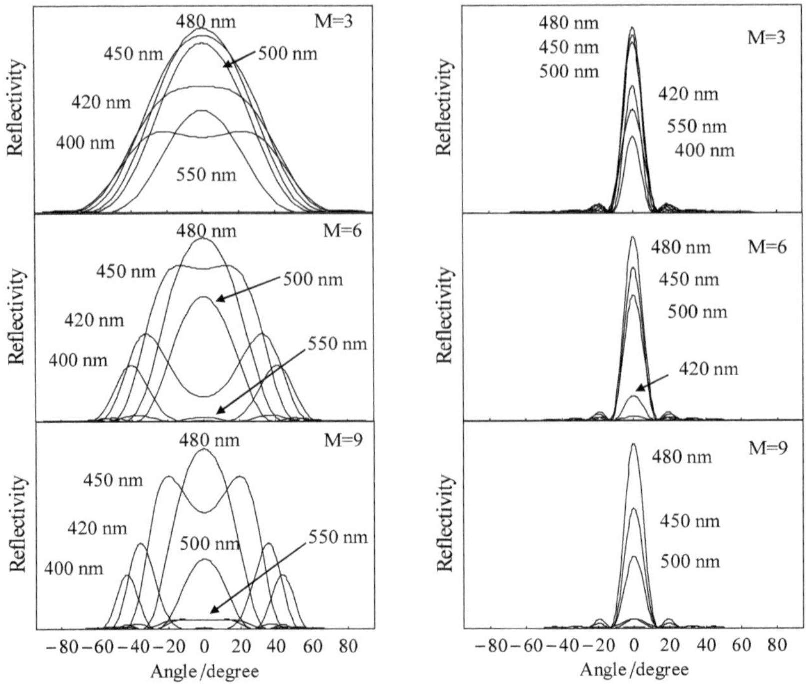

Fig. 9.10 Angular dependence of the reflected light intensity under normal incidence on three, six, and nine shelves, calculated using Eqs. (9.64) and (9.67). Left-hand side shows the angular dependence along shorter side of the shelf, while right-hand side shows that along a longer side. The width and length of a shelf are set at 300 nm and 2 μm, respectively, while the separation of the nearest shelves are set at 235 nm.

emanating from each ridge does not interfere coherently so that the wavelength and angular dependence of the light reflected and diffracted from a scale can be considered to originate essentially from shelves within each ridge. Thus, the total diffraction results in the incoherent sum of single-ridge diffraction.

In Figs. 9.10 and 9.11, we show the simulated data due to a shelf model. The detailed analysis of this result is already described in Sec. 3.2.5.3. Here, we briefly comment on the anisotropic reflection and the reflection spectrum, where we assume the length of a shelf to be l, and apply the same relation only by replacing a with l. As shown in Fig. 9.10, the reflection due to a slender shape of shelf shows large anisotropic reflection. Namely, along the longer side of shelf, the angular dependence is quite sharp and essentially no particular reflection pattern appears. The shape of reflection pattern does

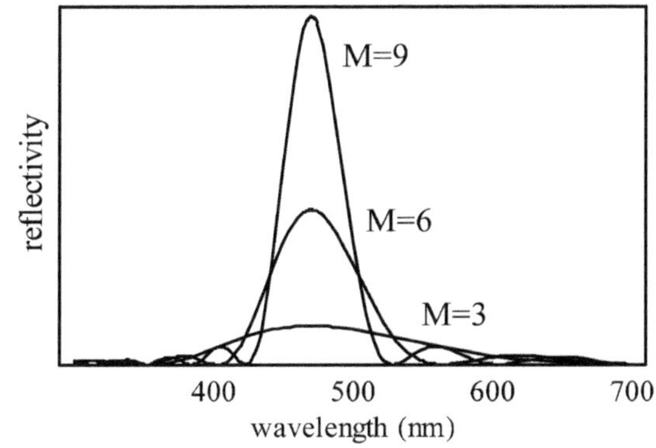

Fig. 9.11 Reflection spectrum under normal incidence for three, six and nine shelves, calculated using Eqs. (9.64) and (9.67) with the same parameters as in Fig. 9.10.

not depend on the number of shelves and only the wavelength-dependent intensity profiles vary remarkably. Thus, the peculiar wevelength-dependent reflection pattern observed in *Morpho* butterfly wing is solely originated from the narrow width of the ridge structure. On the other hand, the intensity and width of the reflection spectrum from shelf structures depend strongly on the number of shelves, as shown in Fig. 9.11. With increasing number of shelves, the reflection peak increases remarkably and its width narrows, which is easily expected from the interference effect. The reflection spectrum of six shelves somewhat resembles the actual reflection spectrum from the *M. didius* of Fig. 3.19, which anyway differs considerably from a simple multilayer calculation.

9.2.2 *Effect of Alternately Sticking Shelf Structure and Ridge Inclination*

The alternately sticking shelf structure, shown in Fig. 9.12(a), is easily taken into account by considering each contribution of the left and right shelf separately,

$$E \propto \sum_{m=0}^{M-1} \left\{ \int_{x}^{x+a/2} \mathrm{e}^{\mathrm{i}k\{x'u-(y+\Delta y-md)v\}} \mathrm{d}x' + \int_{x-a/2}^{x} \mathrm{e}^{\mathrm{i}k\{x'u-(y-md)v\}} \mathrm{d}x' \right\}$$
$$\times B E_{\mathrm{in}}$$

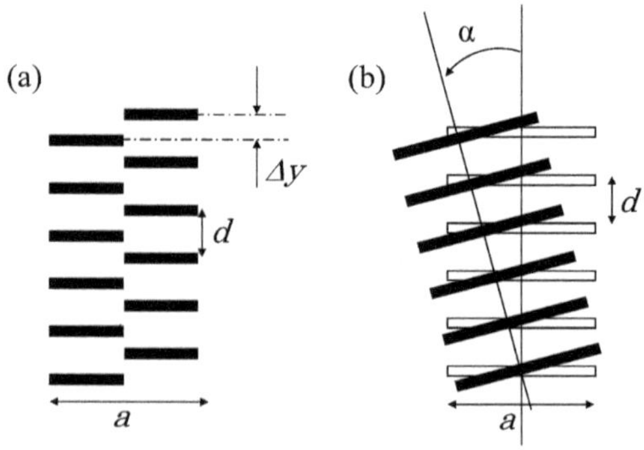

Fig. 9.12 Geometries for (a) alternately sticking shelf structure and (b) ridge inclination.

$$= \frac{\sin(kdvM/2)}{\sin(kdv/2)} \cdot \frac{(a/2)\sin(kua/4)}{kua/4} \cdot 2\cos\{k(ua - 2v\Delta y)/4\}$$

$$\times \mathrm{e}^{ik(xu-yv)-ikv\Delta y/2+ikdv(M-1)/2} BE_{\mathrm{in}}, \tag{9.68}$$

where Δy is the height difference of the left and right shelves. Thus, the reflection intensity in the direction of ϕ becomes

$$I_\phi = \frac{a^2}{2} \cdot \frac{\sin^2(kdvM/2)}{\sin^2(kdv/2)} \cdot \frac{\sin^2(kau/4)}{(kau/4)^2}$$

$$\times \cos^2\{k(ua - 2v\Delta y)/4\} \cdot F_R \cdot I_0 v^2/4\lambda^2, \tag{9.69}$$

which differs from that of the ordinary shelf structure by the presence of the factor $\cos^2\{k(ua - 2v\Delta y)/4\}$ and the diffraction due to the shelf of a half size.

In Fig. 9.13, we show the calculated result for six shelves under normal incidence. With increasing the difference between heights of left and right shelves, the angular dependence becomes strongly asymmetric and the enhancement of reflected light with shorter wavelengths becomes remarkable. Finally the right and left shelves are exactly in an alternate position, the angular dependence is almost symmetric with large reflection components at large reflection angles and at the same time, the reflected light with longer wavelengths almost disappears. This is because blue light, which will be reflected toward normal direction, interferes destructively owing to the phase difference of 180° reflected at right and left shelves. On the other hand, the violet light, which will be reflected obliquely, is enhanced because alternating right and left shelves have oblique reflection planes at

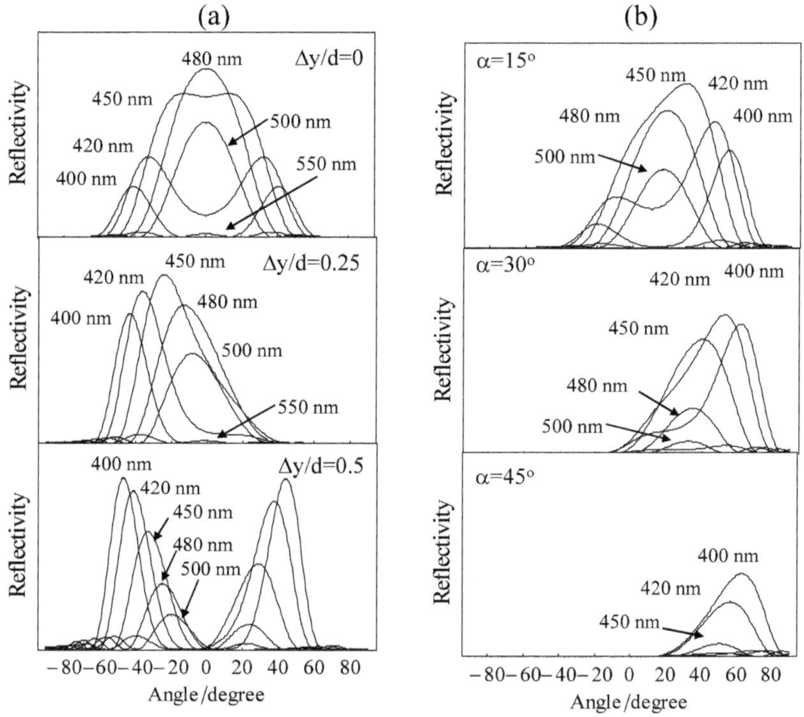

Fig. 9.13 (a) Angular dependence of the reflected light intensity from six alternately sticking shelf structure under normal incidence. The position of the right shelf relative to the left one is changed as indicated. (b) Angular dependence of the reflected light intensity from an obliquely standing ridge having six shelves under normal incidence. The inclination angles measured from the normal direction are indicated in the figure.

both sides. Thus alternately sticked shelf structure hinders the reflection toward the normal direction and plays a role of broadening the reflection angle. Further it contributes the enhancement of light reflection at shorter wavelengths.

The effect of ridge inclination is easily taken into account in the model of "shelf structure." Consider the case that the ridge inclines by α from the normal, as shown in Fig. 9.13(b). Then, under the assumption that the heights of the ridges are randomly distributed, the expression Eq. (9.64) is applicable by replacing

$$\begin{aligned} u \to u_\alpha &= \sin(\theta - \alpha) + \sin(\phi - \alpha) \\ v \to v_\alpha &= \cos(\theta - \alpha) + \cos(\phi - \alpha) \end{aligned} \tag{9.70}$$

with $F_R = 1$. Then,

$$I_\phi = \frac{a^2}{2} \cdot \frac{\sin^2(kdv_\alpha M/2)}{\sin^2(kdv_\alpha/2)} \cdot \frac{\sin^2(kau_\alpha/2)}{(kau_\alpha/2)} \cdot I_0 v_\alpha^2/4\lambda^2. \tag{9.71}$$

The calculated results for various values of α are shown in Fig. 9.13.

The ridge inclination shows a similar effect on the angular dependence as in the alternately sticking shelf structure for small value of α, because the reflection pattern becomes antisymmeteric with respect to the normal direction, and the light of shorter wavelengths becomes enhanced. However, at extremely oblique positions, the plane of the shelf deviates considerably from an illuminated plane and hence the reflected light intensity tends to decrease prominently.

9.2.3 *Effect of Spatial Correlation in the Ridge Height Distribution*

In order to explain a peculiar intensity modulation found in the reflection pattern from a wing of *M. didius* without cover scales (Fig. 3.21(b)), we consider a case where a small amount of the spatial correlation of the ridge height distribution is present. The presence of the spatial correlation implied that the ridges are not completely independent and sometimes have grown from a common root, although a clear correlation longer than the ridge separation has not been found in the experiment (Fig. 3.16).

For this purpose, the factor F_R is reformulated as

$$\begin{aligned} F_R &= \sum_{mn} \exp[\mathrm{i}kbu(m-n) - \mathrm{i}(\psi_m - \psi_n)] \\ &\approx N \sum_{n=-N}^{N} \mathrm{e}^{\mathrm{i}nkbu} \langle \mathrm{e}^{-\mathrm{i}\Delta\psi(nb)} \rangle, \end{aligned} \tag{9.72}$$

where we put

$$\sum_{m=n+1}^{N} \mathrm{e}^{-\mathrm{i}(\psi_m - \psi_{m-n})} \equiv (N-n)\langle \mathrm{e}^{-\mathrm{i}\Delta\psi(nb)} \rangle \approx N \langle \mathrm{e}^{-\mathrm{i}\Delta\psi(nb)} \rangle, \tag{9.73}$$

under the assumption that the correlation length of the ridge height distribution is sufficiently small as compared with the total length Nb. $\Delta\psi(x)$ is a random function expressing the phase difference between ridges separating by x. In order to estimate roughly the effect of spatial correlation, we consider the discrete ridge structure as a continuous function of space and obtain the following result:

$$\langle \mathrm{e}^{-\mathrm{i}\Delta\psi(x)} \rangle = \exp\left(-\frac{1}{2}\langle \Delta\psi(x)^2 \rangle\right), \tag{9.74}$$

where we assume, for simplicity, that the distribution of the ridge height obeys Gaussian random distribution and use the cumulant expansion. Further, if we consider only a leading term of a spatial variable, x, the above equation becomes

$$\exp(-\langle \Delta\psi(x)^2 \rangle/2) \approx \exp(-x^2/l_\mathrm{c}^2). \tag{9.75}$$

where l_c corresponds to a correlation length. Thus,

$$F_R \approx N \sum_{n=-N}^{N} \mathrm{e}^{inkbu} \exp(-(nb)^2/l_\mathrm{c}^2). \tag{9.76}$$

In Fig. 9.14, we show the simulated data for the angular dependence of the reflectivity in the case where a small amount of spatial correlation is present among the height distribution of ridges. It is intuitively understood that the incorporation of spatial correlation causes the modulation corresponding to the diffraction spots due to the interference between adjacent ridges, because the complete spatial correlation makes the arrangement of ridges behave just as a perfect diffraction grating. In fact, at the correlation length of $0.7\,\mu$m, which is equal to the ridge interval, clear diffraction spots are observed. The agreement between the experiment and the simulation becomes better around $l_\mathrm{c} = 0.5\,\mu$m, since it shows both a peak at the normal direction and the shoulders of ± 30–$40°$ at 480 nm. The presence of small spatial correlation implies that of small cooperativity in the lamellar formation on an early stage of the development.

9.2.4 *2D Fourier Analysis of the Shelf Structure*

The problem of how the actual shape of the complex structure affects the light reflection/diffraction properties is visually answered by performing 2D Fourier transform (FT). This method was introduced by Prum *et al.* to analyze the optical properties of ordered collagen arrays in avian skins and blue feather barbs in birds (Prum *et al.*, 1998, 1999b; Prum and Torres, 2003b). The relation between the 2D FT method and the color produced by the periodic structure is understandable from a simple electromagnetic

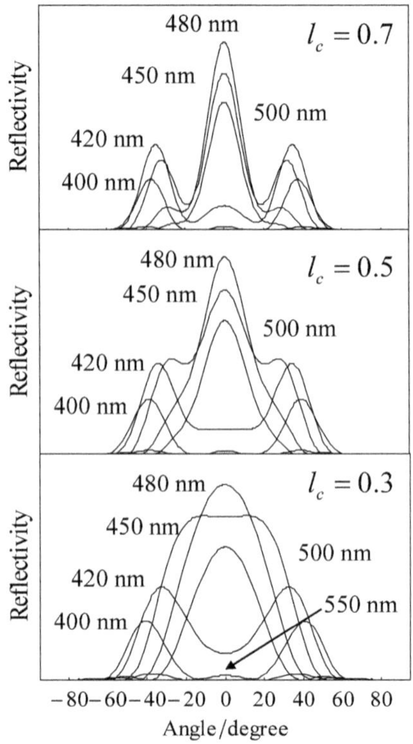

Fig. 9.14 Angular dependence of the reflectivity under normal incidence, calculated under the assumption that a plane wave normally incident on ridges having six shelves of 300 nm in width and 235 nm in interval. The ridge interval is set at 0.7 μm. The spatial correlation of the ridge height is taken into account using Eq. (9.76), where the correlation length is expressed as l_c.

consideration and the 2D FT is the first-order solution of the wave equation under the presence of the complex dielectric material. Although this model does not take into account the propagation delay due to the refractive index of the cuticle and hence the quantitative comparison with the experiment is difficult, it offers the intuitive insight into the scattering problem of the complex material.

We have applied this method to elucidate the essential features of the complex ridge-lamella structure found in a scale of the *Morpho* butterflies (Kinoshita *et al.*, 2002a; Kinoshita and Yoshioka, 2005a). To compare 2D Fourier image with the angular and frequency dependence of the *Morpho* scale/wing, we consider a geometry illustrated in Fig. 9.15, where a plane

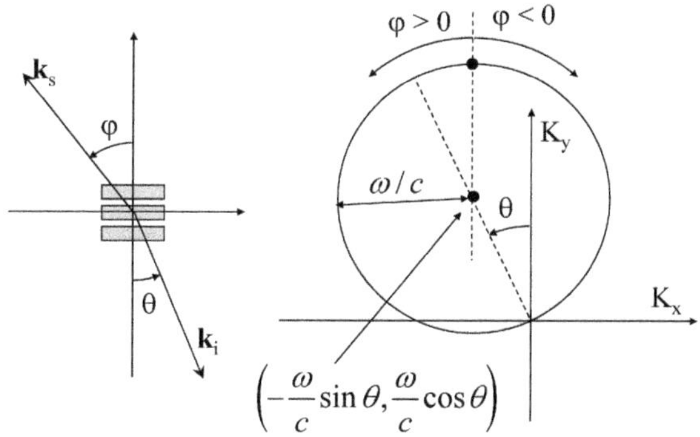

Fig. 9.15 Geometry of light scattering (left) and that in momentum space (right).

wave having the wave vector $\mathbf{k}_\mathrm{i}$ is incident and produces a scattering field with the wave vector $\mathbf{k}_\mathrm{s}$. The scattering vector $\mathbf{K}$ is

$$\mathbf{K} = \mathbf{k}_\mathrm{s} - \mathbf{k}_\mathrm{i} = \frac{\omega}{c}(\mathbf{s}_\mathrm{s} - \mathbf{s}_\mathrm{i}), \tag{9.77}$$

where ω, c, $\mathbf{s}_\mathrm{s}$ and $\mathbf{s}_\mathrm{i}$ are the angular frequency, light speed in vacuum, and unit vectors in the directions of $\mathbf{k}_\mathrm{s}$ and $\mathbf{k}_\mathrm{i}$, respectively. From this relation, we obtain

$$K_x = \frac{\omega}{c}(-\sin\varphi - \sin\theta)$$
$$K_y = \frac{\omega}{c}(\cos\varphi + \cos\theta),$$

where K_x and K_y are two Cartesian components of the scattering vector $\mathbf{K}$ in a momentum space. Eliminating φ from these relations, we obtain

$$\left(K_x + \frac{\omega}{c}\sin\theta\right)^2 + \left(K_y - \frac{\omega}{c}\cos\theta\right)^2 = \left(\frac{\omega}{c}\right)^2. \tag{9.78}$$

Using this relation, we can illustrate the angular and wavelength dependence of the light scattering in a momentum space as shown in Fig. 9.15. First, a circle of a radius $\omega/c = 2\pi/\lambda$ is drawn to pass through an origin, whose center is located at a point of the coordinate $(-(\omega/c)\sin\theta, (\omega/c)\cos\theta) = (-2(\pi/\lambda)\sin\theta, (2\pi/\lambda)\cos\theta)$. Then, the angular dependence of the reflection/diffraction intensity at a wavelength λ is directly connected with the absolute value of the 2D Fourier components, when one rotates on the circumference clockwise for $\varphi < 0$ and anticlockwise for $\varphi > 0$ from the top of the circle. The wavelength dependence is obtained by changing the radius.

First, we show the calculated result for the shape obtained from TEM images of the cross-section of the lamella in Fig. 9.16. The 2D FT image is characterized by (1) a periodic structure along the K_y axis, (2) repeated stick-like structure along K_x axis, and (3) a node observed in the upper and lower sticks on the K_y axis. The periodic structure along K_y axis is solely determined by the periodicity of the lamellar layers and is understandable in terms of multi-slit interference of light. The repeated stick-like structure along K_x direction is connected with the Fourier transform of the finite width of the ridge.

To elucidate the essential part of the lamellar structure, we have calculated 2D Fourier images for various model structures, which are simplified part by part: (1) regularly arranged layers with equal lengths (model 1), (2) model 1 with a stem (model 2), and (3) model 2 with the heights of the right and left sides of the layers shifted by half of the layer interval (model 3). The results are also shown in Fig. 9.16. Comparing the FT image of actual ridge structure with those of models 1 and 2, we have found that the essential part of the FT image comes from the regularly arranged lamella and not from the stem within the present momentum range. From the comparison with model 3, it is concluded that the lamella with alternately shifted layer heights is important for the appearance of the node. The experimental observation was reported previously (Vukusic *et al.*, 1999), who reported the presence of clear node for *M. rhetenor*, but not for *M. didius*. We have also investigated scales from various *Morpho* butterflies and have found that ground scales of *M. rhetenor* and *M. cypris*, and ground and cover scales in *M. sulkowskyi* show a clear split of the scattering direction. However, those of *M. didius* do not show such a node in spite of having the similar lamellar structure as those of *M. rhetenor*. Since Fourier transform of a single ridge of *M. didius* shows a clear node as shown in Fig. 9.16, the tilt-angle distribution of the ridges seems to be the most probable cause.

9.3 Antireflection Effect

9.3.1 *Monolayer and Multilayer Antireflectors*

The antireflection at a surface is basically accomplished by the elimination of reflected light through interference. There are three methods to accomplish it: One is attained to use a discrete film of monolayer and multilayer coating. The second is to accomplish it by preparing a surface having

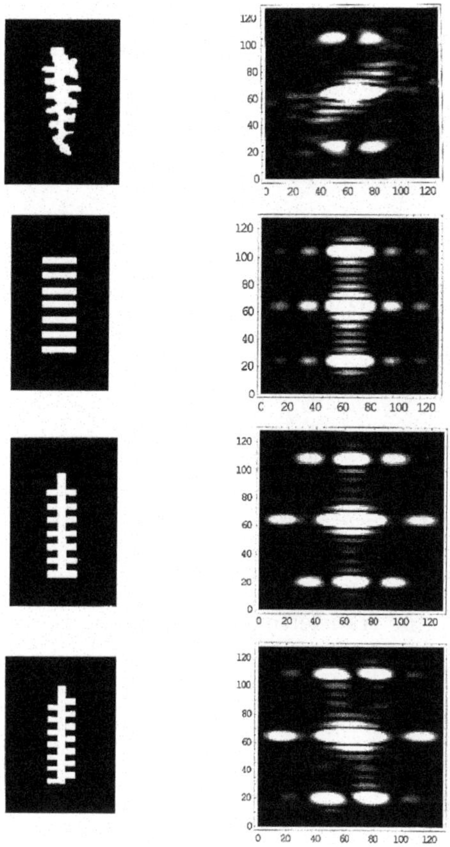

Fig. 9.16 Two-valued real (left) and FT (right) images for a ridge of ground scale of *Morpho didius* (top) and model structures (lower three).

continuously changing refractive index. The last one is moth-eye type, where discrete elements are arranged regularly on a surface.

Consider first the case of monolayer and multilayer coatings (Rabinovitch and Pagis, 1975; Braun and Jungerman, 1995). In a monolayer coating on a substrate of refractive index n_t, the transfer matrix introduced in Sec. 9.1.1 results in

$$\begin{pmatrix} A & B \\ C & D \end{pmatrix} = \begin{pmatrix} \cos n_1 k_0 d_1 & -\mathrm{i}(\eta_0/n_1) \sin n_1 k_0 d_1 \\ -\mathrm{i}(n_1/\eta_0) \sin n_1 k_0 d_1 & \cos n_1 k_0 d_1 \end{pmatrix}, \tag{9.79}$$

where d_1 and n_1 are the thickness and refractive index of a coating material, and the other parameters are defined as before. If the thickness of the coating layer satisfies the relation $n_1 k_0 d_1 = \pi/2$, that is, $n_1 d_1 = \lambda_{\rm min}/4$, where $\lambda_{\rm min}$ is the wavelength giving the minimum reflectivity, then the transfer matrix and the amplitude reflectivity of this system become

$$\begin{pmatrix} A & B \\ C & D \end{pmatrix} = \begin{pmatrix} 0 & -\mathrm{i}(\eta_0/n_1) \\ -\mathrm{i}(n_1/\eta_0) & 0 \end{pmatrix}, \tag{9.80}$$

and

$$\begin{aligned} r &= \frac{n_{\rm i}A + (n_{\rm i}n_{\rm t}/\eta_0)B - \eta_0 C - n_{\rm t}D}{n_{\rm i}A + (n_{\rm i}n_{\rm t}/\eta_0)B + \eta_0 C + n_{\rm t}D} \\ &= \frac{n_{\rm i}n_{\rm t}/n_1 - n_1}{n_{\rm i}n_{\rm t}/n_1 + n_1} = \frac{n_{\rm i}n_{\rm t} - n_1^2}{n_{\rm i}n_{\rm t} + n_1^2}. \end{aligned} \tag{9.81}$$

Thus, the antireflection is attained under the condition that

$$r = \frac{n_{\rm i}n_{\rm t} - n_1^2}{n_{\rm i}n_{\rm t} - n_1^2} = 0 \rightarrow n_1 = \sqrt{n_{\rm i}n_{\rm t}}. \tag{9.82}$$

In case of two-layer coating satisfying the conditions of $n_1 d_1 = \lambda_{\rm min}/4$ and $n_2 d_2 = \lambda_{\rm min}/4$, the transfer matrix becomes

$$\begin{aligned} \begin{pmatrix} A & B \\ C & D \end{pmatrix} &= \begin{pmatrix} 0 & -\mathrm{i}(\eta_0/n_1) \\ -\mathrm{i}(n_1/\eta_0) & 0 \end{pmatrix} \begin{pmatrix} 0 & -\mathrm{i}(\eta_0/n_2) \\ -\mathrm{i}(n_2/\eta_0) & 0 \end{pmatrix} \\ &= \begin{pmatrix} -n_2/n_1 & 0 \\ 0 & -n_1/n_2 \end{pmatrix}, \end{aligned} \tag{9.83}$$

which leads to (Braun and Jungerman, 1995):

$$r = \frac{n_{\rm i}n_2^2 - n_{\rm t}n_1^2}{n_{\rm i}n_2^2 + n_{\rm t}n_1^2} = 0 \rightarrow n_2 = n_1\sqrt{n_{\rm t}/n_{\rm i}}. \tag{9.84}$$

In three-layer coating with $n_1 k_0 d_1 = \pi/2$, $n_2 k_0 d_2 = \pi$, and $n_3 k_0 d_3 = \pi/2$, the relation becomes

$$\begin{aligned} \begin{pmatrix} A & B \\ C & D \end{pmatrix} &= \begin{pmatrix} 0 & -\mathrm{i}(\eta_0/n_1) \\ -\mathrm{i}(n_1/\eta_0) & 0 \end{pmatrix} \begin{pmatrix} -1 & 0 \\ 0 & -1 \end{pmatrix} \begin{pmatrix} 0 & -\mathrm{i}(\eta_0/n_3) \\ -\mathrm{i}(n_3/\eta_0) & 0 \end{pmatrix} \\ &= \begin{pmatrix} n_3/n_1 & 0 \\ 0 & n_1/n_3 \end{pmatrix}, \end{aligned} \tag{9.85}$$

and the same relation as Eq. (9.84) is obtained, though the reflection spectrum will be changed due to the presence of the second layer which increases

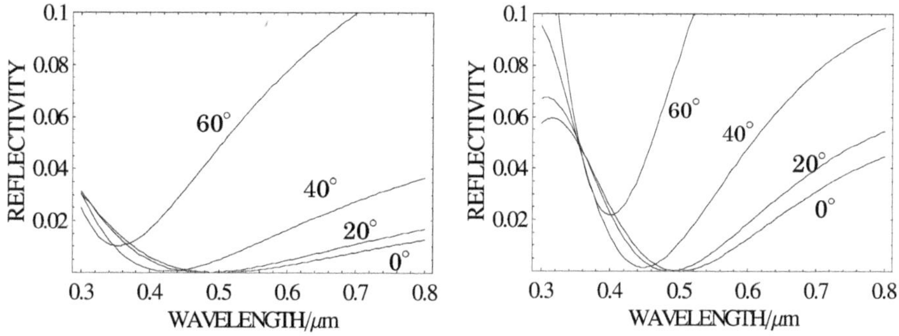

Fig. 9.17 Left: incident angle dependence of monolayer antireflector under s-polarized light illumination with the refractive indices of the layer and substrate as 1.225 and 1.50. The width of the layer is set to satisfy the condition that their optical path lengths are equal to $\lambda_{\min}/4$, where $\lambda_{\min} = 500\,\text{nm}$. Right: incident angle dependence of two-layer antireflector under s-polarized light illumination with the refractive indices of the first and second layers, and substrate as 1.38, 1.69, and 1.50.

the degree of freedom to adjust the parameters (Rabinovitch and Pagis, 1975).

We calculate the reflection spectra for various antireflectors according to the transfer matrix method. The typical results for the antireflection around 500 nm with the refractive index of the substrate 1.50 are shown in Figs. 9.17 and 9.18. It is clear that the monolayer antireflector gives a rather good result with respect to the wavelength range and to the incidence-angle variation. However, in order to satisfy the relation Eq. (9.82), it is necessary to prepare an unrealistic material with the refractive index of 1.225 in the present case of the substrate of 1.50. Thus, the multilayer antireflector is superior in this respect. If we compare the two- and three-layer antireflectors, the latter is clearly superior to the former especially when the appropriate value of refractive index is selected for the intermediate layer.

9.3.2 *Moth-Eye-Type Antireflector*

In animal world, small projections of several tens or a hundred nanometer sometimes cover the whole surface. It is generally believed that these projections have a function of antireflection at the surface. In the present subsection, we will briefly describe the physical basis for such antireflective structures.

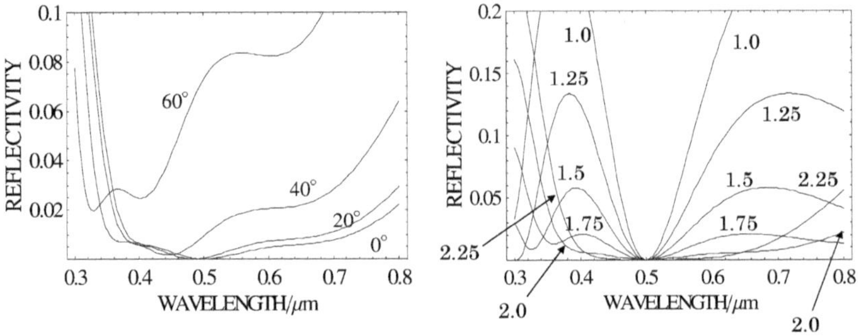

Fig. 9.18 Left: incident angle dependence of three-layer antireflective coating under s-polarized light illumination, where the refractive indices of the first, second, and third layers, and substrate are set constant at 1.38, 2.00, 1.69, and 1.50, respectively. The width of the layers are set to satisfy the condition that their optical path lengths are equal to $\lambda_{\min}/4$, $\lambda_{\min}/2$, and $\lambda_{\min}/4$, where $\lambda_{\min} = 500\,\text{nm}$. Right: reflection spectrum for three-layer multilayer under normal incidence, where the refractive indices of the first and third layers, and substrate are set constant at 1.38, 1.69, and 1.50, respectively. The index of the second layer is varied as 1.0, 1.25, 1.5, 1.75, 2.0, and 2.25.

Regularly aligned projections are considered to be a sort of two-dimensional diffraction grating, as shown in Fig. 2.19. In usual cases of antireflective projections, the distance between the adjacent projections is on the order of 100 nm and much smaller than the wavelength of light. Thus, this grating just corresponds to the zeroth-order grating so that only specular reflection should occur (Sec. 2.4).

As will be described in the following section, the optical response due to periodic structure whose pitch is much smaller than the wavelength of light is generally analyzed in terms of average refractive index and considered as an ensemble of thin slices of layers whose refractive indices gradually changed with the depth from the surface. The reflectivity of light from gradually changing surface was first treated mathematically by Lord Rayleigh in 1879, who considered a case where refractive index changed in a quadratic manner as a function of depth (Rayleigh, 1879). We calculate the cases where the average refractive indices are varied linearly or along a Gaussian function from 1.5 to 1.0 within a range of given thicknesses. For this calculation, we have divided the region into 100 slices and have calculated the reflectivity using the iterative method described in Sec. 9.1.2. The thickness-dependent reflectivity thus calculated is shown in Fig. 9.19, which indicates a clear decrease of reflectivity around 0.1–0.2 μm ($\lambda = 500\,\text{nm}$), irrespectively of the interfacial function employed. This is because the amplitude of the reflected light at each interface of slice, whose phase varies little by little

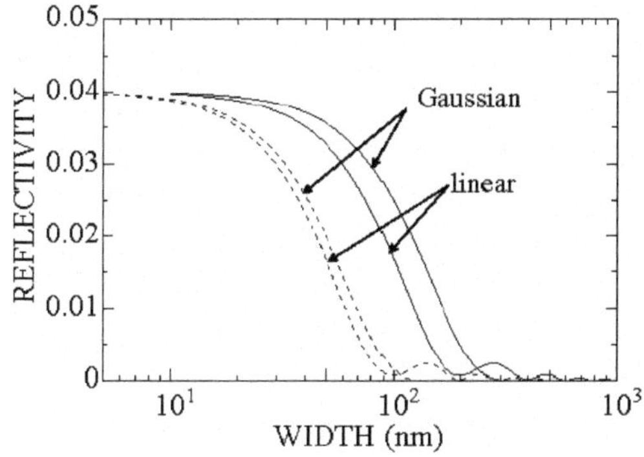

Fig. 9.19 Thickness dependence of the reflectivity at a surface with continuously changing refractive indices at a wavelength of 500 nm. The index changes obeying a Gaussian and linear function are considered. Solid lines indicate the total range of the refractive index variation, while dashed lines indicate the thickness evaluated at half of the variation. The reflectivity is calculated using an iterative method.

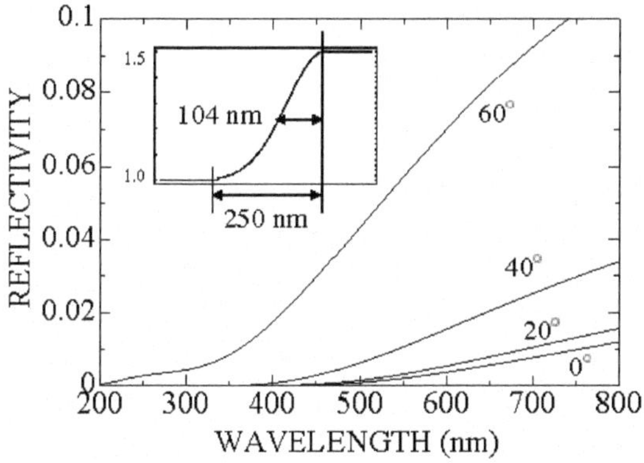

Fig. 9.20 Incidence-angle dependence of reflection at an interface between a substrate of $n = 1.5$ and air with continuously varying refractive index, which obeys a Gaussian function with the half width of 104 nm (shown as inset). The region of 250 nm in thickness is divided into 100 layers and the reflectivity is calculated using an iterative method.

according to the depth, is added destructively and cancels the reflection as a whole. In fact, it is found that the reflectivity decreases and approaches zero with increasing thickness over the wavelength of light. The difference of the interfacial functions only appears as the apparent cutoff position and also the presence of oscillation at the tail.

In Fig. 9.20, we show the reflection spectrum for a Gaussian interfacial profile of the width of 104 nm. The reflection spectrum differs considerably from the monolayer or multilayer coating as shown in Figs. 9.17 and 9.18. Namely, the reflectivity monotonically decreases toward shorter wavelengths and seems to have a cutoff wavelength. With increasing incident angle, the cutoff wavelength shifts gradually to shorter wavelengths, but the absolute reflectivity is by far smaller than the multilayer case. Thus, moth-eye type antireflector is characterized by the presence of the cutoff behaviors both in the thickness of the layer and in the reflection spectrum, and further, it is noticed that the deterioration due to the oblique incidence seems to be considerably reduced.

Thus, the gradual change of the refractive index at the interface is effective to reduce the interfacial reflection particularly at shorter wavelengths. However, to produce an ideal gradation at the surface is quite difficult, because the refraction index should be down to 1.0 for use in air. Although the effort to fabricate such material has been in progress (Xi *et al.*, 2007), the moth-eye structure is superior in this sense, because at the tip of protuberance, the average refractive index eventually becomes 1.0.

9.4 Average Refractive Index

When complicated structures are embedded in a host material, it is sometimes difficult to treat these systems strictly in a mathematical way. Although numerical methods still work in such cases, the physical meaning is usually difficult to derive from the calculated results. In such a case, mean-field treatment will fully exercise its power. Here, we will briefly review such treatments.

First, we will derive the effective permittivity, ϵ, known as Clausius–Mossotti, Lorentz–Lorenz, Maxwell–Garnett, and Bruggeman relations. Let an electric field E_0 be applied perpendicularly to the surface of an isotropic dielectric plate with the thickness of d. We consider a canal field within this plate, as shown in Fig. 9.21(a), where we put the surface charge induced by the applied field as σ. Then, the electric field within the canal E is expressed as the sum of the externally applied field and the induced field by the surface charge, $E = E_0 - \sigma/\epsilon_0$, where ϵ_0 is the permittivity in vacuum. The magnitude of polarization, P, per unit volume gives the relation between P and σ such that $PSd = \sigma Sd$, where S is an arbitrary area on the surface. Hence, the relation $P = \sigma$ generally holds, which leads to a well-known relation:

$$E = E_0 - P/\epsilon_0 \quad \text{or} \quad \epsilon_0 E_0 = \epsilon_0 E + P. \tag{9.86}$$

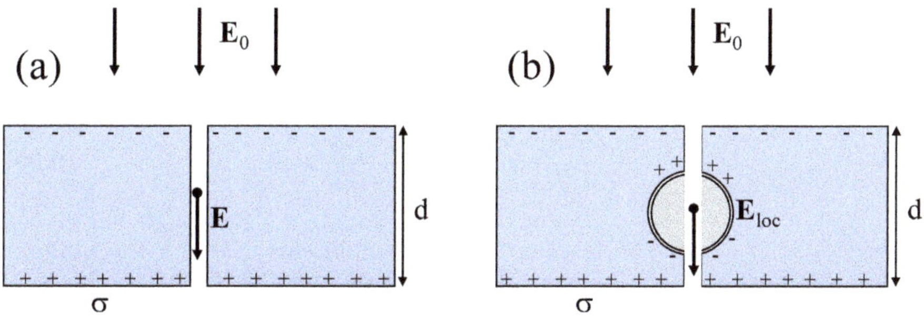

Fig. 9.21 Explanations for canal and local field under the application of external electric field.

Next, we consider a spherical hole created virtually within a dielectric plate, which is connected with a canal field, as shown in Fig. 9.21(b). Within the hole, atomic dipoles are considered to be distributed uniformly, which make the effects of individual dipoles on a dipole located at the center of the sphere to cancel with each other. Then, the local electric field E_{loc} within the hole becomes the sum of the canal field and induced field due to the surface charge on the sphere: $E_{loc} = E_0 - P/\epsilon_0 + P/(3\epsilon_0) = E + P/(3\epsilon_0)$. The last term in the right-hand side is the induced field and the factor 1/3 comes from the depolarization factor of a sphere (Table 9.1). We consider a microscopic dipole, μ, induced on atom within a spherical hole, which is connected by the atomic polarizability, α, as $\mu = \alpha E_{loc} = \alpha\{E + P/(3\epsilon_0)\}$. When the macroscopic polarization is expressed by an ensemble of such microscopic dipoles, the relation $P = N\alpha E_{loc} = N\alpha\{E+P/(3\epsilon_0)\}$ generally holds, where the number of atoms in a unit volume is designated as N. From this relation, the following relation is easily derived:

$$P = \frac{N\alpha E}{1 - N\alpha/(3\epsilon_0)}. \tag{9.87}$$

Inserting Eq. (9.87) into Eq. (9.86) gives the Clausius–Mossotti relation expressed as

$$\frac{\epsilon - \epsilon_0}{\epsilon + 2\epsilon_0} = \frac{N\alpha}{3\epsilon_0}. \tag{9.88}$$

If the dielectric plate consists of two kinds of atom, say a and b, then $P = N_a\mu_a + N_b\mu_b$, which leads to

$$(\epsilon - \epsilon_0)/(\epsilon + 2\epsilon_0) = (N_a\alpha_a + N_b\alpha_b)/(3\epsilon_0). \tag{9.89}$$

If the plate would consist of only one kind of atom, the Clausius–Mossotti relation should be $(\epsilon_j - \epsilon_0)/(\epsilon_j + 2\epsilon_0) = N_j\alpha_j/(3\epsilon_0)$, where $j = a$ or b. Inserting this into Eq. (9.89) results in well-known Lorentz–Lorenz relation:

$$\frac{\epsilon - \epsilon_0}{\epsilon + 2\epsilon_0} = \sum_j f_j \frac{\epsilon_j - \epsilon_0}{\epsilon_j + 2\epsilon_0}, \tag{9.90}$$

where $f_j \equiv N_j/N$ is the fraction of the component j. When the amount of one of these components is overwhelmingly large, e.g. a, as compared with the other, it is regarded that the guest material b is embedded in host material a, which exists as a background material. Then, we replace $\epsilon_0 \to \epsilon_a$. The first term in the right-hand side of Eq. (9.90) vanishes, which leads to

$$\frac{\epsilon - \epsilon_a}{\epsilon + 2\epsilon_a} = f_b \frac{\epsilon_b - \epsilon_a}{\epsilon_b + 2\epsilon_a}, \tag{9.91}$$

or in a well-known form as

$$\epsilon = \epsilon_a + \frac{3f_b(\epsilon_b - \epsilon_a)}{(\epsilon_b + 2\epsilon_a) - f_b(\epsilon_b - \epsilon_a)}. \tag{9.92}$$

This relation is called Maxwell–Garnett model, which has been widely used to estimate the average refractive index in composite materials. Further, when the amounts of the two components are comparable with each other and the mixture of these components can be regarded as a background material, ϵ_0 can be replaced by ϵ in Eq. (9.90), which leads to the Bruggeman's relation:

$$0 = \sum_j f_j \frac{\epsilon_j - \epsilon}{\epsilon_j + 2\epsilon}. \tag{9.93}$$

In order to calculate the average refractive index from the effective permittivity ϵ thus obtained, the relation $n^2 = \epsilon/\epsilon_0$ should be used.

These relations are based on the fact that the material consists of two or more components whose sizes are on the order of atoms or molecules. Thus, a spherical cavity to calculate the local electric field is anyway virtual, and we are able to consider such a sphere anytime with a dipole placed at the center. On the other hand, the above relations have been often extended to a system consisting of components of a much larger size with a real shape. For example, spheres of radius comparable to the wavelength (guest) are randomly or regularly dispersed in a homogeneous medium (host). These cases are obviously different from the situation employed to derive original versions of the relations. It is clear that the sphere is not virtual but is real. Hence, it is not always possible to place the dipole at the center of the sphere, and further the electric flux lines both inside and outside the sphere are anyway deformed by the presence of the sphere with different permittivity. Thus it seems essentially impossible to apply the above relations as it is. At least, we can say that the following conditions are generally necessary to be considered: (1) the difference of the permittivity between the dielectric matter and surrounding medium is small not to deform the electric flux lines considerably, (2) the shape of dielectric matter belongs to a rotational ellipsoid in 3D, an elliptic cylinder in 2D or a flat sheet in 1D to ensure the homogeneous electric field within the matter, and (3) the dielectric matters are sparsely distributed to reduce the effect of the deformation of electric flux lines outside the matter. In spite of these strict restrictions, however, the above relations, particularly Maxwell–Garnett model, have been frequently applied to composite materials, within which dielectric materials of various shapes are densely distributed, and have attained a seemingly good result.

Thus, the Maxwell–Garnett model applied to the composite system should be anyway regarded as an empirical formula. Furthermore, in this case, the above relations should be slightly modified to accommodate the actual shape of guest material:

$$\frac{\epsilon - \epsilon_a}{L\epsilon + (1-L)\epsilon_a} = f_b \frac{\epsilon_b - \epsilon_a}{L\epsilon_b + (1-L)\epsilon_a}, \tag{9.94}$$

$$\sum_j f_j \frac{\epsilon - \epsilon_j}{L\epsilon + (1-L)\epsilon_j} = 0, \tag{9.95}$$

where L is the depolarization factor listed in Table 9.1.

Table 9.1 Depolarization factor L.

Shape	Polarization[a]	Depolarization factor[b]
Sphere (3D)		1/3
Circular cylinder (2D)	s	0
	p	1/2
Sheet (1D)		1

[a] s- and p-polarizations are defined as the electric vectors parallel and perpendicular to an axis of the cylinder.
[b] Depolarization factor is sometimes defined as L/ϵ_0.

In the following, we will investigate the justification and limitation of their applications to the composite systems by means of band calculation for 1D or 2D photonic crystal. So far, the processes of averaging the refractive indices or permittivity of materials have variously appeared in literatures such that

$$n_{av} = \sum_j f_j n_j, \tag{9.96}$$

$$n_{av} = \frac{1}{\sum_j f_j (1/n_j)}, \tag{9.97}$$

$$n_{av} = \sqrt{\sum_j f_j \epsilon_j}, \tag{9.98}$$

$$n_{av} = \frac{1}{\sqrt{\sum_j f_j (1/\epsilon_j)}}, \tag{9.99}$$

where the first two are the volume average of the refractive index or its inverse, while the last two are that of the permittivity or its inverse. In order to evaluate these quantities in addition to the above empirical relations, two criteria should always be considered: One is whether the mean-field approximation itself is adequate or not, and the other is whether the calculated result through averaging is adequate or not. Both of them are dependent on the shape and arrangement of particles, volume fraction, and also difference of the permittivity. In any case, the actual evaluation is quite difficult in general cases except for particular situations such as regularly arranged particles or extremely dilute condition.

It is meaningful to evaluate the above two criteria using the results of photonic band calculations. In the band calculation, a linear increase in the dispersion curve is always found in the low wave number region and shows wave-like character of propagating light in a composite system. In this region, the light regards the system as a uniform material, and thus it gives the average refractive index, which is directly calculated from the gradient of this line. The deviation from a straight line indicates the limitation of the validity of such averaging. Actually the range of the linearity depends on the volume fraction of the guest material. The linearity is well maintained for low and high fractions, while it is worse in the intermediate case, where the amounts of the guest and host materials are comparable. This tendency is more remarkable when the difference of the permittivity is increased.

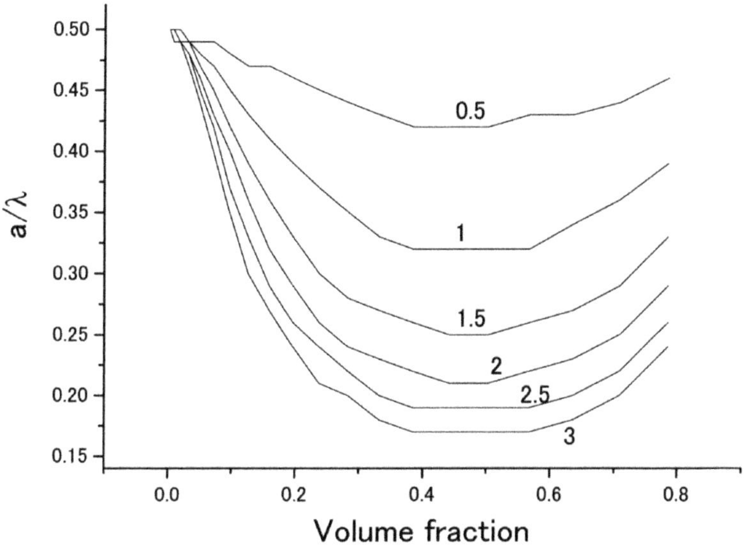

Fig. 9.22 Upper limit of average nearest-neighbor particle distance, a, for various permittivity differences, below which average refractive index is safely applicable. These curves are obtained from a linear range of low wavenumber region of the dispersion curve calculated for two-dimensional band structure of a square lattice consisting of cylinder. The volume fraction is varied by changing the radius of the cylinders, while the difference of the permittivity between cylinder and medium is varied as indicated (Fujimura, 2008).

In a realistic case of the permittivity range of $\Delta\epsilon \leq 1.5$, corresponding to a material of $n \leq 1.6$ in air, the linearity is roughly maintained within the wave number range of $a/\lambda \leq 0.25$, as shown in Fig. 9.22, where a is the lattice constant. This means that the average particle–particle distance should be at least one-fourth of the wavelength of light, $a \leq \lambda/4$, when the average refractive index is safely applicable. This limitation is much more relaxed when the particle size is smaller or larger.

The problem which methods are actually adequate to approximate the refractive index in complex systems, will be answered by comparing the gradient of the straight line with the above approximate calculations. The result is shown in Fig. 9.23 for p-polarization. The above four approximations, Eqs. (9.96)–(9.99), generally show large deviations in a whole region of volume fraction and particularly they are worse around the middle region. They are only valid in the smallest and largest volume fractions, where most of the volume is occupied by either component and thus regarded as being homogeneous. On the other hand, Maxwell–Garnett and

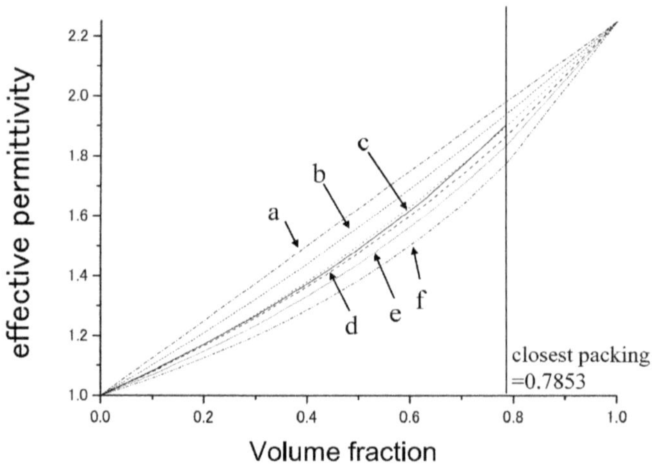

Fig. 9.23 Comparison of methods to obtain the effective permittivity using a model system of two-dimensional square lattice of cylinders with $\epsilon = 2.25$ ($n = 1.5$) embedded in a medium of $\epsilon = 1$ ($n = 1$). The effective permittivity is calculated through (a) volume average of the permittivity (Eq. (9.98)), (b) that of refractive index (Eq. (9.96)), (c) Bruggeman model (Eq. (9.93)), (d) Maxwell–Garnett model (Eq. (9.91)), (e) volume average of the inverse of refractive index (Eq. (9.97)) and (f) that of the reciprocal permittivity (Eq. (9.99)). Solid curve is obtained from the gradient of a linear dispersion curve in low wave number region of the photonic band structure (p-polarization). The volume fraction is changed by increasing the radius of the cylinder up to the closest packing (Fujimura, 2008).

Bruggeman models are in fairly good agreement with the result of the band calculation in a whole region. Particularly, Maxwell–Garnett model is better for the low volume fraction, while Bruggeman model is better for a whole region, which may come from their definitions. On the other hand, under s-polarization, the effective permittivity obtained from band calculation is strictly given by the volume average of the permittivity. In Maxwell–Garnett and Bruggeman models, the same situation is attained by putting $L = 0$, as indicated in Table 9.1. Thus, these models give a fairly good result for the approximate expression of the average permittivity. These results are found to be only weakly dependent on the shape of the cylinder and on the arrangement within the range of the permittivity given above. Thus, these two models will offer the average quantities in an empirical sense to evaluate the optical response, as long as the average particle–particle distance lies within the limitation described above.

However, the above models are often applied beyond the criterion given above, because the criterion essentially assures that the system is completely homogeneous for propagating light so that it behaves like a transparent material. We have confirmed the validity of such a procedure by comparing the angular frequency obtained by extrapolating to Brillouin zone boundary with that obtained by photonic band calculation. One of such a calculation is shown in Fig. 9.8, where the band center is calculated using Maxwell–Garnett model through the relation $\omega = \pi c/(\bar{n}a)$, where $\bar{n}$ is the average refractive index. The agreement is fairly good for multilayer with the difference of refractive index of 0.5. However, in two-dimensional photonic crystal, the agreement is found to be especially worse in a triangle lattice of cylinders even with small difference of refractive index under p-polarization, while it is quite good for s-polarization and fairly good in a square lattice for p-polarization along Γ–X direction (Fujimura, 2008). Further studies are clearly needed to confirm the validity and limitation of average refractive index in these composite systems.

9.5 Cholesteric Liquid Crystal

9.5.1 *Parabolic Patterns*

Parabolic patterns have been often found in the oblique-cut cross-section of the cuticle layer of the crustaceans and insects. The first analysis of this pattern was performed by Bouligand, who considered that the pattern came from fibrous elements uniformly arranged in a plane parallel to the surface (Bouligand, 1965). Further, the direction of the elements gradually changed as a function of the depth from the surface. He explained the origin of the parabolic pattern from geometrical considerations. Here, we will follow his consideration.

We put xy plane parallel to the surface and z-axis normal to it (Fig. 9.24). Consider that xs plane, an oblique-cut plane, makes an angle of α with respect to the xy plane along the y-axis. We consider a point P on the xs plane, whose coordinates are expressed as (x, y, z), corresponding to (x, s) in the xs coordinate, where z is simply expressed as $z = s \sin \alpha$. Bouligand considered that the angle of the preferred direction of the fibrous elements changed linearly with the depth from the surface. Thus, we put the preferred direction in a plane parallel to the surface as a unit vector designated as $\overrightarrow{\mathrm{PQ}}$ with an angle measured from the x-axis as $\beta = az$, where a is a constant. Since, the x and y components of the vector $\overrightarrow{\mathrm{PQ}}$ are expressed as $\cos az$ and $\sin az$, the components of the corresponding vector $\overrightarrow{\mathrm{PQ}'}$ projected onto the xs plane become $\cos az$ and $\sin az \cos \alpha$, respectively. Thus,

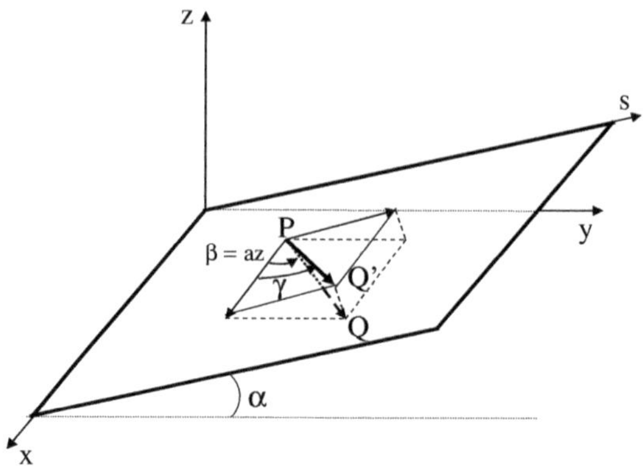

Fig. 9.24 Geometry to explain parabolic pattern.

the gradient of the vector $\overrightarrow{PQ'}$ changed from $\tan az$ in the xy plane into $\tan az \cos\alpha$ in the xs plane, which is then dependent on the height z.

We will calculate the trajectory of this gradient by solving the following differential equation:

$$\frac{\mathrm{d}s}{\mathrm{d}x} = \tan az \cdot \cos\alpha = \tan(as\sin\alpha) \cdot \cos\alpha. \tag{9.100}$$

For convenience, we take an inverse of the above equation as follows:

$$\frac{\mathrm{d}x}{\mathrm{d}s} = \frac{1}{\cos\alpha}\cot(as\sin\alpha), \tag{9.101}$$

which is then solved easily as

$$x = \frac{1}{a\sin\alpha\cos\alpha}\ln|\sin(as\sin\alpha)| + x_0, \tag{9.102}$$

where x_0 is a constant.

We have simulated the above trajectories using Eq. (9.102) with x_0 varying discretely. The obtained results are shown in Fig. 9.25. The overall features of parabolic pattern are clearly reproduced. The positions of the linear linings are easily obtained by satisfying the relation $\sin(as\sin\alpha) = 0$, which results in $s = m\pi/(a\sin\alpha)$ with m an integer. Thus, if α is known, it is easy to estimate a, a rotation angle per unit length normal to the surface. The cause of the parabolic pattern is thus due to the gradual change of the preferred direction and is intuitively understood through the illustration shown in Fig. 9.26.

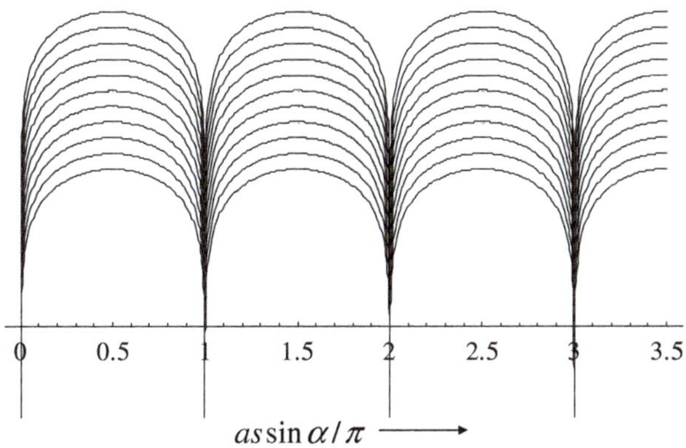

Fig. 9.25 Parabolic pattern simulated using Eq. (9.102) with changing x_0 discretely.

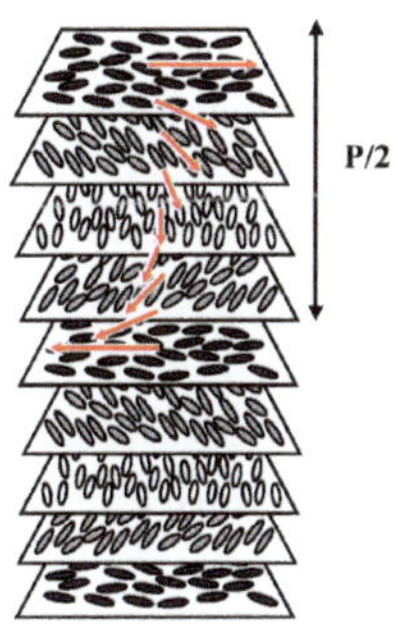

Fig. 9.26 Schematic illustration of cholesteric liquid crystal (illustrated by N. Okamoto). If the orientation of the molecules in a plane is isotropic with respect to the orientational direction, a half of the helicoidal pitch $P/2$ determines the periodicity.

9.5.2 *Dispersion Relations and Optical Responses*

Cholesteric liquid crystal takes a layered structure, within which molecules take a preferred direction that gradually changes from layer to layer so that it takes a form of helicoidal structure as shown in Fig. 9.26. The helicoidal structure shows a periodicity determined by the pitch of the helix, P, and if the molecules take two opposite directions randomly within the preferred direction, then the orientations of the molecules in a plane become symmetric as a whole, and a half of the helicoidal pitch $P/2$ determines the periodicity. In the present subsection, we will show the general properties of optical response of cholesteric liquid crystal. Although a lot of reviews have

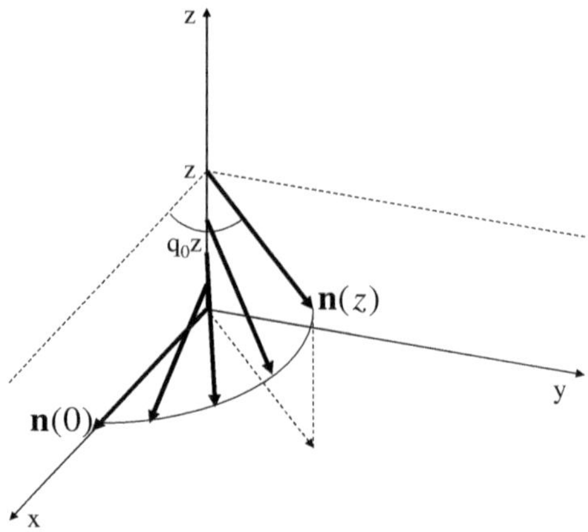

Fig. 9.27 The average orientation of molecules in a plane is designated as **n** that is dependent on the height z. The molecules are assumed to be randomly oriented along **n**, and thus **n** and $-$**n** are equivalent. The azimuth angle of **n** is assumed to be linearly dependent on z, which confirms the helicoidal structure along the z-axis.

been concerned with this problem, here we will follow the books written by De Gennes (1974) and by Orihara (2004).

Let us first consider that anisotropic molecules are aligned unidirectionally within a xy plane. The average orientation of molecules in a plane is designated as **n** that is dependent on the height z of the plane, as shown in Fig. 9.27. We assume that the molecules are randomly oriented along **n**, and thus **n** and $-$**n** are equivalent, which is the usual case for nematic liquid crystal. The azimuth angle of **n** is assumed to be linearly dependent on z, which confirms the helicoidal structure along the z-axis. We thus put $\mathbf{n}(z) = (\cos q_0 z, \sin q_0 z, 0)$, where $q_0 = \pm 2\pi/P$ gives a wave vector concerning the periodicity of the helical structure. Hereafter, we take $q_0 > 0$ for simplicity, indicating a clockwise helix.

If the relative permittivities parallel and perpendicular to $\mathbf{n}(z)$ are denoted as $\epsilon_\parallel$ and $\epsilon_\perp$, respectively, the dielectric tensor in (x, y, z) coordinates can be calculated using the rotational operation of the dielectric matrix along the z-axis:

$$\epsilon = R_z^{-1} \begin{pmatrix} \epsilon_\parallel & 0 & 0 \\ 0 & \epsilon_\perp & 0 \\ 0 & 0 & \epsilon_\perp \end{pmatrix} R_z \tag{9.103}$$

with

$$R_z = \begin{pmatrix} \cos q_0 z & \sin q_0 z & 0 \\ -\sin q_0 z & \cos q_0 z & 0 \\ 0 & 0 & 1 \end{pmatrix}. \tag{9.104}$$

Thus, the matrix calculation results in

$$\epsilon = \begin{pmatrix} \bar{\epsilon} & 0 & 0 \\ 0 & \bar{\epsilon} & 0 \\ 0 & 0 & \epsilon_\perp \end{pmatrix} + \frac{\epsilon_a}{2} \begin{pmatrix} \cos 2q_0 z & \sin 2q_0 z & 0 \\ \sin 2q_0 z & -\cos 2q_0 z & 0 \\ 0 & 0 & 0 \end{pmatrix}, \tag{9.105}$$

where $\bar{\epsilon} = (\epsilon_\| + \epsilon_\perp)/2$ is the average value of the relative permittivity and $\epsilon_a = \epsilon_\| - \epsilon_\perp$ indicates the magnitude of the anisotropy.

Consider z-propagating light satisfying the wave equation

$$\mathrm{d}^2\mathbf{E}(z)/\mathrm{d}z^2 = -(\omega/c)^2\epsilon(z)\mathbf{E}(z), \tag{9.106}$$

where its electric field is assumed to be of a form of $\mathbf{E}(z,t) = \mathbf{E}(z)\exp[\mathrm{i}(kz-\omega t)]$ with k, ω, and c are the wave number, angular frequency, and light velocity in vacuum, respectively. Since the permittivity within a xy plane solely affects the z-propagating light, we consider, hereafter, the xy components of the dielectric tensor. Thus,

$$\begin{aligned} \epsilon(z) &= \bar{\epsilon} I + \delta\epsilon(z) \\ &= \bar{\epsilon} \begin{pmatrix} 1 & 0 \\ 0 & 1 \end{pmatrix} + \frac{\epsilon_a}{2} \begin{pmatrix} \cos 2q_0 z & \sin 2q_0 z \\ \sin 2q_0 z & -\cos 2q_0 z \end{pmatrix}, \end{aligned} \tag{9.107}$$

where I is a unit matrix. In a helicoidal system, it is preferable to take right- (+ sign) and left-circularly (− sign) polarized light as a basis:

$$E^\pm(z) = E_x(z) \pm \mathrm{i}E_y(z), \tag{9.108}$$

which results in

$$\begin{aligned} -\frac{\mathrm{d}^2}{\mathrm{d}z^2}\begin{pmatrix} E^+(z) \\ E^-(z) \end{pmatrix} &= -\begin{pmatrix} 1 & \mathrm{i} \\ 1 & -\mathrm{i} \end{pmatrix} \frac{\mathrm{d}^2}{\mathrm{d}z^2} \begin{pmatrix} E_x(z) \\ E_y(z) \end{pmatrix} \\ &= \left(\frac{\omega}{c}\right)^2 \begin{pmatrix} 1 & \mathrm{i} \\ 1 & -\mathrm{i} \end{pmatrix} \epsilon(z) \begin{pmatrix} E_x(z) \\ E_y(z) \end{pmatrix} \\ &= \begin{pmatrix} k_0^2 & k_1^2 \mathrm{e}^{2\mathrm{i}q_0 z} \\ k_1^2 \mathrm{e}^{-2\mathrm{i}q_0 z} & k_0^2 \end{pmatrix} \begin{pmatrix} E_x(z) \\ E_y(z) \end{pmatrix}, \end{aligned} \tag{9.109}$$

where $k_0^2 = (\omega/c)^2\bar{\epsilon}$ and $k_1^2 = (\omega/c)^2\epsilon_a/2$. It is convenient to consider the following form for each electric field to make closed equations:

$$E^+(z) = a e^{\mathrm{i}(l+q_0)z}$$
$$E^-(z) = b e^{\mathrm{i}(l-q_0)z}. \tag{9.110}$$

Inserting Eq. (9.110) into Eq. (9.109), we obtain

$$\begin{pmatrix} (l+q_0)^2 - k_0^2 & -k_1^2 \\ -k_1^2 & (l-q_0)^2 - k_0^2 \end{pmatrix} \begin{pmatrix} a \\ b \end{pmatrix} = 0. \tag{9.111}$$

Nonzero solutions for the electric fields are obtained to solve the following equation with respect to ω or l:

$$\{(l+q_0)^2 - k_0^2\}\{(l-q_0)^2 - k_0^2\} - k_1^4 = 0, \tag{9.112}$$

which leads to

$$(l^2 + q_o^2 - k_0^2)^2 - 4q_0^2 l^2 - k_1^4 = 0. \tag{9.113}$$

Solving this with respect to $(\omega/c)^2$, we obtain

$$\left(\frac{\omega}{c}\right)^2 = \frac{1}{\bar{\epsilon}} \cdot \frac{(l^2+q_0^2) \pm \sqrt{4q_0^2 l^2 + K^2(l^2-q_0^2)^2}}{1-K^2}, \tag{9.114}$$

which gives the dispersion relation of ω for l. Here, we put $K \equiv \epsilon_a/(2\bar{\epsilon}) = (n_\parallel^2 - n_\perp^2)/(n_\parallel^2 + n_\perp^2)$. We plot the dispersion relation using Eq. (9.114), as shown in Fig. 9.28. It is easily verified that for an arbitrary value of l, ω takes two positive values. The dispersion curves divide into two according to the sign in Eq. (9.114), which we call $\omega_+(l)$ and $\omega_-(l)$ branch corresponding to the $\pm$ sign. It is easily seen in the figure that a band gap is present at $l = 0$, whose edges are obtained as $\omega_+(0) = cq_0/n_\perp$ and $\omega_-(0) = cq_0/n_\parallel$. It is also noticed that at the same frequency range, the propagating modes in $\pm z$ directions exist for $\omega_-(l)$ mode.

On the contrary, Eq. (9.113) can be solved with respect to l and the following result is obtained:

$$l = \pm\sqrt{q_0^2 + k_0^2 \pm \sqrt{4q_0^2 k_0^2 + k_1^2}}. \tag{9.115}$$

This relation generally gives four real values except for the case

$$q_0^2 + k_0^2 - \sqrt{4q_0^2 k_0^2 + k_1^2} < 0, \tag{9.116}$$

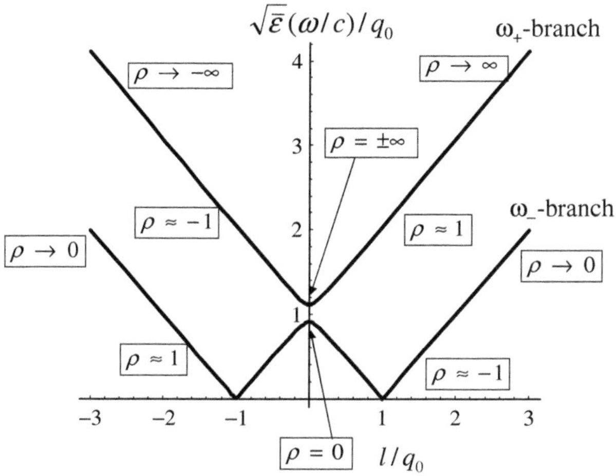

Fig. 9.28 Dispersion relation of light in cholesteric liquid crystal, calculated using Eq. (9.114) with $\epsilon_a/2\epsilon = 0.2$.

or equivalently

$$\omega_-(0) < \omega < \omega_+(0), \tag{9.117}$$

which corresponds to the band gap. Within this region, two of the four solutions become imaginary, indicating that only the light with particular states can selectively propagate within this frequency region, while the light with the other states cannot propagate and decays at a surface.

When Eq. (9.116) holds, the wave vector l takes a pure imaginary value of

$$\begin{aligned} l &= \pm \mathrm{i}\sqrt{\sqrt{4q_0^2k_0^2 + k_1^2} - (q_0^2 + k_0^2)} \\ &\approx \pm \mathrm{i}\frac{k_1^2}{\sqrt{4k_0q_0}}\left(1 - \frac{2q_0^2(k_0 - q_0)^2}{k_1^4} + \cdots\right), \end{aligned} \tag{9.118}$$

which expressed the penetration length into a liquid crystal. We plot in Fig. 9.29, the frequency dependence of the penetration length defined as $L = 2/|l|$. Under the condition of $k_0 = q_0$, the penetration length becomes minimum with $L_{\min} = 2\sqrt{4k_0q_0}/k_1^2 = (2P/\pi)(n_\parallel^2 + n_\perp^2)/(n_\parallel^2 - n_\perp^2)$ and becomes divergent at the band edge. Thus, the minimum penetration length becomes smaller, when the anisotropy of the refractive index is larger.

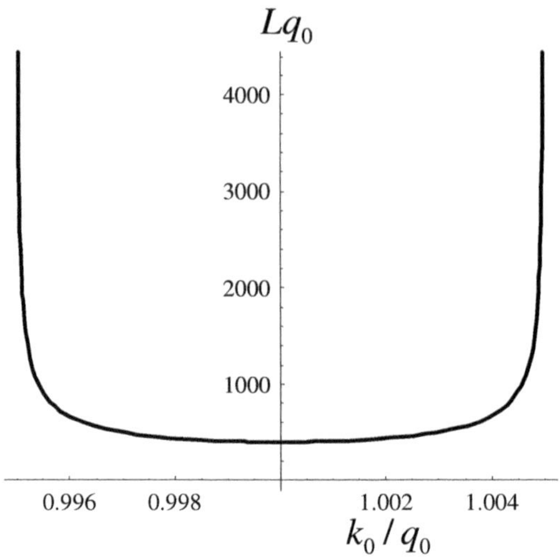

Fig. 9.29 Skin depth of cholesteric liquid crystal calculated using Eq. (9.118) for $k_1/q_0 = 0.1$.

Next, we consider the polarization state within a cholesteric liquid crystal. Since the polarization state of the z-propagating light in a complex plane is expressed as

$$\begin{aligned}\tilde{E}(z,t) &= E_x(z,t) + \mathrm{i}E_y(z,t) \\ &= \frac{1}{2}\left(E^+(z)\mathrm{e}^{-\mathrm{i}\omega t} + E^{-*}(z)\mathrm{e}^{\mathrm{i}\omega t}\right) \\ &= \frac{1}{2}\left(a\mathrm{e}^{\mathrm{i}\{(l+q_0)z-\omega t\}} + b\mathrm{e}^{-\mathrm{i}\{(l-q_0)z+\omega t\}}\right) \\ &\equiv \mathrm{e}^{\mathrm{i}\alpha}(r_1\cos\theta + \mathrm{i}r_2\sin\theta),\end{aligned} \tag{9.119}$$

where $\alpha = q_0 z$, $\theta = lz - \omega t$, $r_1 = (a+b)/2$, and $r_2 = (a-b)/2$. The last expression of Eq. (9.119), indicates that the electric field draws an elliptic trajectory in a complex plane as shown in Fig. 9.30. We define a parameter to express the ellipticity as

$$\rho = \frac{r_1}{r_2} = \frac{a+b}{a-b}, \tag{9.120}$$

where $\rho > 0$ and $\rho < 0$ mean the electric field obeys the clockwise and anticlockwise rotation along the elliptic trajectory when seen from the positive z-axis. Particularly, $\rho = 1$ and -1 indicate right- and left-circular polarizations, while $\rho = 0$ and $\pm\infty$ indicate the linear polarizations.

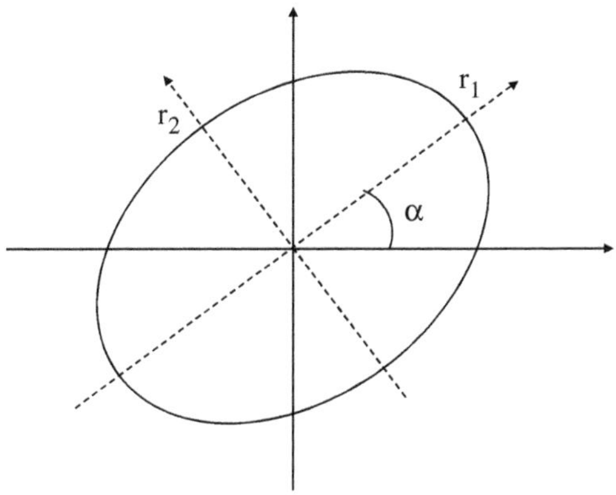

Fig. 9.30 Trajectory of the electric field in a complex plane.

Equation (9.111) gives the ratio of a/b for given ω and l. Thus,

$$\frac{a}{b} = \frac{k_1^2}{\{(l+q_0)^2 - k_0^2\}}, \tag{9.121}$$

or

$$\frac{b}{a} = \frac{k_1^2}{\{(l-q_0)^2 - k_0^2\}}. \tag{9.122}$$

For the ω_- branch, for example, the relation of Eq. (9.121) is employed and

$$\rho = \frac{k_1^2 + k_0^2 - (l+q_0)^2}{k_1^2 - k_0^2 + (l+q_0)^2} = \frac{(\omega/c)^2\epsilon_{\parallel} - (l+q_0)^2}{(\omega/c)^2\epsilon_{\perp} + (l+q_0)^2}, \tag{9.123}$$

which gives $\rho = 0$ for $l = 0$ and $l \to \pm\infty$, while $\rho \approx -1$ for $l \approx q_0$, as summarized in Fig. 9.28. On the other hand, for the ω_+ branch, $\rho \to \pm\infty$ for $l = 0$ and $l \to \pm\infty$, while for $l \approx q_0$, $\rho \approx +1$. Essentially, $+z$-propagating light consists of the mixture of light with right- and left-circular polarizations. However, in the frequency lower than the band gap, both light waves are originated from ω_- branch, while in the frequency higher than the band gap, light wave with right-circular polarization is originated from ω_+ branch, which means a crossover occurs around a band gap. It is also noticed that at a large frequency, both branches tend to give linear polarizations perpendicular to each other. On the other hand, at the band edge, one of the light wave tends to have a linear polarization, but does not show dispersion, indicating the stationary wave is generated through the interference of $+z$ and $-z$-propagating light waves due to the strong Bragg diffraction.

When the angular frequency of the incident light lies within a range of a band gap, in case of a clockwise helix, only the light with left-circular polarization is transmitted, while the light with right-circular polarization is completely reflected within a penetration length. For light propagating toward a positive z direction, the wave vectors l_{R} and l_{L} can be calculated from Eq. (9.115) for the right- and left-circular polarizations as

$$l_{\mathrm{R}} = \begin{cases} \sqrt{k_0^2 + q_0^2 - \sqrt{4k_0^2 q_0^2 + k_1^4}} & \text{for } \omega \geq \omega_+(0) \\ -\sqrt{k_0^2 + q_0^2 - \sqrt{4k_0^2 q_0^2 + k_1^4}} & \text{for } \omega \leq \omega_-(0) \end{cases},$$
$$l_{\mathrm{L}} = \sqrt{k_0^2 + q_0^2 + \sqrt{4k_0^2 q_0^2 + k_1^4}}. \tag{9.124}$$

Thus, the wave vectors $k_{\mathrm{R}} = l_{\mathrm{R}} + q_0$ and $k_{\mathrm{L}} = l_{\mathrm{L}} + q_0$ for the light waves with right- and left-circular polarizations, propagating within a helicoidal system is approximated to the lowest order of k_1 as

$$k_{\mathrm{R}} \approx k_0 + \frac{k_1^4}{\{8k_0 q_0 (q_0 - k_0)\}},$$
$$k_{\mathrm{L}} \approx k_0 + \frac{k_1^4}{\{8k_0 q_0 (q_0 + k_0)\}}. \tag{9.125}$$

If linearly polarized light with its polarization direction inclined by φ from the x-axis is incident on an optically active material with the thickness of d, then for $z < 0$, before entering the material, the electric fields are to be expressed as

$$E_x(z,t) = E_0 \cos\varphi \exp[\mathrm{i}(kz - \omega t)],$$
$$E_y(z,t) = E_0 \sin\varphi \exp[\mathrm{i}(kz - \omega t)], \tag{9.126}$$

where k is the wave vector of light outside the material. In an optically active material, the electric fields of light are well expressed in a form of circular polarizations and the outgoing light at the end of the material is expressed as

$$E^+(d,t) = E_0 \mathrm{e}^{\mathrm{i}\varphi} \mathrm{e}^{\mathrm{i}(k_{\mathrm{R}} d - \omega t)},$$
$$E^-(d,t) = E_0 \mathrm{e}^{-\mathrm{i}\varphi} \mathrm{e}^{\mathrm{i}(k_{\mathrm{R}} d - \omega t)}, \tag{9.127}$$

which results in

$$E_x(d,t) = E_0 \cos[\varphi + (k_\mathrm{R} - k_\mathrm{L})d/2] \mathrm{e}^{\mathrm{i}\{(k_\mathrm{R}+k_\mathrm{L})d/2-\omega t\}},$$
$$E_y(d,t) = E_0 \sin[\varphi + (k_\mathrm{R} - k_\mathrm{L})d/2] \mathrm{e}^{\mathrm{i}\{(k_\mathrm{R}+k_\mathrm{L})d/2-\omega t\}}. \tag{9.128}$$

Thus, the transmitted light is linearly polarized with the angle of the polarization direction at $\varphi+(k_\mathrm{R}-k_\mathrm{L})d/2$. The rotatory power is generally defined as $(k_\mathrm{R} - k_\mathrm{L})/2$, which is calculated as

$$\begin{aligned}(k_\mathrm{R} - k_\mathrm{L})/2 &\approx -k_1^4/\{8q_0(k_0^2 - q_0^2)\}\\ &= -\frac{q_0}{32}\left(\frac{n_\parallel^2 - n_\perp^2}{n_\parallel^2 + n_\perp^2}\right)^2 \frac{1}{\lambda'^2(1-\lambda'^2)}\end{aligned} \tag{9.129}$$

with $\lambda' = q_0/k_0 = \lambda/P\sqrt{\bar{\epsilon}}$. This result was first derived by de Vries (1951). Thus, the optically rotatory power changes its sign, when the wavelength cross the helicoidal pitch.

Summarizing, when linearly polarized light is incident on a cholesteric crystal of a clockwise helix, the right-circular polarized light is selectively reflected, while the left-circular polarized light is transmitted, when the frequency range lies within the band gap of $cq_0/n_\parallel < \omega < cq_0/n_\perp$, which corresponds to $n_\parallel P > \lambda > n_\perp P$ in wavelength. Near the band gap, the light with left- and right-circular polarizations propagate but with different refractive indices, which results in that the transmitted light is linearly polarized but with different polarization direction, which is known as optical rotation. At the band edge, the linearly polarized light appears due to the interference with strongly Bragg-reflected light. On the other hand, for sufficiently higher frequency region, the linear polarization appears again, whose directions are restricted with the principal axes of the dielectric constant in each plane. Thus, the direction of the oscillation of the electric field changes gradually with space, but does not change with time, which is often called waveguide effect.

Bibliography

Ackery, P. R., de Jong, R. and Vane-Wright, R. I. (1999). In: N. P. Kristensen (ed.), *Handbook of Zoology* Vol. IV *Arthropoda: Insecta* Part 35 Lepidoptera, Moths and Butterflies Vol. 1, *Evolution, Systematics, and Biogeography*, Walter de Gruyter, Berlin, pp. 263–300.

Allyn, A. C. and Downey, J. C. (1976). Diffraction structures in the wing scales of *Callophrys* (*Mitoura*) *siva siva* (Lycaenidae). *Bull Allyn Museum* **40**: 1–6.

Allyn, A. C., Jr. and Downey, J. C. (1977). Observations on male U V reflectance and scale ultrastructure in *Phoebis* (Pieridae). *Bull Allyn Museum* **42**: 1–20.

Anderson, T. F. and Richards, A. G., Jr. (1942). An electron microscope study of some structural colors of insects. *J Appl Phys* **13**: 748–758.

Andersson, S. and Amundsen, T. (1997). Ultraviolet colour vision and ornamentation in bluethroats. *Proc R Soc Lond B* **264**: 1587–1591.

Andersson, S. (1996). Bright ultraviolet colouration in the Asian whistling-thrushes (*Myiophonus* spp.). *Proc R Soc Lond B* **263**: 843–848.

Andersson, S. (1999). Morphology of UV reflectance in a whistling-thrush: Implications for the study of structural colour signalling in birds. *J Avian Biol* **30**: 193–204.

Argyros, A., Manos, S., Large, M. C. J., McKenzie, D. R., Cox, G. C. and Dwarte, D. M. (2002). Electron tomography and computer visualisation of a three-dimensional 'photonic' crystal in a butterfly wing-scale. *Micron* **33**: 483–487.

Arnold, J. M., Young, R. E. and King, M. V. (1974). Ultrastructure of a cephalopod photophore. II. Iridophores as reflectors and transmitters. *Biol Bull* **147**: 522–534.

Ashcroft, N. R. and Lythgoe, J. N. (1991). Ionic basis for color changes in the iridescent cornea of the sand goby (*Pomatoschistus minutus*). *Pigment Cell Res* **4**: 209–215.

Auber, L. (1957). The structures producing "non-iridescent" blue colour in bird feathers. *Proc Zool Soc Lond* **129**: 455–486.

Auber, L. (1971/1972). Formation of 'polyhedral' cell cavities in cloudy media of bird feathers. *Proc R Soc Edinb B* **74**: 27–41.

Bancroft, W. D. (1924). Newton and the peacock. *J Phys Chem* **28**: 351–359.

Banerjee, S., Cole, J. B. and Yatagai, T. (2007). Colour characterization of a *Morpho* butterfly wing-scale using a high accuracy nonstandard finite-difference time-domain method. *Micron* **38**: 97–103.

Bernard, G. D. and Miller, W. H. (1968). Interference filters in the corneas of Diptera. *Invest Ophtalmol* **7**: 416–434.

Bernhard, C. G., Miller, W. H. and Møller, A. R. (1965). The insect corneal ripple array. A biological broad-band impedance transformer acts as an antireflection coating. *Acta Physiol Scand* **63**(Suppl. 243): 1–79.

Bernhard, C. G., Gemne, G. and Møller, A. R. (1968). Modification of specular reflexion and light transmission by biological surface structure — To see, to be seen or not to be seen. *Quart Rev Biophys* **1**: 89–105.

Bernhard, C. G., Gemne, G. and Sällström, J. (1970). Comparative ultrastructure of corneal surface topography in insects with aspects on phylogenesis and function. *Z Vergl Physiol* **67**: 1–25.

Bernhard, C. G. and Miller, W. H. (1962). A corneal nipple pattern in insect compound eyes. *Acta Physiol Scand* **56**: 385–386.

Berthier, S., Charron, E. and Da Silva, A. (2003). Determination of the cuticle index of the scales of the iridescent butterfly *Morpho menelaus*. *Opt Commun* **228**: 349–356.

Berthier, S., Charron, E. and Boulenguez, J. (2006). Morphological structure and optical properties of the wings of Morphidae. *Insect Sci* **13**: 145–157.

Berthier, S., Boulenguez, J. and Bálint, Z. (2007). Multiscaled polarization effects in *Suneve coronata* (Lepidoptera) and other insect: Application to anti-counterfeiting of banknotes. *Appl Phys A* **86**: 123–130.

Berthier, S. (2007). *Iridescences — The Physical Colors of Insects*, Springer Science + Business Media, New York.

Bingham, L., Bingham, I., Geary, S., Tanner, J., Driscoll, C., Cluff, B. and Gardner, J. S. (1995). SEM comparison of *Morpho* butterfly dorsal and ventral scales. *Microsc Res Tech* **31**: 93–94.

Biró, L. P., Bálint, Zs., Kertész, K., Vértesy, Z., Márk, G. I., Horváth, Z. E., Balázs, J., Méhn, D., Kiricsi, I., Lousse, V. and Vigneron, J.-P. (2003). Role of photonic-crystal-type structures in the thermal regulation of a Lycaenid butterfly sister species pair. *Phys Rev E* **67**: 021907.

Blandin, P. (1988). *The genus Morpho, Lepidoptera Nymphalidae*, Part 1. *The subgenera Iphimedeia and Schwartzia*, Sciences Nat, Compiegne.

Blandin, P. (1993). *The genus Morpho, Lepidoptera Nymphalidae*, Part 2. *The subgenera Iphixibia, Cytheritis, Balachowskyna and Cypritis*, Sciences Nat, Venette.

Bone, R. A., Lee. D. W. and Norman, J. M. (1985). Epidermal cells functioning as lenses in leaves of tropical rain-forest shade plants. *Appl Opt* **24**: 1408–1412.

Born, M. and Wolf, E. (1975). *Principles of Optics*, 5th ed. Pergamon Press, Oxford.

Bouligand, Y. (1965). Sur une disposition fibrillaire torsadée commune à plusieurs structures biologiques. *C R Acad Sc Paris* **261**: 4864–4867.

Bouligand, Y. (1967). Comparaison de certains matériels biologiques à la structure des cristaux liquides "cholestériques". *J Mircosc* **6**: 41a.

Braun, D. M. and Jungerman, R. L. (1995). Broadband multilayer antireflection coating for semiconductor laser facets. *Opt Lett* **20**: 1154–1156.

Brewster, D. (1854). *Treatise on Optics*, a new edition by A. D. Bache, Blanchard and Lea, Philadelphia.

Brink, D. J., Smit, J. E., Lee, M. E. and Møller, A. (1995). Optical diffraction by the microstructure of the wing of a moth. *Appl Opt* **34**: 6049–6057.

Brink, D. J., van der Berg, N. G. and Botha, A. J. (2002). Iridescent colors on seashells: An optical and structural investigation of *Helcion pruinosus*. *Appl Opt* **41**: 717–722.

Brink, D. J. and Lee, M. E. (1996). Ellipsometry of diffractive insect reflectors. *Appl Opt* **35**: 1950–1955.

Brink, D. J. and Lee, M. E. (1999). Confined blue iridescence by a diffracting microstructure: An optical investigation of the *Cynandra opis* butterfly. *Appl Opt* **38**: 5282–5289.

Brink, D. J. and van der Berg, N. G. (2004). Structural colours from the feathers of the bird *Bostrychia hagedash*. *J Phys D: Appl Phys* **37**: 813–818.

Caveney, S. (1971). Cuticle reflectivity and optical activity in scarab beetles: The rôle of uric acid. *Proc R Soc Lond B* **178**: 205–225.

Chae, J., Kita-Tsukamoto, K., Nishida, S. and Ohwada, K. (1996). Chemical composition of the integumental reflecting platelets in the iridescent copepods of the family Sapphirinidae (Poecilostomatoida). *J Crustacean Biol* **16**: 20–23.

Chae, J. and Nishida, S. (1994). Integumental ultrastructure and color patterns in the iridescent copepods of the family Sapphirinidae (Copepoda: Poecilostomatoida). *Marine Biol* **119**: 205–210.

Chae, J. and Nishida, S. (1999). Spectral patterns of the iridescence in the males of *Sapphirina* (Copepoda: Poecilostomatoida). *J Mar Biol Ass UK* **79**: 437–443.

Clapham, P. B. and Hutley, M. C. (1973). Reduction of lens reflexion by the "moth eye" principle. *Nature* **244**: 281–282.

Clothier, J. and Lythgoe, J. N. (1987). Light-induced colour changes by the iridophores of the Neon tetra, *Paracheirodon innesi*. *J Cell Sci* **88**: 663–668.

Cook, G., Timms, P. L. and Göltner-Spickermann, C. (2003). Exact replication of biological structures by chemical vapor deposition of silica. *Angew Chem Int Ed* **42**: 557–559.

Cooper, K. M., Hanlon, R. T. and Budelmann, B. U. (1990). Physiological color change in squid iridophores II. Ultrastructural mechanisms in *Lolliguncula brevis*. *Cell Tissue Res* **259**: 15–24.

Cooper, K. M. and Hanlon, R. T. (1986). Correlation of iridescence with changes in iridophore platelet ultrastructure in the squid *Lolliguncula brevis*. *J Exp Biol* **121**: 451–455.

Cowan, J. J. (1990). Aztec surface-relief volume diffractive structure. *J Opt Soc Am A* **7**: 1529–1544.

Craig, A. J. F. K. and Hartley, A. H. (1985). The arrangement and structure of feather melanin granules as a taxonomic character in African starlings (Sturnidae). *Auk* **102**: 629–632.

de Genne, P.-G. (1974). *The Physics of Liquid Crystals*, Oxford Univ. Press, London.

Denton, E. J. and Land, M. F. (1967). Optical properties of the lamellae causing interference colours in animal reflectors. *J Physiol* **191**: 23–24P.

Denton, E. J. and Land, M. F. (1971). Mechanism of reflexion in silvery layers of fish and cephalopods. *Proc R Soc Lond A* **178**: 43–61.

Denton, E. J. and Nicol, J. A. C. (1965a). Reflexion of light by external surfaces of the herring, *Clupea harengus*. *J Mar Biol Ass UK* **45**: 711–738.

Denton, E. J. and Nicol, J. A. C. (1965b). Polarization of light reflected from the silvery exterior of the bleak, *Alburnus alburnus*. *J Mar Biol Ass UK* **45**: 705–709.

Denton, E. J. and Nicol, J. A. C. (1965c). Studies on reflexion of light from silvery surfaces of fishes, with special reference to the bleak, *Alburnus alburnus*. *J Mar Biol Ass UK* **45**: 683–703.

Denton, E. J. and Nicol, J. A. C. (1966). A survey of reflectivity in silvery teleosts. *J Mar Biol Ass UK* **46**: 685–722.

Denton, E. J. (1970). On the organization of reflecting surfaces in some marine animals. *Phil Trans Roy Soc Lond B* **258**: 285–313.

Denton, E. (1971). Reflectors in fishes. *Sci Am* **224**: 64–72.

de Vries, H. L. (1951). Rotatory power and other optical properties of certain liquid crystals. *Acta Cryst* **4**: 219–226.

Dey, S., Hooroo, R. N. K. and Bhattacharjee, C. R. (1998). Electron microscopy and spectroscopical studies on the coloured patches on the wing of a butterfly, *Graphium sarpedon* (Lepidoptera: Papillionidae) with reference to their photobiological and electrical properties. *Pigment Cell Res* **11**: 1–11.

Dorst, J., Gastaldi, G., Hagège, R. and Jacquemart, J. (1974). Différents aspects des barbules de quelques Paradisaeidés observés sur coupes en microscopie électronique. Relations avec les phénomènes d'interférences *C R Acad Sc Paris* **278**: 285–290.

Doucet, S. M., Shawkey, M. D., Rathburn, M. K., Mays, H. L., Jr. and Montgomerie, R. (2004). Concordant evolution of plumage colour, feather microstructure and a melanocortin receptor gene between mainland and island population of a fairy-wren. *Proc R Soc Lond B* **271**: 1663–1670.

Doucet, S. M., Shawkey, M. D., Hill, G. E. and Montgomerie, R. (2006). Iridescent plumage in satin bowerbirds: Structure, mechanisms and nanostructural predictors of individual variation in colour. *J Exp Biol* **209**: 380–390.

Dresp, B., Jouventin, P. and Langley, K. (2005). Ultraviolet reflecting photonic microstructures in the king penguin beak. *Biol Lett* **1**: 310–313.

Durrer, H. and Villiger, W. (1962). Schillerfarben der Nektarvögel (Nectariniidae). *Rev Suisse Zool* **69**: 801–813.

Durrer, H. and Villiger, W. (1966). Schillerfarben der Trogoniden — Eine elektronenmikroskopische Untersuchung. *J Orn* **107**: 1–26.

Durrer, H. and Villiger, W. (1967). Bildung der Schillerstruktur beim Glanzstar — Elektronenmikroskoipsche Untersuchungen der Entstehung gasgefüllter Melaninkörner. *Z Zellforsch* **81**: 445–456.

Durrer, H. and Villiger, W. (1970a). Schillerfarben der Stare (Sturnidae) — Elektronenmikroskopisher Beitrag zur Färbung und Systematik der Farmilie. *J Orn* **111**: 133–153.

Durrer, H. and Villiger, W. (1970b). Schillerradien des Goldkuckucks (*Chrysococcyx cupreus* (Shaw)) im Elektronenmikroskop. *Z Zellforsch* **109**: 407–413.

Durrer, H. and Villiger, W. (1972). Schillerfarben von *Euchroma gigantea* (L.): (Coleoptera: Buprestidae): Elektronenmikroskopische Untersuchung der Elytra. *Int J Insect Morphol Embryol* **1**: 233–240.

Durrer, H. and Villiger, W. (1975). Schillerstruktur des Kongopfaus (*Afropavo congensis* Chapin, 1936) im Elektronenmikroskop. *J Orn* **116**: 94–102.

Durrer, H. (1962). Schillerfarben beim Pfau (*Pavo cristantus* L.). *Verhand Naturf Ges Basel* **73**: 204–224.

Durrer, H. (1965). Bau und Bildung der Augfeder des Pfau. *Revue Suiss Zool* **72**: 263–411.

Durrer, H. (1977). Schillerfarben der Vogelfeder als Evolutionsproblem. *Denkschr Schweiz Naturforsch Ges* **91**: 1–127.

Durrer, H. (1986). Colouration. In: J. Bereiter-Hahn, A. G. Matoltsy, and K. S. Richards (eds.), *Biology of the Integument* Vol. 2 *Vertebrates* V. *The Skin of Birds*, Chap. 12, Springer-Verlag, Berlin, pp. 239–247.

Dyck, J. (1971a). Structure and spectral reflectance of green and blue feathers of the rose-faced lovebird (*Agapornis roseicollis*). *Biol Skr Dan Vid Selsk* **18**: 1–67.

Dyck, J. (1971b). Structure and colour-production of the blue barbs of *Agapornis roseicollis* and *Cotinga maynana*. *Z Zellforsch* **115**: 17–29.

Dyck, J. (1973). Feather structure: The sufrace of barbs and barbules. *Zool Jb Anat* **90**: S.550–S.566.

Dyck, J. (1976). Structural colours. *Proc Int Ornithol Congr* **16**: 426–437.

Dyck, J. (1979). Winter plumage of the Rock Ptarmigan: Structure of the air-filled barbules and function of the white colour. *Dansk orn Foren Tidsskr* **73**: 41–58.

Dyck, J. (1987). Structure and light reflection of green feathers of fruit doves (*Ptilinopus* spp.) and an imperial pigeon (*Ducula concinna*). *Biol Skr Dan Vid Selsk* **30**: 1–43.

Dyck, J. (1999). Feather morphology at the ultrastructural level. *Acta Ornithol* **34**: 131–134.

Eguchi, E. and Meyer-Rochow, V. B. (1983). Ultraviolet photography of forty-three species of Lepidoptera representing ten families. *Ann Zool Japon* **56**: 10–18.

Elasäfser, T. (1925). Die Struktur schillernder Federn. *J Orn* **3**: 337–389.

Finger, E. and Burkhardt, D. (1992). Avian plumage colors — Origin of UV reflection in a black parrot. *Naturwiss* **79**: 187–188.

Finger, E. (1995). Visible and UV coloration in birds: Mie scattering as the basis of color in many bird feathers. *Naturwiss* **82**: 570–573.

Foster, K. W. (1933). Color changes in Fundulus with special reference to the color changes of the iridosomes. *Proc Natl Acad Sci* **19**: 535–540.

Fox, D. L. (1976). *Animal Biochromes and Structural Colours*, Univ. California Press, Berkeley.

Frank, F. and Ruska, H. (1939). Übermikroskopische Untersuchung der Blaustruktur der Vogelfeder. *Naturwiss* **27**: 229–230.

Frank, F. (1939). Die Färbung der Vogelfeder durch Pigment und Struktur. *J Orn Lpz* **87**: 426–523.

Fruhstorfer, H. (1913). In: A. Seitz (ed.), *Macrolepidoptera of the World*, Vol. 5 *Family: Morphidae*, Alfred Kernen, Stuttgart, pp. 333–356.

Fujii, R. (1993a). Cytophysiology of fish chromatophores. *Int Rev Cyto* **143**: 191–255.

Fujii, R. (1993b). Coloration and chromatophores. In: D. H. Evans (ed.), *The Physiology of Fish*, CRC Press, Boca Raton, pp. 535–562.

Fujimura, Y. (2008). Photonic structure and average permittivity, Thesis, Osaka University, Osaka.

Fung, K. K. (2005). Photonic iridescence of a blue-banded bee. *Microsc Microanal* **11**(Suppl 2): 1202–1203.

Gemne, G. (1966). Ultrastructural ontogenesis of cornea and corneal nipples in the compound eye of insects. *Acta Physiol Scand* **66**: 511–512.

Gemne, G. (1971). Ontogenesis of corneal surface ultrastructure in nocturnal Lepidoptera. *Phil Trans R Soc B* **262**: 343–363.

Gentil, K. (1941). Beiträge zur Kenntnis schillernder Schmetterlingsschuppen auf Grund polarisationsoptischer Untersuchung. *Z Morph Ökol Tiere* **37**: 591–612.

Gentil, K. (1942). Elektronenmikroskopische Untersuchung des Feinbaues schillernder Leisten von Morpho-Schuppen. *Z Morph Ökol Tiere* **38**: 344–355.

Gentil, K. (1943). Beitrage zur Morphologie und Optik der Schillerschuppen von *Hoplia coerulea* Drury und *Hoplia farinosa* Linné (col.). *Z Morph Ökol Tiere* **40**: 299–313.

Gerwick, W. H. and Lang, N. J. (1977). Structural, chemical and ecological studies on iridescence in *Iridaea* (Rhodophyta). *J Phycol* **13**: 121–127.

Ghiradella, H., Aneshansley, D., Eisner, T., Silberglied, R. E. and Hinton, H. E. (1972). Ultraviolet reflection of a male butterfly: Interference color caused by thin-layer elaboration of wing scales. *Science* **178**: 1214–1217.

Ghiradella, H. and Radican, W. (1976). Development of butterfly scales II. Struts, lattices and surface tension. *J Morph* **150**: 279–298.

Ghiradella, H. (1978). Development of ultraviolet-reflecting butterfly scales: How to make an interference filter. *J Morph* **142**: 395–410.

Ghiradella, H. (1984). Structure of iridescent Lepidopteran scales: Variations on several themes. *Ann Entomol Soc Am* **77**: 637–645.

Ghiradella, H. (1985). Structure and development of iridescent Lepidopteran scales: The Papilionidae as a showcase family. *Ann Entomol Soc Am* **78**: 252–264.

Ghiradella, H. (1989). Structure and development of iridescent butterfly scales: Lattices and laminae. *J Morph* **202**: 69–88.

Ghiradella, H. (1991). Light and color on the wing: Structural colors in butterflies and moths. *Appl Opt* **30**: 3492–3500.

Ghiradella, H. (1994). Structure of butterfly scales: Patterning in an insect cuticle. *Microsc Res Tech* **27**: 429–438.

Ghiradella, H. (1998). Hairs, bristles, and scales. In: F. N. Harrison and M. Locke (eds.), *Microscopic Anatomy of Invertebrates*, Vol. 11A: *Insecta*, Wiley-Liss, New York, pp. 257–287.

Ghiradella, H. (1999). Shining armor: Structural colors in insects. *Opt Photonics News* **March**: 46–48.

Ghiradella, H. (2005). Fine structure of basic Lepidopteran scales. In: S. Kinoshita and S. Yoshioka (eds.), *Structural Colors in Biological Systems — Principles and Applications*, Osaka University Press, Osaka, pp. 75–93.

Goda, M. and Fujii, R. (1998). The blue coloration of the common surgeonfish, *Paracanthurus hepatus* — II. Color revelation and color changes. *Zool Sci* **15**: 323–333.

Goda, M. and Fujii, R. (2001). Coloration and chromatophores of the domino damsel, *Dascyllus trimaculatus. Zool Sci* **18**: 165–174.

Gonzato, J.-C. and Pont, B. (2003). A phenomenological representation of iridescent colors in butterfly wings. *WSCG Short Comm Papers Proc*, Feb. 2–6.

Gould, K. S. and Lee, D. W. (1996). Physical and ultrastructural basis of blue leaf iridescence in four Malaysian understory plants. *Am J Bot* **83**: 45–50.

Graham, R. M., Lee, D. W. and Norstog, K. (1993). Physical and ultrastructural basis of blue leaf iridescence in two neotropical ferns. *Am J Bot* **80**: 198–203.

Gralak, B., Tayeb, G. and Enoch, S. (2001). *Morpho* butterflies wings color modeled with lamellar grating theory. *Opt Express* **9**: 567–578.

Greenewalt, C. H., Brandt, W. and Friel, D. D. (1960a). Iridescent colors of hummingbird feathers. *J Opt Soc Am* **50**: 1005–1013.

Greenewalt, C. H., Brandt, W. and Friel, D. D. (1960b). The iridescent colors of hummingbird feathers. *Proc Am Phil Soc* **104**: 249–253.

Greenewalt, C. H. (1960). *Hummingbirds*, Doubleday & Company, New York, republicated by Dover in 1990, New York.

Gundersen, R. E. and Rivera, E. R. (1982). An ultrastructural study of the development of the dermal iridophores and structural pigmentation in *Poecillia reticulata* (Peters). *J Morph* **172**: 349–359.

Hébant, C. and Lee, D. W. (1984). Ultrastructural basis and developmental control of blue iridescence in Selaginella leaves. *Am J Bot* **71**: 216–219.

Hanlon, R. T., Cooper, K. T., Budelmann, B. U. and Pappas, T. C. (1990). Physiological color change in squid iridophores I. Behavior, morphology and pharmacology in *Lolliguncula brevis. Cell Tissue Res* **259**: 3–14.

Hariyama, T., Hironaka, M., Horiguchi, H. and Stavenga, D. G. (2005). The leaf beetle, the jewel beetle, and the damselfly; Insects with a multilayered show case. In: S. Kinoshita and S. Yoshioka (eds.), *Structural Colors in Biological Systems — Principles and Applications*, Osaka University Press, Osaka, pp. 153–176.

Hecht, E. (2002). *Optics*, 4th ed. Addison-Wesley, New York.

Hemsley, A. R., Collinson, M. E., Kovach, W. L., Vincent, B. and Williams, T. (1994). The role of self-assembly in biological systems: Evidence from iridescent colloidal sporopollenin in *Selaginella* megaspore walls. *Phil Trans R Soc Lond B* **345**: 163–173.

Hinton, H. E. (1969). Some structures of insects as seen with the scanning electron microscope. *Micron* **1**: 84–108.

Hinton, H. E. (1973). Some recent work on the colours of insects and their likely significance. *Proc Brit Ent Nat Hist Soc* **6**: 43–54.

Hinton, H. E. (1976). Recent work on physical colours of insect cuticle. In: H. R. Hepburn (ed.), *The Insect Integument*, Chap. 25, Elsevier Scientific Pub., North Holland, pp. 475–496.

Hinton, H. E. and Jarman, G. M. (1972). Physiological colour change in the Hercules beetle. *Nature* **238**: 160–161.

Honda, K. and Kato, Y. (ed.) (2005). *Biology of Butterflies*, Univ. Tokyo Press, Tokyo, in Japanese.

Hooke, R. (1665). *Micrographia and Some Physiological Descriptions of Minute Bodies Made by Magnifying Glasses with Observations and Inquiries thereupon*, Dover Publications, Mineola, NY.

Hoshino, T., Watanabe, K., Kometani, R., Morita, T., Kanda, K., Haruyama, Y., Kaito, T., Fujita, J., Ishida, M., Ochiai, Y. and Matsui, S. (2003). Development of three-dimensional pattern-generating system for focused-ion-beam chemical-vapor deposition. *J Vac Sci Technol B* **21**: 2732–2736.

Huang, J., Wang, X. and Wang, Z. L. (2006). Controlled replication of butterfly wings for achieving tunable photonic properties. *Nano Lett* **6**: 2325–2331.

Huxley, A. F. (1968). A theoretical treatment of the reflexion of light by multilayer structures. *J Exp Biol* **48**: 227–245.

Huxley, J. (1975). The basis of structural colour variation in two species of *Papilio*. *J Ent (A)* **50**: 9–22.

Huxley, J. (1976). The coloration of *Papilio zalmoxis* and *P. antimachus*, and the discovery of Tyndall blue in butterflies. *Proc R Soc Lond B* **193**: 441–453.

Iga, T., Takabatake, I. and Watanabe, S. (1987). Nervous regulation of motile iridophores of a freshwater goby, *Odontobutis obscura*. *Comp Biochem Physiol* **88C**: 319–324.

Iga, T. and Matsuno, A. (1986). Motile iridophores of a freshwater goby, *Odontobutis obscura*. *Cell Tissue Res* **244**: 165–171.

Imafuku, M., Hirose, Y. and Takeuchi, T. (2002). Wing colors of *Chrysozephyrus* butterflies (Lepidoptera; Lycaenidae): Ultraviolet reflection by males. *Zool Sci* **19**: 175–183.

Iohara, K., Yoshimura, M., Tabata, H. and Shimizu, S. (2000). Structurally colored fibers. *Chem Fibers Int* **50**: 38–39.

Iwase, E., Matsumoto, K. and Shimoyama, I. (2004). The structural-color based on the mechanism of butterfly wing coloring for wide viewing angle reflective display. *Micro Electro Mechanical Systems 2004*, pp. 105–108.

Järvilehto, M. and Meyer-Rochow, V. B. (1996). Ultraviolet colours in Nordic butterflies. *Acta Univ Ouluensis A* **286**: 110–115.

James, R. W. (1965). *The Optical Principles of the Diffraction of X-Rays*, Cornell Univ Press, Ithaca.

Jin, C., Meng, X., Cheng, B., Li, Z. and Zhang, D. (2001). Photonic gap in amorphous photonic materials. *Phys Rev B* **63**: 195107.

Kühn, A. (1955). *Vorlesungen über Entwicklungsphysiologie*, Springer-Verlag, Berlin.

Kasukawa, H., Oshima, N. and Fujii, R. (1987). Mechanism of light reflection in blue damselfish motile iridophore. *Zool Sci* **4**: 243–257.

Kawaguti, S. and Kamishima, Y. (1964). Electron microscopic study on the iridophore of the Japanese porgy. *Biol J Okayama Univ* **10**: 75–81.

Kawaguti, S. and Kamishima, Y. (1966). Electron microscopy on the blue back of a clupeoid fish, *Harengula zunasi*. *Proc Japan Acad* **42**: 389–393.

Kawaguti, S. (1965). Electron microscopy on iridophores in the scale of the blue wrasse. *Proc Japan Acad* **41**: 610–613.

Kawaguti, S. (1966). A supplementary note on the iridophore of the Japanese porgy. *Biol J Okayama Univ* **12**: 57–60.

Kawaguti, S. (1967). Electron microscopy on iridophores from fish and amphibians. *Zool Mag (in Japanese)* **76**: 281–287.

Kemp, D. J., Vukusic, P. and Rutowski, R. L. (2006). Stress-mediated covariance between nano-structural architecture and ultraviolet butterfly coloration. *Func Ecol* **20**: 282–289.

Kenchington, W. and Flower, N. E. (1969). Studies of insect fibrous proteins: The structural protein of the ootheca in the praying mantis, *Sphodromantis centralis* Rehn. *J Microsc* **89**: 263–281.

Kertésa, K., Bálint, Zs., Vértesy, Z., Márk, G. I., Lousse, V., Vigneron, J.-P. and Biró, L. P. (2006). Photonic crystal type structures of biological origins: Structural and spectral characterization. *Curr Appl Phys* **6**: 252–258.

Keyser, A. J. and Hill, G. E. (1999). Condition-dependent variation in the blue-ultraviolet coloration of a structurally based plumage ornament. *Proc R Soc Lond B* **266**: 771–777.

Keyser, A. J. and Hill, G. E. (2000). Structurally based plumage coloration is an honest signal of quality in male blue grosbeaks. *Behav Ecol* **11**: 202–209.

Kinoshita, S., Yoshioka, S., Fujii, Y. and Okamoto, N. (2002a). Photophysics of structural color in the *Morpho* butterflies. *Forma* **17**: 103–121.

Kinoshita, S., Yoshioka, S. and Kawagoe, K. (2002b). Mechanisms of structural colour in the *Morpho* butterfly: Cooperation of regularity and irregularity in an iridescent scale. *Proc R Soc Lond B* **269**: 1417–1421.

Kinoshita, S. and Yoshioka, S. (2005a). Phtophysical approach to blue coloring in the *Morpho* butterflies. In: S. Kinoshita and S. Yoshioka (eds.), *Structural Colors in Biological Systems — Principles and Applications*, Osaka University Press, Osaka, pp. 113–140.

Kinoshita, S. and Yoshioka, S. (2005b). Structural colors in nature: The role of regularity and irregularity in the structure. *Chem Phys Chem* **6**: 1442–1459.

Kinoshita, S. and Yoshioka, S. (2005c). Fundamental optical processes underlying the structural colors. In: S. Kinoshita and S. Yoshioka (eds.),

Structural Colors in Biological Systems — Principles and Applications, Osaka University Press, Osaka, pp. 3–26.

Kinoshita, S. (2005). *Blue Brilliancy of Morpho Butterflies*, Kagaku Dojin, Kyoto, in Japanese.

Kinoshita, S., Yoshioka, S. and Miyazaki, J. (2008). Physics of structural colors. *Rep Prog Phys* (in press).

Kirchhoff, G. R. (1882). Zur Theorie der Lichtstrahlen *Berlin Ber* In: *Sitzungsberichte der Kniglich Preussischen Akademie der Wissenschaften zu Berlin* II, pp. 641–669.

Kirchhoff, G. R. (1883). Zur Theorie der Lichtstrahlen. *Ann Physik* **18**: 663–695.

Knopp, M. C. N. and Krenn, H. W. (2003). Efficiency of fruit juice feeding in *Morpho peleides* (Nymphalidae, Lepidoptera). *J Insect Behavior* **16**: 67–77.

Koon, D. W. and Crawford, A. B. (2000). Insect thin films as sun blocks, not solar collectors. *Appl Opt* **39**: 2496–2498.

Kosaku, A. and Miyamoto, K. (1994). Art of structural coloration in insects. *Forma* **9**: 297–302.

Kristensen, N. P. and Skalski, A. W. (1999). Phylogeny and palaeontology. In: N. P. Kristensen (ed.), *Handbook of Zoology*, Vol. IV, *Arthropoda: Insecta*, Part 35, Lepidoptera, Moths and Butterflies, Vol. 1, *Evolution, Systematics, and Biogeography*, Walter de Gruyter, Berlin, pp. 7–25.

Kristensen, N. P. (1999). In: N. P. Kristensen (ed.), *Handbook of Zoology*, Vol. IV, *Arthropoda: Insecta*, Part 35, Lepidoptera, Moths and Butterflies, Vol. 1, *Evolution, Systematics, and Biogeography*, Walter de Gruyter, Berlin.

Kristensen, N. P. (2003). In: N. P. Kristensen (ed.), *Handbook of Zoology*, Vol. IV, *Arthropoda: Insecta*, Part 36, Lepidoptera, Moths and Butterflies, Vol. 2, *Morphology, Physiology, and Development*, Walter de Gruyter, Berlin.

Kumazawa, K., Tanaka, S., Negita, K. and Tabata, H. (1994). Fluorescence from wing of *Morpho sulkowskyi* butterfly. *Jpn J Appl Phys* **33**: 2119–2122.

Kumazawa, K., Negita, K., Hasegawa, T. and Tabata, H. (1996). Fluorescence from cover and basal scales of *Morpho sulkowskyi* and *Papilio xuthus* butterflies. *J Exp Zool* **275**: 15–19.

Kurachi, M., Takaku, Y., Komiya, Y. and Hariyama, T. (2002). The origin of extensive colour polymorphism in *Plateumaris sericea* (Chrysomelidae, Coleoptera). *Naturwiss* **89**: 295–298.

LaBastille, A., Allen, D. G. and Durrell, L. W. (1972). Behavior and feather structure of the Quetzal. *Auk* **89**: 339–348.

Large, M. C. J., McKenzie, D. R., Parker, A. R., Steel, B. C., Ho, K., Bosi, S. G., Nicorovici, N. and McPhedran, R. C. (2001). The mechanism of light reflectance in silverfish. *Proc R Soc Lond A* **457**: 511–518.

Lawrence, C., Vukusic, P. and Sambles, R. (2002). Grazing-incidence iridescence from a butterfly wing. *Appl Opt* **41**: 437–441.

Le Moult, E. and Réal, P. (1962). *Les Morpho d'Amérique du Sud et Centrale. Historique — Morphologie — Systématique*, Éditions du Cabinet Entomologique, E. Le Moult, Paris.

Le Moult, E. and Réal, P. (1963). *Les Morpho d'Amérique du Sud et Centrale* Tome II *Planches en couleurs et en noir*, Éditions du Cabinet Entomologique, E. Le Moult, Paris.
Lee, D. W., Taylor, G. T. and Irvine, A. K. (2000). Structural fruit coloration in *Delarbrea michieana* (Araliaceae). *Int J Plant Sci* **161**: 297–300.
Lee, D. W. and Lowry, J. B. (1975). Physical basis and ecological significance of iridescence in blue plants. *Nature* **254**: 50–51.
Lee, D. W. (1977). On iridescent plants. *Gardens' Bull Singapore* **30**: 21–29.
Lee, D. W. (1991). Ultrastructural basis and function of iridescent blue colour of fruits in *Elaeocarpus*. *Nature* **349**: 260–262.
Lee, D. W. (1997). Iridescence blue plants. *Am Sci* **85**: 56–63.
Li, Y., Lu, Z., Yin, H., Yu, X., Liu, X. and Zi, J. (2005). Structural origin of the brown color of barbules in male peacock tail feathers. *Phys Rev E* **72**: 010902.
Lippert, W. and Gentil, K. (1959). Über lamellare Feinstrukturen bei den Schillerschuppen der Schmetterlinge vom Urania- und Morpho-Typ. *Z Morph Ökol Tiere* **48**: 115–122.
Lippert, W. and Gentil, K. (1963). Über den Feinbau der Schillerhaare des Polychaeten *Aphrodite aculeata* L. *Z Morph Ökol Tiere* **53**: 22–28.
Liu, Y., Shigley, J. E. and Hurwit, K. N. (1999). Iridescence color of a shell of the mollusk *Pinctada Margaritifera* caused by diffraction. *Opt Express* **4**: 177–182.
Locket, N. A. (1972). The reflecting structure in the iridescent cornea of the serranid teleost *Nemanthias carberryi*. *Proc R Soc Lond B* **182**: 249–254.
Lythgoe, J. N., Shand, J. and Foster, R. G. (1984). Visual pigment in fish iridocytes. *Nature* **308**: 83–84.
Lythgoe, J. N. and Shand, J. (1982). Changes in spectral reflexions from the irodophores of the neon tetra. *J Physiol* **325**: 23–34.
Lythgoe, J. N. and Shand, J. (1989). The structural basis for iridescent colour changes in dermal and corneal iridophores in fish. *J Exp Biol* **141**: 313–325.
Lythgoe, J. N. (1971). Iridescent corneas in fishes. *Nature* **233**: 205–207.
Lythgoe, J. N. (1975). The structure and function of iridescent corneas in teleost fishes. *Proc R Soc Lond B* **188**: 437–457.
Mäthger, L. M., Land, M. F., Siebeck, U. E. and Marshall, N. J. (2003). Rapid colour changes in multilayer reflecting stripes in the paradise whiptail, *Pentapodus paradiseus*. *J Exp Biol* **206**: 3607–3613.
Maeno, N. and Iga, T. (1992). Adrenergic mechanisms associated with the movement of platelets in iridophores from the freshwater goby. *Odontobutis obscura Comp Biochem Physiol* **102C**: 233–237.
Mallock, A. (1911). Note on the iridescent colours of birds and insects. *Proc R Soc Lond* **85**: 598–605.
Mann, S. E., Miaoulis, I. N. and Wong, P. Y. (2001). Spectral imaging, reflectivity measurements, and modeling of iridescent butterfly scale structures. *Opt Eng* **40**: 2061–2068.
Mason, C. W. (1923a). Structural colors in feathers. I. *J Phys Chem* **27**: 201–251.
Mason, C. W. (1923b). Structural colors in feathers. II. *J Phys Chem* **27**: 401–447.

Mason, C. W. (1926). Structural colors in insects. I. *J Phys Chem* **30**: 383–395.
Mason, C. W. (1927a). Structural colors in insects. II. *J Phys Chem* **31**: 321–354.
Mason, C. W. (1927b). Structural colors in insects. III. *J Phys Chem* **31**: 1856–1872.
Matsui, S., Kaito, T., Fujita, J., Komuro, M., Kanda, K. and Haruyama, Y. (2000). Three-dimensional nanostructure fabrication by focused-ion-beam chemical vapor deposition. *J Vac Sci Technol B* **18**: 3181–3184.
McGraw, K. J. (2004). Multiple UV reflectance peaks in the iridescent neck feathers of pigeons. *Naturwiss* **91**: 125–129.
McPhedran, R. C., Nicorovici, N. A., McKenzie, D. R., Botten, L. C., Parker, A. R. and Rouse, G. W. (2001). The sea mouse and the photonic crystal. *Aust J Chem* **54**: 241–244.
McPhedran, R. C., Nicorovici, N. A., McKenzie, D. R., Rouse, G. W., Botten, L. C., Weilch, V., Parker, A. R., Wohlgennant, M. and Vardeny, V. (2003). Structural colours through photonic crystals. *Physica B* **338**: 182–185.
Menter, D. G., Obika, M., Tchen, T. T. and Taylor, J. D. (1979). Leucophores and iridophores of *Fundulus heteroclitus*: Biophysical and ultrastructural properties. *J Morph* **160**: 103–120.
Merritt, E. (1925). A spectrophotometric study of certain cases of structural color. *J Opt Soc Am Rev Sci Instrum* **11**: 93–98.
Meyer-Rochow, V. B. and Eguchi, E. (1983). "Flügelfarben, wie sie die Falter sehen" — A study of UV- and other colour patterns in Lepidoptera. *Ann Zool Japon* **56**: 85–99.
Meyer-Rochow, V. B. and Järvilehto, M. (1997). Ultraviolet colours in *Pieris napi* from northern and southern Finland: Arctic females are the brightest! *Naturwiss* **84**: 165–168.
Meyer-Rochow, V. B. (1991). Differences in ultraviolet wing patterns in the New Zealand lycaenid butterflies *Lycaena salustius*, *L. rauparaha*, and *L. feredayi* as a likely isolating mechanism. *J Roy Soc New Zealand* **21**: 169–177.
Michelson, A. A. (1911). On metallic colouring in birds and insects. *Phil Mag* **21**: 554–566.
Mie, G. (1908). Beiträge zur Optik truber Medien, speziell kolloidaler Metallösungen. *Ann Physik* **25**: 377–445.
Millott, N. and Manly, B. M. (1961). The iridophores of the Echinoid *Diadema antillarum*. *Quart J Microsc Sci* **102**: 181–194.
Miyamoto, K. and Kosaku, A. (2005). Structural color of shield bugs (Insecta, Heteroptera) in Japan. In: S. Kinoshita and S. Yoshioka (eds.), *Structural Colors in Biological Systems — Principles and Applications*, Osaka University Press, Osaka, pp. 177–191.
Morimoto, G., Yamaguchi, N. and Ueda, K. (2005). Plumage color as a status signal in male-male interaction in the red-flanked bushrobin, *Tarsiger cyanurus*. *J Ethol* **24**: 261–266.
Morris, R. B. (1975). Iridescence from diffraction structures in the wing scales of *Callophrys rubi*, the Green Hairstreak. *J. Ent. (A)* **49**: 149–154.
Mossakowski, D. (1979). Reflection measurements used in the analysis of structural colours of beetles. *J Microsc* **116**: 351–364.

Nagaishi, H., Oshima, N. and Fujii, R. (1990). Light-reflecting properties of the iridophores of the neon tetra, *Paracheirodon innesi*. *Comp Biochem Physiol* **95A**, 337–341.

Nagaishi, H. and Oshima, N. (1989). Neural control of motile activity of light-sensitive iridophores in the neon tetra. *Pigment Cell Res* **2**: 485–492.

Nagaishi, H. and Oshima, N. (1992). Ultrastructure of the motile iridophores of the neon tetra. *Zool Sci* **9**: 65–75.

Nassau, K. (2001). *The Physics and Chemistry of Color*, 2nd edn. Wiley, New York.

Nekrutenko, Y. P. (1965). Gynandromorphic effect and the optical nature of hidden wing-pattern in *Gonepteryx rhamni* L. (Lepidoptera, Pieridae). *Nature* **205**: 417–418.

Neville, A. C., Parry, D. A. D. and Woodhead-Galloway, J. (1976). The chitin crystallite in arthropod cuticle. *J Cell Sci* **21**: 73–82.

Neville, A. C. and Caveney, S. (1969). Scarabaeid beetle exocuticle as an optical analogue of cholesteric liquid crystals. *Biol Rev* **44**: 531–562.

Neville, A. C. and Luke, B. M. (1969). A two-system model for chitin-protein complexes in insect cuticles. *Tissue Cell* **1**: 689–707.

Neville, A. C. (1975). *Biology of the Arthropod Cuticle*, Springer, New York.

Neville, A. C. (1977). Metallic gold and silver colours in some insect cuticles. *J Insect Physiol* **23**: 1267–1274.

Newton, I. (1704). *Opticks: Or a Treatise of the Reflections, Refractions, Inflections & Colours of Light*, Dover, New York.

Nijhout, H. F. (1991). *The Development and Evolution of Butterfly Wing Patterns*, Smithonian Institution Press, Washington and London.

Onslow, H. (1920a). The iridescent colours of insects. *Nature* **106**: 149–152.

Onslow, H. (1920b). The iridescent colours of insects. *Nature* **106**: 181–183.

Onslow, H. (1923). On a periodic structure in many insect scales, and the cause of their iridescent colours. *Phil Trans* **211**: 1–74.

Orihara, H. (2004). *The Physics of Liquid Crystals*, Uchida Rokakuho Pub., Tokyo, in Japanese.

Oshima, N., Sato, M., Kumazawa, T., Okeda, N., Kasukawa, H. and Fujii, R. (1985). Motile iridophores play the leading role in damselfish coloration. In: J. Bagnara, S. N. Klaus, E. Paul and M. Schartl (eds.), *Biological, Molecular and Clinical Aspects of Pigmentation. Proceedings of the XIIth International Pigment Cell Conference*, Univ. Tokyo Press, Tokyo, pp. 241–246.

Oshima, N. and Fujii, R. (1987). Motile mechanism of blue damselfish (*Chrysiptera cyanea*) iridophores. *Cell Motility Cytoskeleton* **8**: 85–90.

Oshima, N. and Nagaishi, H. (1992). Study of the motile mechanism in neon tetra (*Paracheirodon innesi*) iridophores. *Comp Biochem Physiol* **102A**: 273–278.

Oshima, N. (2005). Light reflection in motile iridophores of fish. In: S. Kinoshita and S. Yoshioka (eds.), *Structural Colors in Biological Systems — Principles and Applications*, Osaka University Press, Osaka, pp. 211–229.

Osorio, D. and Ham, A. D. (2002). Spectral reflectance and directional properties of structural coloration in bird plumage. *J Exp Biol* **205**: 2017–2027.

Overton, J. (1966). Microtubules and microfibrils in morphogenesis of the scale cells of *Ephestia kühniella*. *J Cell Biol* **29**: 293–305.

Oxford, G. S. and Gillespie, R. G. (1998). Evolution and ecology of spider coloration. *Ann Rev Entomol* **43**: 619–643.

Parker, A. R., Hegedus, Z. and Watts, R. A. (1998a). Solar-absorber antireflector on the eye of an Eocene fly (45 Ma). *Proc R Soc Lond B* **265**: 811–815.

Parker, A. R., McKenzie, D. R. and Large, M. C. J. (1998b). Multilayer reflectors in animals using green and gold beetles as contrasting examples. *J Exp Biol* **201**: 1307–1313.

Parker, A. R., Mckenzie, D. R. and Ahyong, S. T. (1998c). A unique form of light reflector and the evolution of signalling in *Ovalipes* (Crustacea: Decapoda: Portunidae). *Proc R Soc Lond B* **265**: 861–867.

Parker, A. R., Welch, V. L., Driver, D. and Martini, N. (2003). Opal analogue discovered in a weevil. *Nature* **426**: 786–787.

Parker, A. R. and Hegedus, Z. (2003). Diffractive optics in spiders. *J Opt A: Pure Appl Opt* **5**: S111–S116.

Parker, A. R. and Martini, N. (2006). Structural colour in animals — Simple to complex optics. *Opt Laser Techol* **38**: 315–322.

Parker, A. R. (1995). Discovery of functional iridescence and its coevolution with eyes in the phylogeny of Ostracoda (Crustacea). *Proc R Soc Lond B* **262**: 349–355.

Parker, A. R. (2000). 515 million years of structural colour. *J Opt A: Pure Appl Opt* **2**: R15–R28.

Parker, A. R., McPhedran, R. C., McKenzie, D. R., Botten, L. C. and Nicorovici, N.-A. P. (2001). Aphrodite's iridescence. *Nature* **409**: 36–37.

Parker, A. R. (2002). Natural photonic engineers. *Materialstoday* Sept.: 26–31.

Parker, A. R. (2003). *In the Blink of an Eye: How Vision Kick-Started the Big Bang of Evolution*, Simon & Schuster, Sydney.

Parker, A. R. (2004). A vision for natural photonics. *Phil Trans R Soc Lond A* **362**: 2709–2720.

Parker, A. R. (2005a). A geological history of reflecting optics. *J R Soc Interface* **2**: 1–17.

Parker, A. R. (2005b). Simple classical optics in animals — Reflectors and antireflectors. In: S. Kinoshita and S. Yoshioka (eds.), *Structural Colors in Biological Systems — Principles and Applications*, Osaka University Press, Osaka, pp. 29–52.

Penz, C. M. and DeVries, P. J. (2002). Phylogenetic analysis of *Morpho* butterflies (Nymphalidae, Morphinae): Implications for classification and natural history. *Am Musium Novitates* **3374**: 1–33.

Pillai, P. K. C. (1968). Spectral reflection characteristics of *Morpho* butterfly wing. *J Opt Soc Am* **58**: 1019–1022.

Plattner, L. (2004). Optical properties of the scales of *Morpho rhetenor* butterflies: Theoretical and experimental investigation of the back-scattering of light in the visible spectrum. *J R Soc Interface* **1**: 49–59.

Prum, R. O., Morrison, R. L. and Eyck, G. R. T. (1994). Structural color production by constructive reflection from ordered collagen arrays in a bird (*Philepitta castanea*: Eurylaimidae). *J Morph* **222**: 61–72.

Prum, R. O., Torres, R. H., Williamson, S. and Dyck, J. (1998). Coherent light scattering by blue feather barbs. *Nature* **396**: 28–29.

Prum, R. O., Torres, R., Kovach, C., Williamson, S. and Goodman, S. M. (1999a). Coherent light scattering by nanostructured collagen arrays in the caruncles of the Malagasy asities (Eurylaimidae: Aves). *J Exp Biol* **202**: 3507–3522.

Prum, R. O., Torres, R., Williamson, S. and Dyck, J. (1999b). Two-dimensional Fourier analysis of the spongy medullary keratin of structurally coloured feather barbs. *Proc R Soc Lond B* **266**: 13–22.

Prum, R. O., Andersson, S. and Torres, R. H. (2003). Coherent scattering of ultraviolet light by avian feather barbs. *Auk* **120**: 163–170.

Prum, R. O., Cole, J. A. and Torres, R. H. (2004). Blue integumentary structural colours in dragonflies (Odonata) are not produced by incoherent Tyndall scattering. *J Exp Biol* **207**: 3999–4009.

Prum, R. O., Quinn, T. and Torres, R. H. (2006). Anatomically diverse butterfly scales all produce structural colours by coherent scattering. *J Exp Biol* **209**: 748–765.

Prum, R. O. and Torres, R. H. (2003a). A Fourier tool for the analysis of coherent light scattering by bio-optical nanostructures. *Integr Comp Biol* **43**: 591–602.

Prum, R. O. and Torres, R. (2003b). Structural colouration of avian skin: Convergent evolution of coherently scattering dermal collagen arrays. *J Exp Biol* **206**: 2409–2429.

Prum, R. O. (2006). Anatomy, physics, and evolution of structural colors. In: G. E. Hill and K. J. McGraw (eds.) *Bird Coloration*, Vol. I, *Mechanisms and Measurements*, Harvard Univ. Press., Cambridge, pp. 295–353.

Rabinovitch, K. and Pagis, A. (1975). Multilayer antireflection coatings: Theoretical model and design parameters. *Appl Opt* **14**: 1326–1334.

Ralph, C. L. (1969). The control of color in birds. *Am Zoologist* **9**: 521–530.

Raman, C. V. and Krishnamurti, D. (1954a). The structure and optical behaviour of iridescent shells. *Proc Indian Acad Sci A* **39**: 1–13.

Raman, C. V. and Krishnamurti, D. (1954b). The structure and optical behaviour of pearls. *Proc Indian Acad Sci A* **39**: 215–222.

Raman, C. V. (1934a). The orgin of the colours in the plumage of birds. *Proc Ind Acad Sci* **1**: 1–7.

Raman, C. V. (1934b). On iridescent shells — Part I. Introductory. *Proc Indian Acad Sci A* **1**: 567–573.

Raman, C. V. (1934c). On iridescent shells — Part II. Colours of laminar diffraction. *Proc Indian Acad Sci A* **1**: 574–589.

Raman, C. V. (1934d). On iridescent shells — Part III. Body-colours and diffusion haloes. *Proc Indian Acad Sci A* **1**: 859–870.

Lord Rayleigh (1871a). On the scattering of light by small particles. *Phil Mag* **41**: 447–454.

Lord Rayleigh (1871b). On the light from the sky, its polarization and colour. *Phil Mag* **41**: 87–103.

Lord Rayleigh, F. R. S. (1879). On reflection of vibrations at the confines of two media between which the transition is gradual. *Proc Lond Math Soc* **s1–s11**: 51–56.

Lord Rayleigh, O. M., F. R. S. (1917). On the reflection of light from a regularly stratified medium. *Proc R Soc Lond A* **93**: 565–577.

Lord Rayleigh, O. M., F. R. S. (1919). VII. On the optical character of some brilliant animal colours. *Phil Mag* **37**: 98–111.

Lord Rayleigh, F. R. S. (1923). Studies of iridescent colour, and the structure producing it. — II. Mother-of-pearl. *Proc R Soc Lond A* **102**: 674–677.

Lord Rayleigh, For. Sec. R. S. (1930). The iridescent colours of birds and insects. *Proc R Soc Lond A* **128**: 624–641.

Richards, A. G. (1951). *The Integument of Arthropods*, Univ. Minnesota Press, Minneapois.

Rohrlich, S. T. (1974). Fine structural demonstration of ordered arrays of cytoplasmic filaments in vertebrate iridophores. A comparative survey. *J Cell Biol* **62**: 295–304.

Rothschild, M., Gardiner, B. and Mummery, R. (1978). The role of carotenoids in the "golden glance" of danaid pupae (Insecta: Lepidoptera). *J Zool Lond* **186**: 351–358.

Rutowski, R. L., Macedonia, J. M., Morehouse, N. and Taylor-Taft, L. (2005). Pterin pigments amplify iridescent ultraviolet signal in males of the orange sulphur butterfly, *Colias eurytheme*. *Proc R Soc Lond B* **272**: 2329–2335.

Rutschke, E. (1966a). Die submikroskopische Struktur schillender Federn von Entenvögeln. *Z Zellforsch* **73**: 432–443.

Rutschke, E. (1966b). Über den bau und die Färbung der Vogelfeder. *Monats Ornithol Vivarienk* **13**: 292–299.

Süffert, F. (1924). Morphologie und Optik der Schmetterlingsschuppen, insbesondere die Schillerfarben der Schmetterlinge. *Z f Morph Ökol D Tiere* **1**: 171–308.

Saito, A., Yoshioka, S. and Kinoshita, S. (2004). Reproduction of the *Morpho* butterfly's blue: Arbitration of contradicting factors. *Proc SPIE* **5526**: 188–194.

Saito, A., Miyamura, Y., Nakajima, M., Ishikawa, Y., Sogo, K., Kuwahara, Y. and Hirai, Y. (2006). Reproduction of the *Morpho* blue by nanocasting lithography. *J Vac Sci Technol B* **24**: 3248–3251.

Sakoda, K. (2001). *Optical Properties of Photonic Crystals*, Springer Verlag, Berlin.

Schmidt, W. J. and Ruska, H. (1961). Elektronenmikroskopische Untersuchung der Pigmentganula in den schillernden Federstrahlen der Taube *Columba trocaz* H. *Z Zellforsch* **55**: 379–388.

Schmidt, W. J. and Ruska, H. (1962a). Tyndallblau-Struktur von Federn im Elektronenmikroskop. *Z Zellforsch* **56**: 693–708.

Schmidt, W. J. and Ruska, H. (1962b). Über das schillernde Federmelanin bei *Heliangelus* und *Lophophorus*. *Z Zellforsch* **57**: 1–36.

Schmidt, W. J. (1943). Die Mosaikschuppen des *Teinopalpus imperialis* Hope, ein neues Muster schillernder Schmetterlingsschuppen. *Z Morph Ökol Tiere* **39**: 176–216.

Schmidt, W. J. (1952). Wie entstehen die Schillerfarben der Federn? *Naturwiss* **39**: 313–318.

Schultz, T. D. and Rankin, M. A. (1985a). The ultrastructure of the epicuticular interference reflectors of tiger beetles (*Cicindela*). *J Exp Biol* **117**: 87–110.

Schultz, T. D. and Rankin, M. A. (1985b). Developmental changes in the interference reflectors and colorations of tiger beetles (*Cicindela*). *J Exp Biol* **117**: 111–117.

Schultz, T. D. (1986). Role of structural colors in predator avoidance by tiger beetles of the genus *Cincindela* (Coleoptera: Cicindelidae). *Bull ESA* **Fall**: 142–146.

Shand, J. and Lythgoe, J. N. (1987). Light-indcued changes in corneal iridescence in fish. *Vision Res* **27**: 303–305.

Shand, J. (1988). Corneal iridescence in fishes: Light-induced colour changes in relation to structure. *J Fish Biol* **32**: 625–632.

Shashar, N., Borst, D. T., Ament, S. A., Saidel, W. M., Smolowitz, R. M. and Hanlon, R. T. (2001). Polarization reflecting iridophores in the arms of the squid *Loligo pealeii. Biol Bull* **201**: 267–268.

Shawkey, M. D., Estes, A. M., Siefferman, L. M. and Hill, G. E. (2003). Nanostructure predicts intraspecific variation in ultraviolet-blue plumage colour. *Proc R Soc Lond B* **270**: 1455–1460.

Shawkey, M. D. and Hill, G. E. (2004). Feathers at a fine scale. *Auk* **121**: 652–655.

Shawkey, M. D. and Hill, G. E. (2006). Significance of a basal melanin layer to production of non-iridescent structural plumage color: Evidence from an amelanotic Steller's jay (*Cyanocitta stelleri*). *J Exp Biol* **209**: 1245–1250.

Sibley, C. G., Ahlquist, J. E. and Monroe, B. L., Jr. (1988). A classification of the living birds of the world based on DNA-DNA hybridization studies. *Auk* **105**: 409–423.

Sibley, C. G. and Ahlquist, J. E. (1990). *Phylogeny and Classification of Birds. A Study in Molecular Evolution*, Yale Univ. Press, New Haven & London.

Siefferman, L. and Hill, G. E. (2003). Structural and melanin coloration indicate parental effort and reproductive success in male eastern bluebirds. *Behav Ecol* **14**: 855–861.

Silver, J., Withnall, R., Ireland, T. G. and Fern, G. R. (2005). Novel nanostructured phosphor materials cast from natural *Morpho* butterfly scales. *J Mod Opt* **52**: 999–1007.

Simon, H. (1971). *The Splendor of Iridescence — Structural Colors in the Animal World*, Dodd, Mead & Company, New York.

Srinivasarao, M. (1999). Nano-optics in the biological world: Beetles, butterflies, birds, and moths. *Chem Rev* **99**: 1935–1961.

Stavenga, D. G., Stowe, S., Siebke, K., Zeil, J. and Arikawa, K. (2004). Butterfly wing colours: Scale beads make white pierid wings brighter. *Proc R Soc Lond B* **271**: 1577–1584.

Stavenga, D. G., Giraldo, M. A. and Hoenders, B. J. (2006a). Reflectance and transmittance of scattering scales stacked on the wings of pierid butterflies. *Opt Express* **14**: 4880–4890.

Stavenga, D G., Foletti, S., Palasantzas, G. and Arikawa, K. (2006b). Light on the moth-eye corneal nipple array of butterflies. *Proc R Soc B* **273**: 661–667.

Stavenga, D. G. (2002). Colour in the eyes of insects. *J Comp Physiol A* **188**: 337–348.

Steinbrecht, R. A., Mohren, W., Pulker, H. K. and Schneider, D. (1985). Cuticular interference reflectors in the golden pupae of danaine butterflies. *Proc R Soc Lond B* **226**: 367–390.

Steinbrecht, R. A. (1985). Fine structure and development of the silver and golden cuticle in butterfly pupae. *Tissue Cell* **17**: 745–762.

Steinbrecht, R. A. (2005). The golden pupae of butterflies — A special biological multi-layer reflector system. In: S. Kinoshita and S. Yoshioka (eds.), *Structural Colors in Biological Systems — Principles and Applications*, Osaka University Press, Osaka, pp. 233–247.

Sweeney, A., Jiggins, C. and Johnsen, S. (2003). Polarized light as a butterfly mating signal. *Nature* **423**: 31–32.

Tabata, H., Kumazawa, K., Funakawa, M., Takimoto, J. and Akimoto, M. (1996a). Microstructures and optical properties of scales of butterfly wings. *Opt Rev* **3**: 139–145.

Tabata, H., Hasegawa, T., Nakagoshi, M., Takikawa, S. and Tsusue, M. (1996b). Occurrence of biopterin in the wings of *Morpho* butterflies. *Cell Mol Life Sci* **52**: 85–87.

Tada, H., Mann, S. E., Miaoulis, I. N. and Wong, P. Y. (1999). Effects of a butterfly scale microstructure on the iridescent color observed at different angles. *Opt Express* **37**: 1579–1584.

Takeuchi, I. K. (1976). Electron microscopy of two types of reflecting chromatophores (Iridophores and leucophores) in the guppy, *Lebistes reticulatus* Peters. *Cell Tissue Res* **173**: 17–27.

Tan, T. L., Wong, D. and Lee, P. (2004). Iridescence of a shell of mollusk *Haliotis Glabra*. *Opt Express* **12**: 4847–4854.

Tayeb, G., Gralak, B. and Enoch, S. (2003). Structural colors in nature and butterfly-wing modeling. *Opt Photonics News* **Feb.**: 40–49.

Tilley, R. J. D. and Eliot, J. N. (2002). Scale microstructure and its phylogenetic implications in lycaenid butterflies (Lepidoptera, Lycaenidae). *Trans Lepid Soc Japan* **53**: 153–180.

Townsend, V. R., Jr. and Felgenhauer, B. E. (1999). Ultrastructure of the cuticular scales of lynx spiders (Araneae, Oxyopidae) and jumping spiders (Araneae, Salticidae). *J Morph* **240**: 77–92.

Vértesy, Z., Bálint, Zs., Kertész, K., Vigneron, J. P., Lousse, V. and Biró, L. P. (2006). Wing scale microstructures and nanostructures in butterflies — natural photonic crystals. *J Microsc* **224**: 108–110.

van de Hulst, H. C. (1957). *Light Scattering by Small Particles*, Wiley, New York; republicated in 1981 by Dover, New York.

Veron, J. E. N., O'Farrell, A. F. and Dixon, B. (1974). The fine structure of Odonata chromatophores. *Tissue Cell* **6**: 613–626.

Vigneron, J. P., Colomer, J.-F., Rassart, M., Ingram, A. L. and Lousse, V. (2006). Structural origin of the colored reflections from the black-billed magpie feathers. *Phys Rev E* **73**: 021914.

Vogelmann, T. C. (1993). Plant tissue optics. *Annu Rev Plant Physiol Plant Mol Biol* **44**: 231–251.

Vukusic, P., Sambles, J. R., Lawrence, C. R. and Wootton, R. J. (1999). Quantified interference and diffraction in single *Morpho* butterfly scales. *Proc R Soc Lond B* **266**: 1403–1411.

Vukusic, P., Sambles, J. R. and Lawrence, C. R. (2000). Colour mixing in wing scales of a butterfly. *Nature* **404**: 457.

Vukusic, P., Sambles, R., Lawrence, C. and Wakely, G. (2001a). Sculpted-multilayer optical effects in two species of *Papilio* butterfly. *Appl Opt* **40**: 1116–1125.

Vukusic, P., Sambles, J. R., Lawrence, C. R. and Wootton, R. J. (2001b). Now you see it — now you don't. *Nature* **410**: 36.

Vukusic, P., Sambles, J. R., and Lawrence, C. R. (2004a). Structurally assisted blackness in butterfly scales. *Proc R Soc Lond B (Suppl)* **271**: S237–S239.

Vukusic, P., Wootton, R. J. and Sambles, J. R. (2004b). Remarkable iridescence in the hindwings of the damselfly *Neurobasis chinensis chinensis* (Linnaeus) (Zygoptera: Calopterygidae). *Proc R Soc Lond B* **271**: 595–601.

Vukusic, P., Hallam, B. and Noyes, J. (2007). Brilliant whiteness in ultrathin beetle scales. *Science* **315**: 348.

Vukusic, P. and Hooper, I. (2005). Directionally controlled fluorescence emission in butterflies. *Science* **310**: 1151.

Vukusic, P. and Sambles, R. (2000). Colour effects in bright butterflies. *JSDC* **116**: 376–380.

Vukusic, P. and Sambles, J. R. (2003). Photonic structures in biology. *Nature* **424**: 852–855.

Vukusic, P. (2003). Natural coatings. In: N. Kaiser and H. K. Pulker (eds.), *Optical Interference Coatings*, Springer, Berlin, New York, pp. 1–34.

Vukusic, P. (2005). Structural colour effects in Lepidoptera. In: S. Kinoshita and S. Yoshioka (eds.), *Structural Colors in Biological Systems — Principles and Applications*, Osaka University Press, Osaka, pp. 95–112.

Vukusic, P. (2006). Structural colour in Lepidoptera. *Curr Biol* **16**: R621–R623.

Vulinec, K. (1997). Iridescent dung beetles: A different angle. *Florida Entomol* **80**: 132–141.

Watanabe, K., Hoshino, T., Kanda, K., Haruyama, Y., Kaito, T. and Matsui, S. (2005a). Optical measurement and fabrication from a *Morpho*-butterfly-scale quasistructure by focused ion beam chemical vapor deposition. *J Vac Sci Technol B* **23**: 570–574.

Watanabe, K., Hoshino, T., Kanda, K., Haruyama, Y. and Matsui, S. (2005b). Brilliant blue observation from a *Morpho*-butterfly-scale quasi-structure. *Jpn J Appl Phys* **44**: L48–L50.

Watson, G. S. and Watson, J. A. (2004). Natural nano-structures on insects — Possible functions of ordered arrays characterized by atomic force microscopy. *Appl Surf Sci* **235**: 139–144.

Welch, V., Vigneron, J. P., Lousse, V. and Parker, A. R. (2006). Optical properties of the iridescent organ of the comb-jellyfish *Beroë cucumin* (Ctenophora). *Phys Rev E* **73**: 041916.

Welch, V. L. (2005). Photonic crystals in biology. In: S. Kinoshita and S. Yoshioka (eds.), *Structural Colors in Biological Systems — Principles and Applications*, Osaka University Press, Osaka, pp. 53–71.

Wickham, S., Large, M. C. J., Poladian, L. and Jermiin, L. S. (2006). Exaggeration and suppression of iridescence: The evolution of two-dimensional butterfly structural colours. *J R Soc Interface* **3**: 99–109.

Wilson, S. J. and Hutley, M. C. (1982). The optical properties of 'moth eye' antireflection surfaces. *Opt Acta* **29**: 993–1009.

Wong, T.-H., Gupta, M. C., Robins, B. and Levendusky, T. L. (2003). Color generation in butterfly wings and fabrication of such structures. *Opt Lett* **28**: 2342–2344.

Wright, W. D. (1963). The rays are not coloured. *Nature* **4887**: 1239–1244.

Xi, J.-Q., Schubert, M. F., Kim, J. K., Schubert, E. F., Chen, M., Lin, S.-Y., Liu, W. and Smart, J. A. (2007). Optical thin-film materials with low refractive index for broadband elimination of Fresnel reflection. *Nat Photon* **1**: 176–179.

Yin, H., Shi, L., Sha, J., Li, Y., Qin, Y., Dong, B., Meyer, S., Liu, X., Zhao, L. and Zi, J. (2006). Iridescence in the neck feathers of domestic pigeons. *Phys Rev E* **74**: 051916.

Yoshida, A., Motoyama, M., Kosaku, A. and Miyamoto, K. (1996). Nanoprotuberance array in the transparent wing of a hawkmoth, *Cephonodes hylas*. *Zool Sci* **13**: 525–526.

Yoshida, A., Motoyama, M., Kosaku, A. and Miyamoto, K. (1997). Antireflective nanoprotuberance array in the transparent wing of a hawkmoth, *Cephonodes hylas*. *Zool Sci* **14**: 737–741.

Yoshida, A. (2005). Antireflective microstructure on the wings of butterflies and moths. In: S. Kinoshita and S. Yoshioka (eds.), *Structural Colors in Biological Systems — Principles and Applications*, Osaka University Press, Osaka, pp. 249–264.

Yoshioka, S., Kinoshita, S. and Saito, A (2004). Coloration mechanisms of *Morpho* butterflies and production of *Morpho* color plate. *Oyo Butsuri (in Japanese)* **73**: 939–942.

Yoshioka, S., Nakamura, E. and Kinoshita, S. (2007). Origin of two-color iridescence in rock dove's feather. *J Phys Soc Jpn* **76**: 013801.

Yoshioka, S. and Kinoshita, S. (2002). Effect of macroscopic structure in iridescent color of the peacock feathers. *Forma* **17**: 169–181.

Yoshioka, S. and Kinoshita, S. (2004). Wavelength-selective and anisotropic light-diffusing scale on the wing of the *Morpho* butterfly. *Proc R Soc Lond B* **271**: 581–587.

Yoshioka, S. and Kinoshita, S. (2005a). Inter-scale mechanism in blue-light reflection of a *Morpho* butterfly. In: S. Kinoshita and S. Yoshioka (eds.), *Structural Color in Biological Systems — Principles and Applications*, Osaka University Press, Osaka, pp. 141–152.

Yoshioka, S. and Kinoshita, S. (2005b). Structural color of peacock feathers. In: S. Kinoshita and S. Yoshioka (eds.), *Structural Colors in Biological Systems — Principles and Applications*, Osaka University Press, Osaka, pp. 195–208.

Yoshioka, S. and Kinoshita, S. (2006a). Single-scale spectroscopy of structurally colored butterflies: Measurements of quantified reflectance and transmittance. *J Opt Soc Am A* **23**: 134–141.

Yoshioka, S. and Kinoshita, S. (2006b). Structural or pigmentary? Origin of the distinctive white stripe on the blue wing of a *Morpho* butterfly. *Proc R Soc Lond B* **273**: 129–134.

Yoshioka, S. and Kinoshita, S. (2007). Polarization-sensitive color mixing in the wing of the Madagascan sunset moth. *Opt Express* **15**: 2691–2701.

Young, A. M. (1971). Wing coloration and reflectance in *Morpho* butterflies as related to reproductive behavior and escape from avian predators. *Oecologia* **7**: 209–222.

Zarnescu, O. (2007). Ultrastructure of the skin melanophores and iridophores in paddlefish, *Polyodon spathula*. *Micron* **38**: 81–84.

Zi, J., Yu, X., Li, Y., Hu, X., Xu, C., Wang, X., Liu, X. and Fu, R. (2003). Coloration strategies in peacock feathers. *Proc Natl Acad Sci* **100**: 12576–12578.

Appendix A

Table A.1 List of butterfly species quoted in papers concerning structural colors. Scientific names given in the column "Genus/species" are those quoted in papers as they are. Where necessary, we replace them with the proper names employed in recent classification, which are shown in column "Reference." Column "Type/EM" shows the reference numbers, in which electron microscopic data are available, together with the type of microstructure as shown in Fig. 3.35. Note that the type of microstructure is only tentatively shown, because we only know it from the images shown in the paper and also no evidence is usually given whether it truly contributes to the iridescence or not. The reference numbers, in which electron microscopic data for corneal nipple (CN) and pupal cuticle (PC) are available, are also shown. The last column indicates old names that should be replaced owing to synonym, misidentification, transfer, and so on.

Family	Subfamily	Genus/species	Type/EM	Reference	Synonym/ misidentification
Hesperiidae	Pyrrhopyginae	*Elbella polyzona*		*Parelbella polyzona*	
		Parelbella polyzona	13	13,42	*Elbella polyzona*
		Pyrrhopyge venezuela?	13	13	
	Pyrginae	*Astraptes azul*		*Astraptes fulgerator azul*	
		Astraptes fulgerator azul	F1(1)	1	*Astraptes azul*
		Phocides lilea		*Phocides polybius lilea*	
		Phocides polybius lilea	D(13)	13	*Phocides lilea*
		Phocides sp.	13	13	
		Urbanus proteus	G(42),40	40,42	
	Hesperiinae	*Hylephila phylaeus*		*Hylephila phyleus*	
		Hylephila phyleus		178	*Hylephila phylaeus*
		Ochlodes venata	CN(112)	112	
Papilionidae	Parnanassiinae	*Parnassius apollo*	CN(356),356	356	
	Papilioninae	*Atrophaneura aristolochiae*	120	108,120	*Pachliopta aristolochiae*
		Battus belus		6	
		Battus philenor	D(6)	6,50	*Papilio philenor*

(*Continued*)

Table A.1 (*Continued*)

Family	Subfamily	Genus/species	Type/EM	Reference	Synonym/ misidentification
		Graphium agamemnon		6,47	*Papilio agamemnon*
		Graphium cyrnus		6	
		Graphium illyris		6	
		Graphium ridleyanus		6	
		Graphium sarpedon	6,96,D(120),A(120)	6,96,108,120	
		Graphium sp.	13	13	
		Graphium weiskei	356	6,356	
		Lamproptera curius	42	42	
		Ornithoptera arruana		*Troides priamus arruana*	
		Ornithoptera brookiana		*Troides brookiana*	
		Ornithoptera goliath		*Troides goliath*	
		Ornithoptera paradisea		*Troides paradisea*	
		Ornithoptera poseidon		*Troides priamus poseidon*	
		Ornithoptera priamus		*Troides priamus*	
		Ornithoptera urvilliana		*Troides urvillianus*	
		Ornithopteren croesus		*Troides croesus*	
		Pachliopta aristolochiae		*Atrophaneura aristolochiae*	
		Papilio agamemnon		*Graphium agamemnon*	
		Papilio antimachus	B(29)	29	
		Papilio asterias		*Papilio polyxenes asterius*	

(*Continued*)

Table A.1 (*Continued*)

Family	Subfamily	Genus/species	Type/EM	Reference	Synonym/ misidentification
		Papilio blumei	F1(8,356)	8,21,25,26,356	
		Papilio blumei fruhstoferi		38	
		Papilio buddha		21,25	
		Papilio crino		21,25	
		Papilio daedalus		*Papilio palinurus daedalus*	
		Papilio epiphorbas		6	
		Papilio erithalion		*Parides erithalion*	
		Papilio hoppo		6	
		Papilio karna karna	F1(30)	30	
		Papilio lorquinianus	F1(6)	6	
		Papilio machaon	356	48,356	
		Papilio memmon		108	
		Papilio nephalion		*Parides nephalion*	
		Papilio nireus	B(13,27)	6,13,27	
		Papilio oribazus	B(2)	2,6	
		Papilio palinurus	F1(6,14,21,23,25, 32,81,86)	6,14,21,23,24, 25,32,81,85,86	
		Papilio palinurus daedalus	F1(13,30,42),87	13,30,42,87	*Papilio daedalus*
		Papilio parinda		*Papilio polymnestor parinda*	
		Papilio paris	F1(59)	6,8,59	
		Papilio peranthus		8,48,50	
		Papilio philenor		*Battus philenor*	
		Papilio polymnestor parinda		47	*Papilio parinda*

(*Continued*)

Table A.1 (*Continued*)

Family	Subfamily	Genus/species	Type/EM	Reference	Synonym/ misidentification
		Papilio polyxenes asterius		50	*Papilio asterias*
		Papilio ulysses	F1(6,14,21,81,356), B(14,98,109)	6,14,21,47,81,98,109, 171,356	
		Papilio ulysses autocylus	F1(17)	17	
		Papilio xuthus		37,108	
		Papilio zalmoxis	B(1,6,17,29,42)	1,6,17,29,42,50	
		Parides erithalion		48	*Papilio erithalion*
		Parides erithalion polyzelus	A(6)	6,42	*Parides polyzelus*
		Parides nephalion		6	*Papilio nephalion*
		Parides neophilus olivencius		50	
		Parides polyzelus		*Parides erithalion polyzelus*	
		Parides sesostris	G(1,6,13,14,17,32,81,86)	1,2,6,13,14,17,32,81,86	
		Teinopalpus imperialis	G(2,10,42,85)	2,10,42,92	
		Trogonoptera brookianus		*Troides brookiana*	
		Troides aeacus		178	
		Troides brookiana	6,A(17)	6,17,67	*Ornithoptera brookiana, Trogonoptera brookianus*
		Troides croesus		47,48	*Ornithopteren croesus*
		Troides goliath		48	

(*Continued*)

Table A.1 (*Continued*)

Family	Subfamily	Genus/species	Type/EM	Reference	Synonym/ misidentification
		Troides magellanus	A(14)	14,67,91	*Ornithoptera goliath*
		Troides paradisea		171	*Ornithoptera paradisea*
		Troides priamus	A(6),87	6,87	
		Troides priamus arruana	59	59	*Ornithoptera arruana*
		Troides priamus hecuba	F1(17),A(17)	17	
		Troides priamus poseidon	F1(356),A(356)	47,48,50,94,171,190,356	*Ornithoptera poseidon*
		Troides priamus urvillianus	F1(17),A(17)	17,47,48,171	*Troides urvillianus, Ornithoptera urvilliana*
		Troides urvillianus		*Troides priamus urvillianus*	
Pieridae		*Pieridae sp.*		94	
	Pierinae	*Anthocharis cardamines*		108	
		Colotis phlegyas		47	*Teracolus phlegyas*
		Delias nigrina	EG(31)	31	
		Hebomoia glaucippe	EG(356)	47,356	*Hebonia glocippe*
		Hebomoia glaucippe vossi		356	
		Hebonia glocippe		*Hebomoia glaucippe*	
		Pieris brassicae	EG(356)	31,123,178,356	
		Pieris napi	EG(118)	118,131	
		Pieris protodice		*Pontia protodice*	
		Pieris rapae	EG(28,31),102,CN(108)	28,31,102,108,356	

(*Continued*)

Table A.1 (*Continued*)

Family	Subfamily	Genus/species	Type/EM	Reference	Synonym/ misidentification
		Pieris rapae rapae	EG(31)	31	
		Rhabdodryas trite		7	
		Teracolus phlegyas		*Colotis phlegyas*	
	Coliadinae	*Aphrissa boisduvalii*		7	
		Aphrissa statira		7	
		Catopsilia argante		*Phoebis argante?*	
		Catopsilia florella	356	356	
		Catopsilia hersilia		*Phoebis argante larra?*	
		Colias crocea			
		Colias croceus	EG(356),A(356)	356	*Colias crocea*
		Colias eurytheme	A(1,2,3,16,40,41,42),E(1,42)	1,2,3,4,16,40,41,42,95	
		Colias philodice		16,40	
		Eurema lisa	A(4,41),EG(40)	4,40,41	
		Gonepteryx aspasia		*Gonepteryx mahaguru aspasis*	
		Gonepteryx cleopatra	A(356),EG(356)	356	
		Gonepteryx rhamni	CN(112),15	9,15,37,112,120,122	
		Phoebis agarithe		7	
		Phoebis argante		7	*Catopsilia argante?*
		Phoebis avellaneda		7	
		Phoebis neocypris		7	
		Phoebis neocypris rurina		7	*Phoebis rurina*
		Phoebis philea		7	
		Phoebis philea thalestris		7	

(*Continued*)

Table A.1 (*Continued*)

Family	Subfamily	Genus/species	Type/EM	Reference	Synonym/ misidentification
		Phoebis rurina		*Phoebis neocypris rurina*	
		Phoebis sennae marcellina		7	
		Phoebis sp.	A(7),D(7),EG(7)	7	
		Prestonia clarki		7	
Lycaenidae	Riodininae	*Abisara echerius*		101	
		Ancyluris inca miranda		48	*Ancyluris miranda*
		Ancyluris meliboeus	A(14,22,81,91,101)	14,22,81,91,101	
		Ancyluris miranda		*Ancyluris inca miranda*	
		Calliona sp.		*Calospila sp.*	
		Calospila sp.	2	2	*Calliona sp.*
		Diorina periander		*Rhetus periander*	
		Hamearis lucina		101	
		Helicopis cupido		50,94	
		Lasaia narses		*Lasaia sula*	
		Lasaia sula	A(101)	101	*Lasaia narses*
		Lyropteryx lyra	A(101)	101	
		Menander pretus picta		48	*Tharops picta*
		Necyria duellona	87	87	
		Paralaxita telesia		101	
		Rhetus periander	A(101)	101	*Diorina periander*
		Tharops picta		*Menander pretus picta*	
	Poritiinae	*Poritia sumatrae*	F2(101)	101	
	Miletinae	*Allotinus subviolaceus*	101	101	

(*Continued*)

Table A.1 (*Continued*)

Family	Subfamily	Genus/species	Type/EM	Reference	Synonym/ misidentification
	Curetinae	*Curetis bulis stigmata*	101	101	
	Lycaeninae	*Amblypodia eumolphus*		*Arhopala eumolphus*	
		Ancema blanka		101	
		Anthene lycaenina	D(101),F1(101)	101	
		Arawacus meliboeus		101	
		Arcas ducalis		48	*Thecla ducalis*
		Arcas imperialis		48	*Thecla imperialis*
		Arhopala aurea	F1(101)	101	
		Arhopala eumolphus		47,101	*Amblypodia eumolphus*
		Arhopala pseudocentaurus	F1(101)	101	
		Arhopala sp.	F1(6)	6	
		Arhopara japonica japonica	F1(85)	85	*Narathura japonica japonica*
		Artipe eryx	A(101)	101	
		Atlides halesus	F2(101)	101	
		Callophrys (Mitoura) siva siva		*Mitoura gryneus siva*	
		Callophrys avis	G(101)	101	
		Callophrys dumetorum	G(17)	17	
		Callophrys rubi	G(39,40,41,85), F2?(101),14	14,39,40,41,47, 85,87,92,101,172	*Thecla rubi*
		Callophrys siva		*Mitoura gryneus siva*	
		Callophrys sp.		2	
		Candalides margarita	F2(101)	101	
		Catapaecilma major	F2(101)	101	

(*Continued*)

Table A.1 (*Continued*)

Family	Subfamily	Genus/species	Type/EM	Reference	Synonym/ misidentification
		Catapaecilma sp.	G(39)	39	
		Celastrina argiolus	F2(101)	101	
		Celastrina argiolus ladonides	F2(85)	85	
		Celastrina ladon	F2(2,5,17,40),42	2,5,17,40,42	
		Chliaria othona	A(101),F2(101)	101	
		Chrysophanus phlaeas		*Lycaena phlaeas*	
		Chrysophanus virgaureae		*Lycaena virgaureae*	
		Chrysoritis chrysaor	F1(101)	101	*Niphanda tesselata, Poecilmitis chrysaor*
		Chrysozephyrus ataxus		20	
		Chrysozephyrus ataxus kirishimaenis	F1(28)	28	
		Chrysozephyrus aurorinus		*Chrysozephyrus brillantinus*	*Chrysozephyrus aurorinus*
		Chrysozephyrus brillantinus		20	
		Chrysozephyrus hisamatsusanus		20	
		Chrysozephyrus smaragdinus		20	
		Cigaritis nipalicus	A(101)	101	*Spindasis nipalicus*
		Cigaritis syama	A(101)	101	*Spindasis syama*
		Cupido argiades		108	*Everes argiades*

(*Continued*)

Table A.1 (*Continued*)

Family	Subfamily	Genus/species	Type/EM	Reference	Synonym/ misidentification
		Cupido minimus	F2(101)	101	
		Cyanophrys herodotus	42	42	*Thecla herodotus*
		Denivia hemon		*Theritas hemon*	
		Drina maneia	F2(101)	101	
		Drupadia ravindra	F1(101)	101	
		Eumaeus minijas	D(101),F1(101)	101	*Eumaeus minyas*
		Eumaeus minyas		*Eumaeus minijas*	
		Evenus coronata	F1(8)	8	*Thecla coronata, Suneve coronata*
		Evenus regalis		48(?)	*Thecla regalis?*
		Everes argiades		*Cupido argiades*	
		Favonius quercus	F1(101)	94,101	*Quercusia quercus, Zephyrus quercus, Thecla quercus*
		Glaucopsyche alexis	F2(101)	101	
		Heliophorus androcles	F1(101)	101	
		Heliophorus sena	F1(97)	97	
		Heliophorus tamu		97	
		Heodes virgaureae		*Lycaena virgaureae*	
		Horaga syrinx	F2(101)	101	
		Hypochrysops sp.	G(39)	39	
		Hypolycaena erylus	A(101)	101	
		Iraota rochana	F1(101)	101	
		Jalmenus evagoras		47	
		Jamides bochus		47	

(*Continued*)

Table A.1 (*Continued*)

Family	Subfamily	Genus/species	Type/EM	Reference	Synonym/ misidentification
		Jamides caerulea	F1(101)	101	*Jamides caeruleus*
		Jamides caeruleus		*Jamides caerulea*	
		Laeosopis evippus		*Laeosopis roboris*	
		Laeosopis roboris	F1(101)	101	*Laeosopis evippus*
		Lycaeides argyrognomon		*Plebejus argyrognomon*	
		Lycaena alciphron	F2?(101)	97,101	
		Lycaena corydon		*Polyommatus corydonius*	
		Lycaena epixanthe	F2(2,5)	2,5	
		Lycaena heteronea		5,97	
		Lycaena icarus		*Polyommatus icarus*	
		Lycaena phlaeas	356	19,171,356	*Chrysophanus phlaeas*
		Lycaena rauparaha	121	121	
		Lycaena rubida		*Lycaena rubidus*	
		Lycaena rubidus	G(5,42)	5,42	*Lycaena rubida*
		Lycaena salustius	F1(121)	121	
		Lycaena virgaureae	98,101	94,98,101	*Heodes virgaureae, Chrysophanus virgaureae*
		Maculinea teleius		122	
		Mitoura grynea		*Mitoura gryneus*	
		Mitoura gryneus	G(1,2,5,17,42,85)	1,2,5,17,42,85	*Mitoura grynea*
		Mitoura gryneus siva	G(34)	34	*Callophrys (Mitoura) siva siva, Callophrys siva*

(*Continued*)

Table A.1 (*Continued*)

Family	Subfamily	Genus/species	Type/EM	Reference	Synonym/ misidentification
		Mitoura spp.		2	
		Narathura japonica japonica		*Arhopala japonica japonica*	
		Neozephyrus taxila	F1(120)	120	
		Niphanda tesselata		*Chrysoritis chrysaor*	
		Paiwarria telemus		47	*Thecla telemus*
		Palaeochrysophanus candens		*Lycaena candens*	
		Palaeochrysophanus hippothoe		*Lycaena hippothoe*	
		Parrhasius m-album		17	
		Plebejus argyrognomon		122	*Lycaeides argyrognomon*
		Poecilmitis chrysaor		*Chrysoritis chrysaor*	
		Polyommatus bellargus	F2(98)	97,98	
		Polyommatus coridon	F1(101)	50,101	
		Polyommatus corydonius		171	*Lycaena corydon*
		Polyommatus daphnis	F1(101)	33,101	
		Polyommatus daphnis versicolor		97	*Polyommatus versicolor*
		Polyommatus escheri	F2(101)	101	
		Polyommatus icarus	F2(97,98,356)	47,50,97,98,171,356	*Lycaena icarus*
		Polyommatus marcida		33	*Polyommatus marcidus*
		Polyommatus marcidus		*Polyommatus marcida*	
		Polyommatus sp.	F2(33)	33	

(*Continued*)

Table A.1 (*Continued*)

Family	Subfamily	Genus/species	Type/EM	Reference	Synonym/ misidentification
		Polyommatus versicolor		*Polyommatus daphnis versicolor*	
		Pseudolucia plumbea		97	
		Pseudolycaena damo	G(2)	2	*Thecla damo*
		Pseudolycaena marsyas	F2(97)	97	*Thecla marsyas*
		Pseudozizeeria maha	CN(108)	108	
		Quercusia quercus		*Favonius quercus*	
		Rapala manea	A(101)	101	
		Rapala pheritima		101	
		Rapala rhoecus	A(101)	101	
		Remelana jangala	F1(101)	101	
		Siderus tephraeus		*Strephonota tephraeus*	
		Sithon nedymond	A(101)	101	
		Spindasis nipalicus		*Cigaritis nipalicus*	
		Spindasis syama		*Cigaritis syama*	
		Strephonota tephraeus	F1(101)	101	*Siderus tephraeus*
		Suneve coronata		*Evenus coronata*	
		Surendra florimel	F1?(101)	101	
		Surendra vivarna amisena	F1(101)	101	
		Tajuria cippus	F2(101)	101	
		Thecla coronata		*Evenus coronata*	
		Thecla damo		*Pseudolycaena damo*	
		Thecla ducalis		*Arcas ducalis*	
		Thecla gewölbt?		48	
		Thecla hemon		*Theritas hemon*	

(*Continued*)

Table A.1 (*Continued*)

Family	Subfamily	Genus/species	Type/EM	Reference	Synonym/ misidentification
		Thecla herodotus		*Cyanophrys herodotus*	
		Thecla imperialis		*Arcas imperialis*	
		Thecla marsyas		*Pseudolycaena marsyas*	
		Thecla quercus		*Favonius quercus*	
		Thecla regalis		*Evenus regalis?*	
		Thecla rubi		*Callophrys rubi*	
		Thecla telemus		*Paiwarria telemus*	
		Theritas hemon		97	*Thecla hemon, Denivia hemon*
		Theritas paupera		97	
		Tomares ballus	G(13),F2(101)	13,101	
		Udara sp.	G(39)	39	*Vaga sp.*
		Vaga sp.		*Udara sp.*	
		Zephyrus quercus		*Favonius quercus*	
		Zesius chrysomallus		47	
Nymphalidae	Heliconiinae	*Acraea sp.*		94	
		Argynnis latonia		*Issoria lathonia?*	
		Argynnis sp.		50,94	
		Dione juno		47,171	
		Heliconius charithonia	PC(100)	100	*Heliconius charitonius*
		Heliconius charitonius		*Heliconius charithonia*	
		Heliconius cydno		18	
		Heliconius melpomene	31	31,108	

(*Continued*)

Table A.1 (*Continued*)

Family	Subfamily	Genus/species	Type/EM	Reference	Synonym/ misidentification
		Heliconius melpomene malleti		18	
		Issoria lathonia		48	*Argynnis latonia?*
		Philaethria dido	42	42	
	Nymphalinae	*Aglais urticae*		*Nymphalis urticae*	
		Araschnia prorsa		*Araschnia levana f. prorsa*	
		Hypolimnas salmacis		47	
		Inachis io		*Nymphalis io*	
		Junonia lemonias		122	*Precis lemonias*
		Kallima sp.		103	
		Kaniska canace		*Nymphalis canace*	
		Nymphalis c-album	CN(108)	48,108	*Polygonia c-album*
		Nymphalis c-aureum		108	
		Nymphalis io	CN(108)	108	*Inachis io*
		Nymphalis urticae	CN(103),PC(113)	103,113,149	*Aglais urticae, Vanessa urticae*
		Polygonia c-album		*Nymphalis c-album*	
		Precis lemonias		*Junonia lemonias*	
		Pyrameis atalanta		*Vanessa atalanta*	
		Salamis parhassus		47,171	
		Vanessa kershawi	CN(83,110)	83,110	
		Vanessa urticae		*Nymphalis urticae*	
	Limenitidinae	*Ageronia feronia*		*Hamadryas feronia*	
		Basilarchia arthemis		50	
		Batesia prola?		47	
		Callicore aegina		67	

(*Continued*)

Table A.1 (*Continued*)

Family	Subfamily	Genus/species	Type/EM	Reference	Synonym/ misidentification
		Callicore eluina		*Diaethria eluina*	
		Cynandra opis	36	36	
		Diaethria eluina		94	*Callicore eluina*
		Diaethria neglecta		67	
		Dynamine mylitta		47	
		Euphaedra sp.		108	
		Hamadryas feronia	CN(178)	178	*Ageronia feronia*
		Paulogramma peristera		*Paulogramma pyracmon*	
		Paulogramma pyracmon		67	*Paulogramma peristera*
		Perisama humboldtii		67	
	Charaxinae	*Anaea sp.*		50	
		Charaxes fulvescens		108	
		Euxanthe makefieldi?		108	
	Apaturinae	*Apatura ilia*		94	
		Apatura iris		171	
		Apatura seraphina		*Doxocopa seraphina*	
		Chlorippe laurentia		*Doxocopa laurentia*	
		Chlorippe seraphina		*Doxocopa seraphina*	
		Doxocopa laurentia		47,171	*Chlorippe laurentia*
		Doxocopa seraphina		50,94	*Chlorippe laurentia, Apatura seraphina*
		Sasakia charonda		120	

(*Continued*)

Table A.1 (*Continued*)

Family	Subfamily	Genus/species	Type/EM	Reference	Synonym/ misidentification
	Morphinae	*Caligo beltrao*	A(1)	1	
		Caligo braziliensis		*Caligo eurilochus brasiliensis*	
		Caligo eurilochus brasiliensis	CN(178)	178	*Caligo braziliensis*
		Caligo illioneus	D(1)	1,32	
		Caligo martia	1	1,67	
		Caligo memnon	D(1,42),A(2)	1,2,42	
		Caligo sp.	13	13	
		Caligo uranus	13,A(40)	13,40	
		Eryphanis aesacus	C(2,13,42)	2,13,42,67	
		Morpho achilles	A(51),59	47,48,51,59	
		Morpho adonis	A(28,51,55,64,85,356),59	28,51,55,59,61,64, 85,166,356	
		Morpho aega	A(17,51,52,55),59	17,45,48,51, 52,55,59,98	
		Morpho amathonte	A(40,55)	40,53,55	
		Morpho anaxibia	356	356	
		Morpho aurora	55	55	
		Morpho cisseis	55	55	
		Morpho cypris	A(11,63),51,356	11,19,47,48, 51,55,59,63,85, 99,356	
		Morpho deidamia granadensis		53	*Morpho granadensis*
		Morpho didius	A(14,23,55,64,65, 69,75,79,81,85),H(65)	14,23,55,64,65,66,67,69, 73,75,79,80,81,85	

(*Continued*)

Table A.1 (*Continued*)

Family	Subfamily	Genus/species	Type/EM	Reference	Synonym/ misidentification
		Morpho eugenia		166	
		Morpho godarti	356	356	
		Morpho godarti asarpai?	A(55)	55	
		Morpho granadensis		*Morpho deidamia granadensis*	
		Morpho helenor	A(55)	55	
		Morpho hercules	42	42	
		Morpho menelaus	A(1,13,40,42,52,55,56,57, 58,87,356),41	1,3,13,38,40, 41,42,46,47, 48,49,50,52, 55,56,57,58, 59,72,87, 166,171,356	
		Morpho peleides	59,A(74),262	40,53,54,59,74,262	
		Morpho portis	55	55,59	
		Morpho rhetenor	A(14,23,55,64,68, 69,71,72,81,82, 85,86),70,87	14,23,55,59, 64,68,69,70, 71,72,81,82, 85,86,87,166	
		Morpho spp.	A(73,76,88,90)	73,76,77,78, 83,84,88,89, 90,102,152	
		Morpho sulkowskyi	A(28,52,55,60,64,79, 85,356),37,59	28,37,47,48, 50,52,55,59, 60,61,62,64, 65,79,85,166,356	
		Morpho theseus		40	

(*Continued*)

Table A.1 (*Continued*)

Family	Subfamily	Genus/species	Type/EM	Reference	Synonym/ misidentification
		Morpho zephiritis	55,A(356)	55,356	
		Zeuxidia amethystus	A(2)	2	
	Satyrinae	*Bicyclus anynana*	CN(108)	108	
		Callitaera esmeralda		*Cithaerias andromeda*	
		Cercyonis alope		*Cercyonis pegala*	
		Cercyonis pegala		50	*Cercyonis alope*
		Cithaerias andromeda		47,171	*Callitaera esmeralda*
		Cithaerias menander		*Cithaerias pireta pireta*	
		Cithaerias pireta pireta	356	356	*Cithaerias menander*
		Mycalesis francisca		108	
		Ypthima argus		*Ypthima baldus*	
		Ypthima baldus		122	*Ypthima argus*
	Danainae	*Amauris niavius*		35,106	
		Amauris ochlea	PC(100,106)	35,100,106	
		Danaus chrysippus	PC(100)	35,100,123	
		Danaus genutia		122	*Salatura genutia*
		Danaus plexippus	CN(125,178)	35,123,125	
		Euploea algea deione		47,171	*Euploea deione*
		Euploea core	PC(35,100,106)	35,100,106,110	
		Euploea core amymone		123	
		Euploea deione		*Euploea algea deione*	
		Euploea dufresue		*Euploea mulciber dufresne*	
		Euploea midamus		67,123	

(*Continued*)

Table A.1 (*Continued*)

Family	Subfamily	Genus/species	Type/EM	Reference	Synonym/ misidentification
		Euploea mulciber dufresne	A(13)	13	*Euploea midamus*
		Idea leuconoe		85,122	
		Parantica sita	120	120,122	
		Salatura genutia		*Danaus genutia*	
		Edenopsis taxila?	32	32	
		Dixocopa chlotilda?	356	356	
		Nomiades arion?		171	

Table A.2 Avian species whose microstructures have been already clarified through electron microscopy. "Bl", "Bl/H", "B", "S", and "Bk" mean that the structural colors are produced by filled and hollow melanin granules located in barbule, spongy structure in barb, collagen fibers in skin, and lamellar structure in beak, respectively. Alphabetical and numerical expressions for melanin types should be referred to (Durrer, 1977) and (Durrer, 1986).

Infraclass	Order	Family/ subfamily	Genus/species	Common name	Type	EM	Reference
Eoaves	Craciformes	Cracidae	*Oreophasis derbianus*	Horned guan	Bl	StK-type(263) [type 3.2(254)]	245,263
		Megapodidae	*Megapodius freycinet*	Dusky scrubfowl	Bl	OSK-type(263) [type 2.2.1(254)]	245,263
	Galliformes	Phasianidae/ Tetraoninae	*Tetrao tetrix*	Black grouse	Bl	StK-type(263) [type 3.2(254)]	245,263,264
		Phasianidae/ Meleagridinae	*Agriocharis ocellata*	Ocellated turkey	Bl/H	ORK-type(263) [type 5.2(254)]	245,263
			Meleagris gallopavo	Wild turkey	Bl/H	ORK-type(263) [type 5.2(254)]	218,243,245,263
		Phasianidae/ Phasianinae	*Afropavo congensis*	Congo peafowl	Bl	peacock type(221), StG-type(263) [type 3.4(254)]	221,245,263
			Caloperdix oculea	Ferruginous partridge	Bl	OSK-type(263) [type 2.2.1(254)]	245,263
			Chrysolophus pictus	Golden pheasant	Bl	OStK-type(263) [type 3.2(254)]	245,263
			Francolinus francolinus	Black francolin	Bl	OSK-type(263) [type 2.2.1(254)]	245,263
			Gallus gallus	Red junglefowl	Bl/H	no structure in barb(241), ORE-type(263) [type 5.1(254)]	241,245,257,263
			Lophophorus impejanus	Himalayan monal	Bl/H	6–7 layers of hollow tubes(214), RS-type(263) [type 5.3(254)]	81,208,209,214, 243,245,263

(Continued)

Table A.2 (*Continued*)

Infraclass	Order	Family/ subfamily	Genus/species	Common name	Type	EM	Reference
			Lophura erythrophthalmus pyronotus	Crestless fireback pheasant	Bl	StK-type(263) [type 3.2(254)]	263
			Lophura ignita	Crested fireback pheasant	Bl	StK-type(263) [type 3.2(254)]	245,263
			Lophura leucomelanos melanotus	Kalij pheasant	Bl	StK-type(263) [type 3.2(254)]	263
			Melanoperdix nigra	Black partridge	Bl	OSK-type(263) [type 2.2.1(254)]	245,263
			Pavo cristatus	Indian peafowl	Bl	peacock type(194,195,268), lattice constants at various regions of feather(267), StG-type(263) [type 3.4(254)]	85,194,195,208, 218,232,245, 263,267,268
			Pavo muticus	Green peafowl	Bl	peacock type(199)	199,263
			Pavo sp.	peacock	Bl	peacock type(196)	166,196,197,200
			Phasianus versicolor	Green pheasant	Bl	OStK-type(263) [type 3.2(254)]	263
			Polyplectron bicalcaratum	Grey peacock pheasant	Bl	OStK-type(263) [type 3.2(254)]	245,263
			Rollulus roulroul	Crested partridge	Bl	OSK-type(263) [type 2.2.1(254)]	245,263
			Syrmaticus elliot i	Elliot's pheasant	Bl	StK-type(263) [type 3.2(254)]	245,263
			Tragopan caboti	Cabot's tragopan	S	random collagen array in skin(243)	243,245

(*Continued*)

Table A.2 (*Continued*)

Infraclass	Order	Family/ subfamily	Genus/species	Common name	Type	EM	Reference
			Tragopan melanocephalus	Western tragopan	Bl	OSK-type(263) [type 2.2.1(254)]	245,263
		Numididae	*Acryllium vulturinum*	Vulturine guineafowl	Bl	OSK-type(263) [type 2.2.1(254)]	245,263
			Numida meleagris	Helmeted guineafowl	S	random collagen array in skin(243)	218,243,245
	Anseriformes	Anatidae/ Oxyurinae	*Oxyura jamaicensis*	Ruddy duck	S	random collagen array in skin(243)	243,245
		Anatidae/ Anatinae	*Tadorna tadorna*	Common shelduck	Bl	OStK-type(263) [type 3.2(254)]	215,245,263
			Aix galericulata	Mandarin duck	Bl	duck type(215), StK-type(263) [type 3.2(254)]	215,245,263
			Aix sponsa	Wood duck	Bl	StK-type(263) [type 3.2(254)]	245,263
			Amazonetta brasiliensis	Brazilian teal	Bl	duck type(215), StK-type(263) [type 3.2(254)]	215,245,263
			Anas crecca	Common teal	Bl	duck type(268)	215,245,268
			Anas gibberifrons	Sunda teal	Bl	duck type(215), OStK-type(263) [type 3.2.1(254)]	215,245,263
			Anas platrhynchos	Mallard	Bl	duck type(215), StK-type(263) [type 3.2(254)]	215,232,245,263
			Anas specularis	Spectacled duck	Bl	duck type(215), OStK-type(263) [type 3.2.1(254)]	215,245,263
Neoaves	Piciformes	Megalaimidae	*Megalaima chrysopogon*	Gold-whiskered barbet	B	spongy structure in barb(245)	245

(*Continued*)

Table A.2 (*Continued*)

Infraclass	Order	Family/ subfamily	Genus/species	Common name	Type	EM	Reference
		Ramphastidae	*Ramphastos toco*	Toco toucan	S	random collagen array in skin(243)	243,245
	Galbuliformes	Galbulidae	*Galbula ruficauda*	Green-tailed jacamar	Bl/H	RS-type(263) [type 5.3(254)]	216,245,263
	Upupiformes	Phoeniculidae	*Phoeniculus purpureus*	Green woodhoopoe	Bl/H	RK-type(263) [type 5.2(254)]	245,263
	Trogoniformes	Trogonidae/ Apaloder-matinae	*Apaloderma narina*	Narina trogon	Bl/H	several layers of hollow tubes in barbule(216), RS-type(263) [type 5.3(254)]	216,245,263
		Trogonidae/ Trogoninae	*Pharomachrus mocinno*	Resplendent quetzal	Bl/H	several layers of hummingbird-type platelets(245)	152,245
			Pharomachrus mocinno mocinno	Resplendent quetzal	Bl/H	several layers of hummingbird-type platelets(216,258), KS-type(263) [type 6.3(254)]	216,258,263
			Pharomachrus pavoninus helactin	Pavonine quetzal	Bl/H	several layers of hummingbird-type platelets(216), KS-type(263) [type 6.3(254)]	216,263
			Priotelus temnurus temnurus	Cuban trogon	Bl/H	2 layers of hollow tube in barbule(216), RS-type(263) [type 5.3(254)]	216,263
			Trogon massena	Slaty-tailed trogon	Bl/H	10 layers of hollow tube in barbule(245)	245

(*Continued*)

Table A.2 (*Continued*)

Infraclass	Order	Family/subfamily	Genus/species	Common name	Type	EM	Reference
			Trogon rufus sulphureus	Black-throated trogon	Bl/H	3–4 layers of hollow tube in barbule(216), RK-type(263) [type 5.2(254)]	216,263
			Trogon violaceus	Violaceous trogon	Bl/H	6 arrays of close-packed hollow tubes in barbule(268)	245,268
			Trogon violaceus crissalis	Violaceous trogon	Bl/H	6–7 arrays of hollow tubes in barbule(216), RK-type(263)	216,263
			Harpactes reinwardtii mackloti	Blue-tailed trogon	Bl/H	2 layers of hollow tubes in barbule(216), RS-type(263) [type 5.3(254)]	216,263
	Coraciiformes	Alcedinidae	*Alcedo atthis*	Common kingfisher	B	spongy structure in barb(85,224,268)	85,224,232, 245,268
		Dacelonidae	*Halcyon pileata*	Black-capped kingfisher	B	spongy structure in barb(245)	245
		Cerylidae	*Chloroceryle americana cabanisii*	Green kingfisher	Bl/H	ORE-type(263) [type 5.1(254)]	263
		Meropidae	*Nyctyornis amictus*	Blue-bearded bee-eater	B	spongy structure in barb(245)	244,245,280
	Cuculiformes	Cuculidae	*Chrysococcyx cupreus*	African emerald cuckoo	Bl	OStS-type(263) [type 3.3(254)]	245,263
			Chrysococcyx cupreus cupreus	African emerald cuckoo	Bl	duck type(220)	220
	Psittaciformes	Psittacidae	*Agapornis roseicollis*	Rosy-faced lovebird	B	spongy structure in barb(229,235,247,251,253,268)	229,235,245, 251,253,261, 268,247

(*Continued*)

Table A.2 (*Continued*)

Infraclass	Order	Family/ subfamily	Genus/species	Common name	Type	EM	Reference
			Ara ararauna	Blue-and-yellow macaw	B	spongy structure in barb(213)	213,245
			Chalcopsitta atra	Black lory	B	spongy structure in barb(224,247)	224,225,245,247
			Melopsittacus undulatus	Budgerigar	B	spongy structure in barb(224,229)	224,229,244,245
			Trichoglossus haematodus	Rainbow lorikeet	B	spongy structure in barb(247,268)	245,247,268
	Apodiformes	Apodidae	*Collocalia esculenta*	Glossy swiftlet	Bl	StK-type(263) [type 3.2(254)]	245,263
		Hemiprocnidae	*Hemiprocne comata*	Whiskered treeswift	Bl	PK-type(263) [type 4.2(254)]	245,263
	Trochiliformes	Trochilidae/ Trochilinae	*Amazilia cyanura*	Blue-tailed hummingbird	Bl/H	KK-type(263) [type 6.2(254)]	245,263
			Calypte anna	Anna's hummingbird	Bl/H	[KK-type(263)] [type 6.2(254)]	263
			Chlorestes notatus	Blue-chinned sapphire	Bl/H	12 layers of hummingbird-type platelets in barbule(268)	245,268
			Chrysolampis mosquitus	Ruby-topaz hummingbird	Bl/H	hummingbird-type platelets in barbule(210), [KK-type(263)] [type 6.2(254)]	201,210,245,263
			Clytolaema rubricauda	Fawn-breasted brilliant	Bl/H	14 layers of hummingbird-type platelets(201,210), [KK-type(263)] [type 6.2(254)]	87,201,210, 245,263

(*Continued*)

Table A.2 (*Continued*)

Infraclass	Order	Family/ subfamily	Genus/species	Common name	Type	EM	Reference
			Heliangelus strophianus	Gorgeted sunangel	Bl/H	layers of hummingbird-type platelets(214), [KK-type(263)] [type 6.2(254)]	209,214, 245,263
			Heliangelus viola	Purple-throated sunangel	Bl/H	hummingbird-type platelets?(210)	87,201, 210,245
			Lampornis clemenciae	Blue-throated hummingbird	Bl/H	hummingbird-type platelets?(210), [KK-type(263)] [type 6.2(254)]	201,210, 263,245
			Thalurania furcata	Fork-tailed woodnymph	Bl/H	[KK-type(263)] [type 6.2(254)]	201,245,263
			Topaza pyra	Fiery topaz	Bl/H	[KK-type(263)] [type 6.2(254)]	201,263
	Musophagi- formes	Musophagidae/ Musophagi- nae	*Tauraco corythaix emini*	Knysna turaco	Bl	StK-type(263), StS-type(263) [type 3.2(254)]	263
	Columbiformes	Columbidae	*Caloenas nicobarica*	Nicobar pigeon	Bl	StS-type(263) [type 3.3(254)]	245,263
			Chalcophaps indica	Emerald dove	Bl	9 layers of rod-shaped melanin granules in barbule(268) [type 3.3(254)]	245,268
			Columba fasciata albilinea	Band-tailed pigeon	Bl	SK-type(263) [type 2.2(254)]	263

(*Continued*)

Table A.2 (*Continued*)

Infraclass	Order	Family/ subfamily	Genus/species	Common name	Type	EM	Reference
			Columba livia	Rock pigeon	Bl	[type 1(254)], large melanin granules are randomly packed, which are enclosed by thin layer(276)	236,245,276
			Columba livia domestica	Rock pigeon	Bl	thin layer(239)	239
			Columba livia livia	Rock pigeon	Bl	OL-type(263)	263
			Columba trocaz	Trocaz pigeon	Bl	thin layer(211)	211,245,263
			Ducula concinna	Elegant imperial pigeon	Bl	several layers of rod-shaped melanin granules(250), StS-type(263) [type 3.3(254)]	245,250,263
			Phaps chalcoptera	Common bronzewing	Bl	StS-type(263) [type 3.3(254)]	245,263
			Ptilinopus cinctus	Black-backed fruit dove	Bl	3–4 layers of rod-shaped melanin granules in barbule(250)	245,250
			Ptilinopus jambu	Jambu fruit dove	Bl	12 layers of rod-shaped melanin granules in barbule(250)	245,250
			Ptilinopus rivoli	White-bibbed fruit dove	Bl	>10 layers of rod-shaped melanin granules in barbule(245,250,265)	245,250,265
			Ptilinopus superbus	Superb fruit dove	Bl	several layers of rod-shaped melanin granules in barbule(250)	245,250
			Ptilinopus victor	Orange dove	Bl	10 layers of rod-shaped melanin granules in barbule(251)	245,250

(*Continued*)

Table A.2 (*Continued*)

Infraclass	Order	Family/subfamily	Genus/species	Common name	Type	EM	Reference
			Ptilinopus viridis	Claret-breasted fruit dove	Bl	10 layers of rod-shaped melanin granules in barbule(252)	245,250,265
			Zenaida asiatica	White-winged dove	Bl	OL-type(263) [type 1(254)]	245,263
	Ciconiiformes	Charadriidae	*Vanellus vanellus*	Northern lapwing	Bl	OSK-type(263) [type 2.2.1(254)]	245,263
		Ardeidae	*Butorides virescens*	Green heron	Bl	OStK-type(263) [type 3.2(254)]	263
			Syrigma sibilatrix	Whistling heron	S	random collagen array in skin(243)	243,245
		Threskiornithidae	*Bostrychia hagedash*	Hadada ibis	Bl/H	monolayer of hollow tubes in barbule(249,270)	245,249
			Threskiornis spinicollis	Straw-necked ibis	Bl/H	RE-type(263), [type 5.1(254)]	245,263
		Spheniscidae	*Aptenodytes patagonicus*	King penguin	Bk	lamellar structure in beak(238)	238
	Passeriformes	Pittidae	*Pitta maxima*	Ivory-breasted pitta	B	spongy structure in barb?(259,282)	245,259,282
		Eurylaimidae	*Calyptomena viridis*	Green broadbill	B	spongy structure in barb(245)	245
		Philepittidae	*Neodrepanis coruscans*	Sunbird asity	S	slightly random collagen array in skin(228)	228,243,245
			Neodrepanis hypoxantha	Yellow-bellied asity	S	random collagen array in skin(228)	228,243,245
			Philepitta castanea	Velvet asity	S	2D hexagonal crystal of collagen fibers in skin(223,228,235, 243)	223,228,235, 243,245

(*Continued*)

Table A.2 (*Continued*)

Infraclass	Order	Family/subfamily	Genus/species	Common name	Type	EM	Reference
		Tyrannidae/Cotinginae	*Cotinga cotinga*	Purple-breasted cotinga	B	spongy structure in barb(213,245)	213,245
			Cotinga maynana	Plum-throated cotinga	B	spongy structure in barb(253)	227,245,253,261
			Procnias nudicollis	Bare-throated bellbird	S	2D array of seemingly random collagen fibers in skin(235,243)	235,245
		Tyrannidae/Pirinae	*Pipra coronata*	Blue-crowned manakin	B	spongy structure in barb(245)	245
			Pipra nattereri	Snow-capped manakin	B	spongy structure in barb(245)	245
		Thamnophilidae	*Gymnopithys leucaspis*	White-cheeked antbird	S	random collagen array in skin(243)	243,245
		Ptilonor-hynchidae	*Ptilonorhynchus violaceus minor*	Satin bowerbird	Bl	random melanin granules in barbule(242)	242
		Maluridae	*Malurus leucopterus leuconotus*	White-winged fairywren	B	spongy structure in barb(257)	257
		Meliphagidae	*Prosthemadera novaeseelandiae*	Tui	Bl	StE-type(263) [type 3.1(254)]	245,263
		Irenidae	*Irena puella*	Asian fairy-bluebird	B	spongy structure in barb(245)	245
		Corvidae	*Cyanocitta stelleri*	Steller's jay	B	spongy structure in barb(241)	241

(*Continued*)

Table A.2 (*Continued*)

Infraclass	Order	Family/subfamily	Genus/species	Common name	Type	EM	Reference
			Cyanocorax beecheii	Purplish-backed jay	B	spongy structure in barb(245)	245
			Garrulus glandarius	Eurasian jay	B	spongy structure in barb(213)	213,232,244,245
			Pica pica	European magpie	Bl/H	2D air-hole type crystal in the membrane of barbule(240), monolayer of hollow melanin granules under 365 nm surface layer of barbule(268) [type 5.1(254)]	232,240,245,268
			Astrapia rothschildi	Huon astrapia	Bl	St(doppel)-S-type(263) [type 3.3(254)]	245,263
			Astrapia splendidissima	Splendid astrapia	Bl	St(doppel)-S-type(263) [type 3.3(254)]	245,263
			Cicinnurus magnificus	Magnificent bird-of-paradise	Bl	6–7 layers of two-row melanin granules(260), St(doppel)-S-type(263) [type 3.3(254)]	245,260,263
			Cicinnurus regius	King bird-of-paradise	Bl	St(doppel)-S-type(263) [type 3.3(254)]	245,263
			Epimachus fastuosus	Black sicklebill	Bl	StS-type(263) [type 3.3(254)]	245,263
			Lophorina superba	Superb bird-of-paradise	Bl	St(doppel)-S-type(263) [type 3.3(254)]	245,263
			Paradisea raggiana	Raggiana bird-of-paradise	Bl	7 layers of hollow tubes in barbule(245)	245
			Paradisea rubra	Red bird-of-paradise	Bl	SS-type(263) [type 2.3(254)] [type 3.3(254)]	245,263

(*Continued*)

Table A.2 (*Continued*)

Infraclass	Order	Family/ subfamily	Genus/species	Common name	Type	EM	Reference
			Parotia sefilata	Western parotia	Bl	10–20 layers of melanin granules(260), StS-type(263) [type 3.3(254)]	245,260,263
			Ptiloris magnificus	Magnificent riflebird	Bl	10 layers of plate-shaped melanin granules in barbule(245), St(doppel)-S-type(263) [type 3.3(254)]	245,263
			Ptiloris paradiseus	Paradise riflebird	Bl	3 layers of two-low melanin granules(260), S-type(263) [type 3.3(254)]	245,260,263
		Corvidae/ Dicrurinae	*Terpsiphone mutata*	Madagascar paradise flycatcher	S	random collagen array in skin(243)	243,245
		Corvidae/ Aegithininae	*Aegithina tiphia*	Common iora	Bl	SK-type(263) [type 2.2(254)]	245,263
		Corvidae/ Malaconoti-nae	*Telophorus dohertyi*	Doherty's bush shrike	B	spongy structure in barb(265)	265
		Muscicapidae/ Turdinae	*Myiophonus caeruleus*	Blue whistling thrush	B	spongy structure in barb(230,269)	230,245,269
			Sialia sialis	Eastern bluebird	B	spongy structure in barb(234)	233,234,245
		Sturnidae	*Acridotheres cristatellus*	Crested myna	Bl	monolayer of rod-shaped melanin granules(219), OSK-type(263) [type 2.2.1(254)]	219,245,263

(*Continued*)

Table A.2 (*Continued*)

Infraclass	Order	Family/ subfamily	Genus/species	Common name	Type	EM	Reference
			Ampeliceps coronatus	Golden-crested myna	Bl	monolayer of rod-shaped melanin granules in barbule(219), OSK-type(263) [type 2.2.1(254)]	219,245,263
			Aplonis panayensis neglectus	Asian glossy starling	Bl	OSE-type(263) [type 2.1 or 2.2.1(254)]	219,263
			Basilornis celebensis	Sulawesi myna	Bl	monolayer of rod-shaped melanin granules in barbule(219), OSE-type(263) [type 2.1.1(254)]	219,245,263
			Buphagus erythrorhynchus	Red-billed oxpecker	Bl	monolayer of rod-shaped melanin granules in barbule(219), OSK-type(263) [type 2.2.1(254)]	219,245,263
			Cinnyricinclus leucogaster	Violet-backed starling	Bl/H	2 layers of hollow granules in barbule(222), 4 layers of hollow platelet(245), RK-type(263) [type 5.2(254)]	209,222,245,263
			Cinnyricinclus sp.		Bl/H	3–4 layers of hollow tube in barbule(219)	219
			Coccycolius iris	Iris glossy-starling	Bl/H	2–3 layers of hollow tubes in barbule(219,222), KS-type(263) [type 6.3(254)]	219,222,245,263
			Cosmopsarus regius	Golden-breasted starling	Bl/H	1–2 layers of hollow platelets in barbule(219), KE-type(263) [type 6.1(254)]	219,245,263

(*Continued*)

Table A.2 (*Continued*)

Infraclass	Order	Family/subfamily	Genus/species	Common name	Type	EM	Reference
			Creatophora cinerea	Wattled starling	Bl	spherical melanin granules distributed randomly(222), monolayer of melanin granules(219), OSE-type(263) [type 2.1.1(254)]	219,245,263
			Enodes erythrophris erythrophris	Fiery-browed myna	Bl	monolayer of rod-shaped melanin granules in barbule(219), OSE-type(263) [type 2.1.1(254)]	219,263
			Gracula religiosa religiosa	Hill myna	Bl	monolayer of rod-shaped melanin granules in barbule(219), OSE-type(263) [type 2.1.1(254)]	219,263
			Grafisia torquata	White-collared starling	Bl	monolayer of rod-shaped melanin granules in barbule(219), SE-type(263) [type 2.1(254)]	219,222,245,263
			Lamprotornis caudatus	Long-tailed glossy starling	Bl/H	KE-type(263) [type 6.1(254)]	219,245,263
			Lamprotornis chalybaeus	Greater blue-eared glossy starling	Bl/H	monolayer of slender hollow platelets in barbule(222)	209,222,245
			Lamprotornis chloropterus	Lesser blue-eared glossy starling	Bl/H	monolayer of hummingbird-type platelet and its development(217)	216,217,245

(*Continued*)

Table A.2 (*Continued*)

Infraclass	Order	Family/ subfamily	Genus/species	Common name	Type	EM	Reference
			Lamprotornis chloropterus chloropterus	Lesser blue-eared glossy starling	Bl/H	monolayer of hollow tubes in barbule(219), KE-type(263)	219,263
			Lamprotornis corruscus	Black-bellied glossy starling	Bl/H	monolayer of hollow platelets in barbule(222)	222,245
			Lamprotornis purpureiceps	Purple-heated glossy starling	Bl	slender rod-shaped melanin granules in barbule(222)	222,245
			Leucopsar rothschildi	Bali myna	Bl	rod-shaped melanin granules distributed at a surface in barbule(219), OSK-type(263) [type 2.2.1(254)]	219,243,245,263
			Mino dumontii dumontii	Yellow-faced myna	Bl	rod-shaped melanin granules distributed at a surface in barbule(219), OSE-type(263) [type 2.1.1(254)]	219,263
			Neocichla gutturalis gutturalis	Babbling starling	Bl	rod-shaped melanin granules distributed at a surface in barbule(219), OSK-type(263) [type 2.2.1(254)]	219,263
			Onychognathus salvadorii	Bristle-crowned starling	Bl	rod-shaped melanin granules distributed at a surface in barbule(219)	219,245,263
			Onychognathus tenuirostris	Slender-billed starling	Bl	monolayer of rod-shaped melanin granules in barbule(219), SE-type(263)	219,245,263

(*Continued*)

Table A.2 (*Continued*)

Infraclass	Order	Family/ subfamily	Genus/species	Common name	Type	EM	Reference
			Poeoptera kenricki	Kenrick's starling	Bl	slender rod-shaped melanin granules randomly distributed(222)	222,245
			Poeoptera lugubris	Narrow-tailed starling	Bl	monolayer of plate-shaped melanin granules in barbule(219), PK-type(263) [type 4.2(254)]	219,245,263
			Sarcops calvus	Coleto	Bl	monolayer of rod-shaped melanin granules in barbule(219), SE-type(263) [type 2.1(254)]	219,245,263
			Saroglossa aurata	Madagascar starling	Bl	OSE-type(263) [type 2.1 or 2.1.1(254)]	219,245,263
			Scissirostrum dubium dubium	Finch-billed myna	Bl	monolayer of rod-shaped melanin granules in barbule(219), SE-type(263) [type 2.1(254)]	219,263
			Speculipastor bicolor	Magpie starling	Bl	monolayer of rod-shaped melanin granules in barbule(219), SE-type(263) [type 2.1(254)]	219,245,263
			Spreo fischeri	Fischer's starling	Bl/H	hollow granules randomly distributed in barbule(222)	222,245
			Spreo superbus	Superb starling	Bl/H	monolayer of flat hollow granules(219,268), KE-type(263) [type 6.1(254)]	219,245,263,268

(*Continued*)

Table A.2 (*Continued*)

Infraclass	Order	Family/ subfamily	Genus/species	Common name	Type	EM	Reference
			Streptocitta albicollis torquata	White-necked myna	Bl	monolayer of rod-shaped melanin granules(219), SE-type(263) [type 2.1(254)]	219,263
			Sturnus vulgaris	Common starling	Bl	monolayer of melanin granules in barbule(245)	208,232,245
			Sturnus vulgaris vulgaris	Common starling	Bl	OSE-type(263) [type 2.1.1(254)]	219,263
		Hirundinidae/ Hirundininae	*Tachycineta albiventer*	White-winged swallow	Bl	OPE-type(ubergang:PK-type)(263) [type 4.1(254)]	245,263
		Pycnonotidae	*Pycnonotus atriceps*	Black-headed bulbul	Bl	melanin granules densely distributed under keratin layer of 500 nm thickness (thin layer interference)(268) [type 2.2.1(254)]	245,268
		Nectariniidae/ Nectariniinae	*Nectarinia coccinigaster*	Splendid sunbird	Bl	4 layers of plate-like melanin granules in barbule(235,268)	235,245,268
			Nectarinia cuprea septentrion-alis	copper sunbird	Bl	PS-type(263) [type 4.3(254)]	212,263
			Nectarinia sperata brasiliana	Purple-throated sunbird	Bl	several layers of flat melanin granules in barbule(212), PS-type(263) [type 4.3(254)]	212,263

(*Continued*)

Table A.2 (*Continued*)

Infraclass	Order	Family/ subfamily	Genus/species	Common name	Type	EM	Reference
		Passeridae/ Estrildinae	*Taeniopygia guttata*	Chestnut-eared finch	B	spongy structure in barb(229)	229,244,245
		Fringillidae/ Emberizinae	*Cyanerpes cyaneus*	Red-legged honeycreeper	B	spongy structure in barb(245)	244,245
			Euphonia affinis	Scrub euphonia	Bl/H	hollow granules randomly distributed in barbule(245)	245
			Tangara cyanicollis	Blue-necked tanager	B	spongy structure in barb(224)	224,245
			Tangara mexicana	Turquoise tanager	B	spongy structure in barb?(282)	245
			Passerina cyanea	Indigo bunting	B	spongy structure in barb(245)	245
			Molothrus bonariensis	Shiny cowbird	Bl	SE-type(263) [type 2.1(254)]	245,263

References

1) [Ghiradella (1994)]
2) [Ghiradella (1991)]
3) [Ghiradella (1978)]
4) [Ghiradella *et al.* (1972)]
5) [Ghiradella (1989)]
6) [Ghiradella (1985)]
7) [Allyn and Downey (1977)]
8) [Berthier *et al.* (2007)]
9) [Nekrutenko (1965)]
10) [Argyros *et al.* (2002)]
11) [Anderson and Richards (1942)]
13) [Ghiradella (2005)]
14) [Vukusic (2005)]
15) [Stavenga *et al.* (2006a)]
16) [Kemp *et al.* (2006)]
17) [Prum *et al.* (2006)]
18) [Sweeney *et al.* (2003)]
19) [Yoshioka and Kinoshita (2006a)]
20) [Imafuku *et al.* (2002)]
21) [Vukusic *et al.* (2001a)]
22) [Vukusic *et al.* (2001b)]
23) [Vukusic and Sambles (2000)]
24) [Vukusic (2006)]
25) [Vukusic *et al.* (2000)]
26) [Tada *et al.* (1999)]
27) [Vukusic and Hooper (2005)]
28) [Tabata *et al.* (1996a)]
29) [Huxley (1976)]
30) [Huxley (1975)]
31) [Stavenga *et al.* (2004)]
32) [Ghiradella (1999)]
33) [Biró *et al.* (2003)]
34) [Allyn and Downey (1976)]
35) [Steinbrecht *et al.* (1985)]
36) [Brink and Lee (1999)]
37) [Kumazawa *et al.* (1996)]
38) [Koon and Crawford (2000)]
39) [Morris (1975)]
40) [Ghiradella (1984)]
41) [Ghiradella and Radican (1976)]
42) [Ghiradella (1998)]
44) [Parker *et al.* (2001)]
45) [Michelson (1911)]
46) [Rayleigh (1919)]
47) [Onslow (1923)]
48) [Süffert (1924)]
49) [Merrit (1925)]
50) [Mason (1927a)]
51) [Gentil (1942)]
52) [Lippert and Gentil (1959)]
53) [Young (1971)]
54) [Knopp and Krenn (2003)]
55) [Berthier *et al.* (2006)]
56) [Berthier *et al.* (2003)]
57) [Mann *et al.* (2001)]
58) [Gonzato and Pont (2003)]
59) [Gentil (1941)]
60) [Kumazawa *et al.* (1994)]
61) [Tabata *et al.* (1996b)]
62) [Banerjee *et al.* (2007)]
63) [Yoshioka and Kinoshita (2006b)]
64) [Kinoshita *et al.* (2002a)]
65) [Kinoshita *et al.* (2002b)]
66) [Yoshioka and Kinoshita (2004)]
67) [Wickham *et al.* (2006)]
68) [Bingham *et al.* (1995)]
69) [Vukusic *et al.* (1999)]
70) [Plattner (2004)]
71) [Gralak *et al.* (2001)]
72) [Cowan (1990)]
73) [Saito *et al.* (2006)]
74) [Huang *et al.* (2006)]
75) [Iwase *et al.* (2004)]
76) [Watanabe *et al.* (2005a)]
77) [Pillai (1968)]
78) [Wright (1963)]
79) [Kinoshita and Yoshioka (2005a)]
80) [Yoshioka and Kinoshita (2005a)]
81) [Vukusic (2003)]
82) [Tayeb *et al.* (2003)]
83) [Parker (2002)]
84) [Parker and Martini (2006)]
85) [Kinoshita and Yoshioka (2005b)]
86) [Vukusic and Sambles (2003)]
87) [Srinivasarao (1999)]
88) [Watanabe *et al.* (2005b)]
89) [Hoshino *et al.* (2003)]
90) [Wong *et al.* (2003)]
91) [Lawrence *et al.* (2002)]
92) [Schmidt (1943)]
94) [Mason (1926)]
95) [Rutowski *et al.* (2005)]
96) [Dey *et al.* (1998)]
97) [Vértesy *et al.* (2006)]
98) [Kertésa *et al.* (2006)]
99) [Land (1972)]
100) [Steinbrecht (1985)]
101) [Tilley and Eliot (2002)]
102) [Yoshida (2005)]
103) [Bernhard *et al.* (1965)]
106) [Steinbreht (2005)]
107) [Bernhard *et al.* (1968)]

108) [Stavenga *et al.* (2006b)]
109) [Vukusic *et al.* (2004a)]
110) [Parker (2005b)]
112) [Bernhard *et al.* (1970)]
113) [Neville (1977)]
118) [Järvilehto and Meyer-Rochow (1996)]
120) [Meyer-Rochow and Eguchi (1983)]
121) [Meyer-Rochow (1991)]
122) [Eguchi and Meyer-Rochow (1983)]
123) [Rothschild *et al.* (1978)]
125) [Hinton (1976)]
131) [Meyer-Rochow and Järvilehto (1997)]
138) [Gentil (1943)]
139) [Neville and Caveney (1969)]
146) [Vukusic *et al.* (2007)]
149) [Neville *et al.* (1976)]
152) [Vulinec (1997)]
156) [Kosaku and Miyamoto (1994)]
163) [Miyamoto and Kosaku (2005)]
166) [Rayleigh (1930)]
171) [Onslow (1920a)]
172) [Onslow (1920b)]
178) [Hinton (1969)]
190) [Mallock (1911)]
193) [Stavenga (2002)]
194) [Yoshioka and Kinoshita (2005b)]
195) [Durrer (1962)]
196) [Yoshioka and Kinoshita (2002)]
197) [Li *et al.* (2005)]
199) [Zi *et al.* (2003)]
200) [Bancroft (1924)]
201) [Greenewalt *et al.* (1960a)]
208) [Elasäfser (1925)]
209) [Schmidt (1952)]
210) [Greenewalt *et al.* (1960b)]
211) [Schmidt and Ruska (1961)]
212) [Durrer and Villiger (1962)]
213) [Schmidt and Ruska (1962a)]
214) [Schmidt and Ruska (1962b)]
215) [Rutschke (1966a)]
216) [Durrer and Villiger (1966)]
217) [Durrer and Villiger (1967)]
218) [Ralph (1969)]
219) [Durrer and Villiger (1970a)]
220) [Durrer and Villiger (1970b)]
221) [Durrer and Villiger (1975)]
222) [Craig and Hartley (1985)]
223) [Prum *et al.* (1994)]
224) [Finger (1995)]
225) [Andersson (1996)]
226) [Andersson and Amundsen (1997)]
227) [Prum *et al.* (1998)]
228) [Prum *et al.* (1999a)]
229) [Prum *et al.* (1999b)]
230) [Andersson (1999)]
231) [Keyser and Hill (2000)]
232) [Osorio and Ham (2002)]
233) [Siefferman and Hill (2003)]
234) [Shawkey *et al.* (2003)]
235) [Prum and Torres (2003a)]
236) [McGraw (2004)]
237) [Shawkey and Hill (2004)]
238) [Dresp *et al.* (2005)]
239) [Yin *et al.* (2006)]
240) [Vigneron *et al.* (2006)]
241) [Shawkey and Hill (2006)]
242) [Doucet *et al.* (2006)]
243) [Prum and Torres (2003b)]
244) [Auber (1957)]
245) [Prum (2006)]
247) [Finger and Burkhardt (1992)]
248) [Morimoto *et al.* (2005)]
249) [Brink and van der Berg (2004)]
250) [Dyck (1987)]
251) [Dyck (1971a)]
253) [Dyck (1971b)]
254) [Durrer (1986)]
257) [Doucet *et al.* (2004)]
258) [LaBastille *et al.* (1972)]
259) [Frank and Ruska (1939)]
260) [Dorst *et al.* (1974)]
261) [Dyck (1973)]
262) [Silver *et al.* (2005)]
263) [Durrer (1977)]
264) [Rutschke (1966b)]
265) [Dyck (1999)]
266) [Dyck (1979)]
267) [Durrer (1965)]
268) [Dyck (1976)]
269) [Prum *et al.* (2003)]
270) [Keyser and Hill (1999)]
276) [Yoshioka *et al.* (2007)]
282) [Frank (1939)]
356) [Berthier (2007)]

Appendix B

B.1 Fundamental Optical Processes

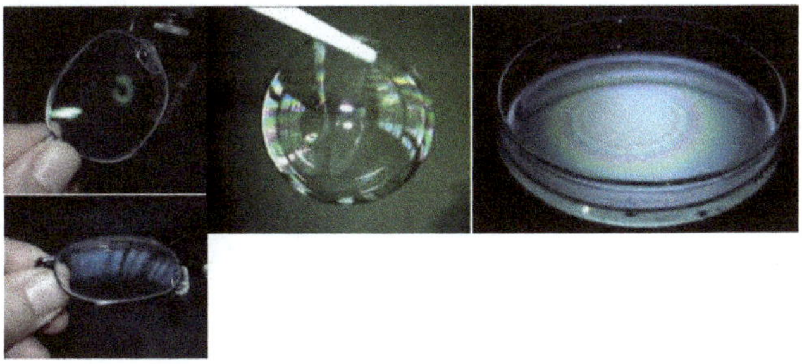

Plate 1 Examples of thin-film interference: antireflection coating (left), soap bubble (center), and oil film (right).

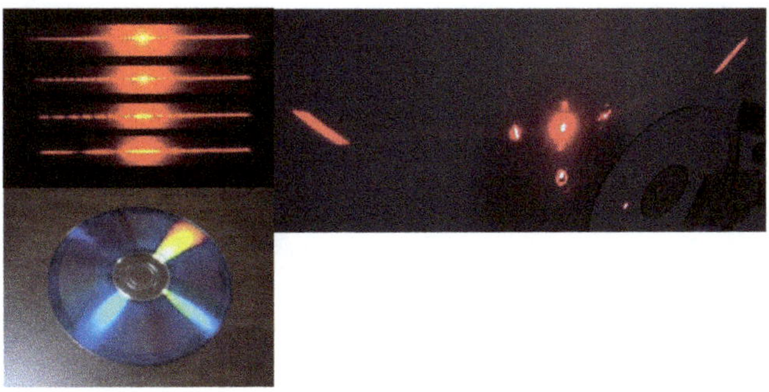

Plate 2 Examples of diffraction of light and diffraction grating: diffraction of light due to a slit with various widths (upper left), CD (lower left), and diffraction spots from CD (right). Laser light is incident from a hole and the diffraction spots are observed up to the second order.

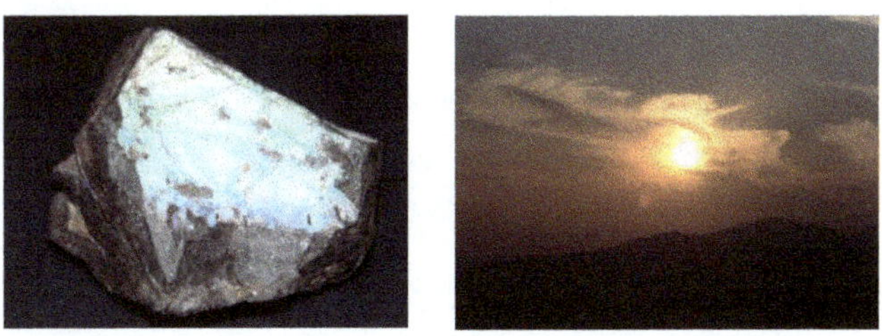

Plate 3 Example of photonic crystal: natural opal in the possession of The Museum of Osaka University (left). Evening sun with a tinge of red caused by Rayleigh scattering of light of shorter wavelengths (right).

B.2 Characteristics of Morpho Butterflies

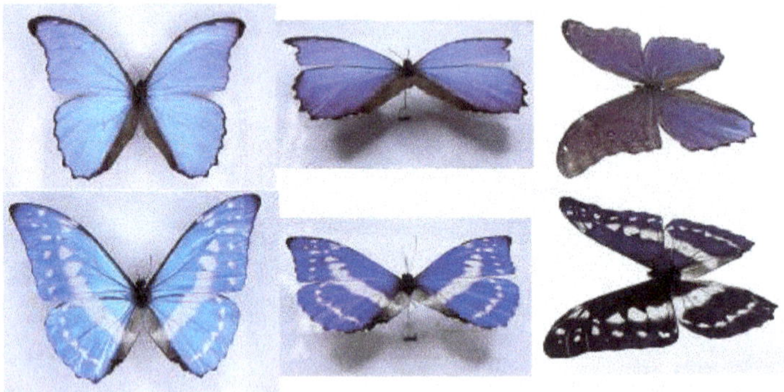

Plate 4 Viewing-angle dependence of the *Morpho* butterflies: *Morpho didius* (upper) and *Morpho cypris* (lower). Photographed by Dr. S. Yoshioka.

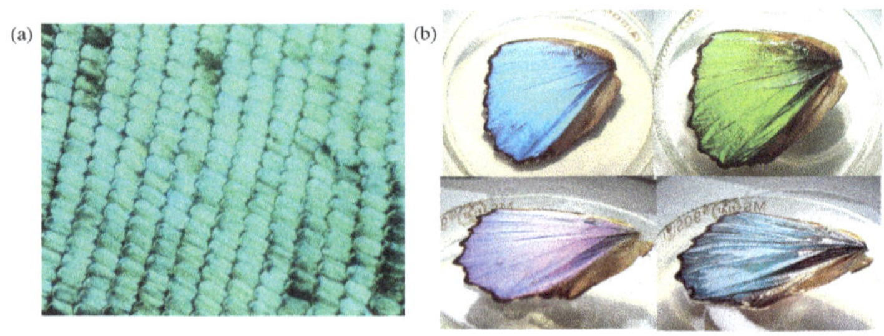

Plate 5 (a) Scales of *Morpho rhetenor*. (b) Wings of *Morpho didius* in air (left) and in ethanol (right) photographed from normal (upper) and oblique directions (lower). Photographed by N. Okamoto.

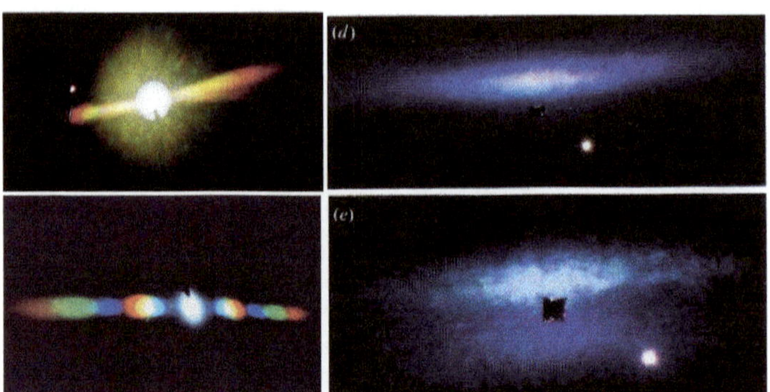

Plate 6 Transmission (left) and reflection (right) patterns from a single ground (upper) and cover (lower) scale of *Morpho didius* (Yoshioka and Kinoshita, 2004).

B.3 Morpho Butterflies

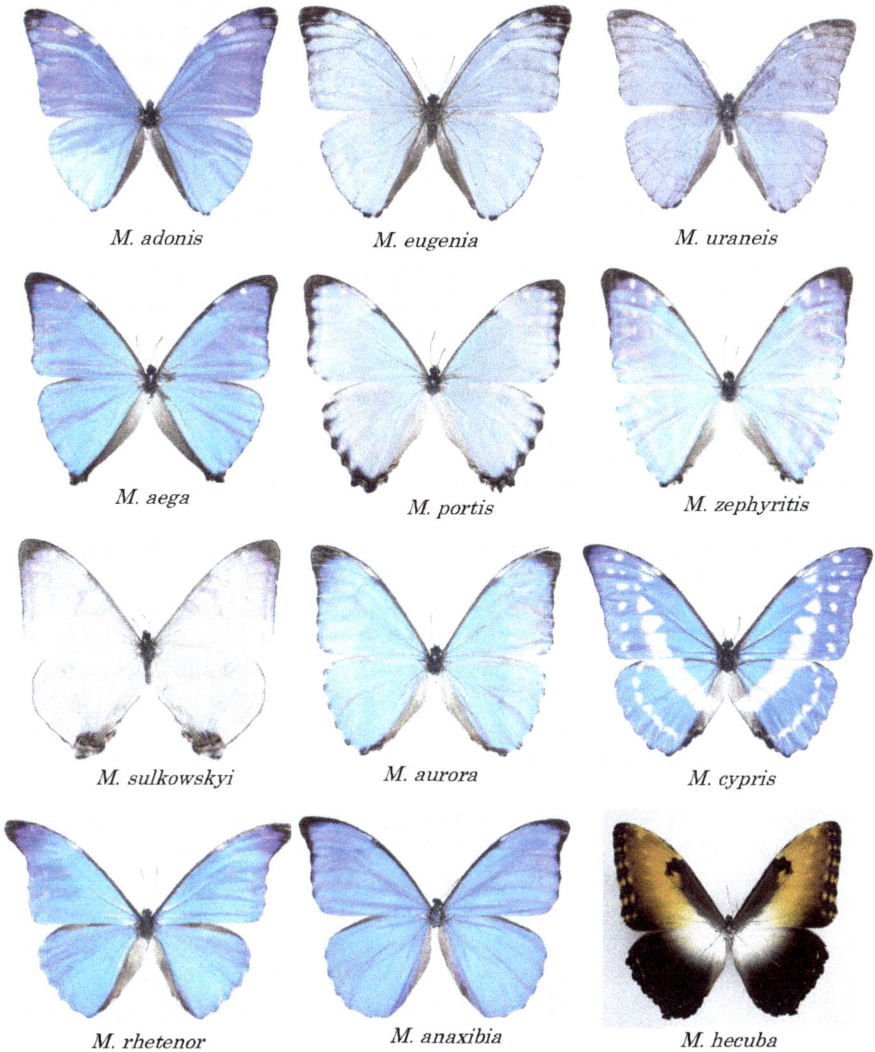

Plate 7 The male *Morpho* butterflies, photographed by N. Okamoto, in the possession of The Nature and Human Activities, Hyogo, Japan.

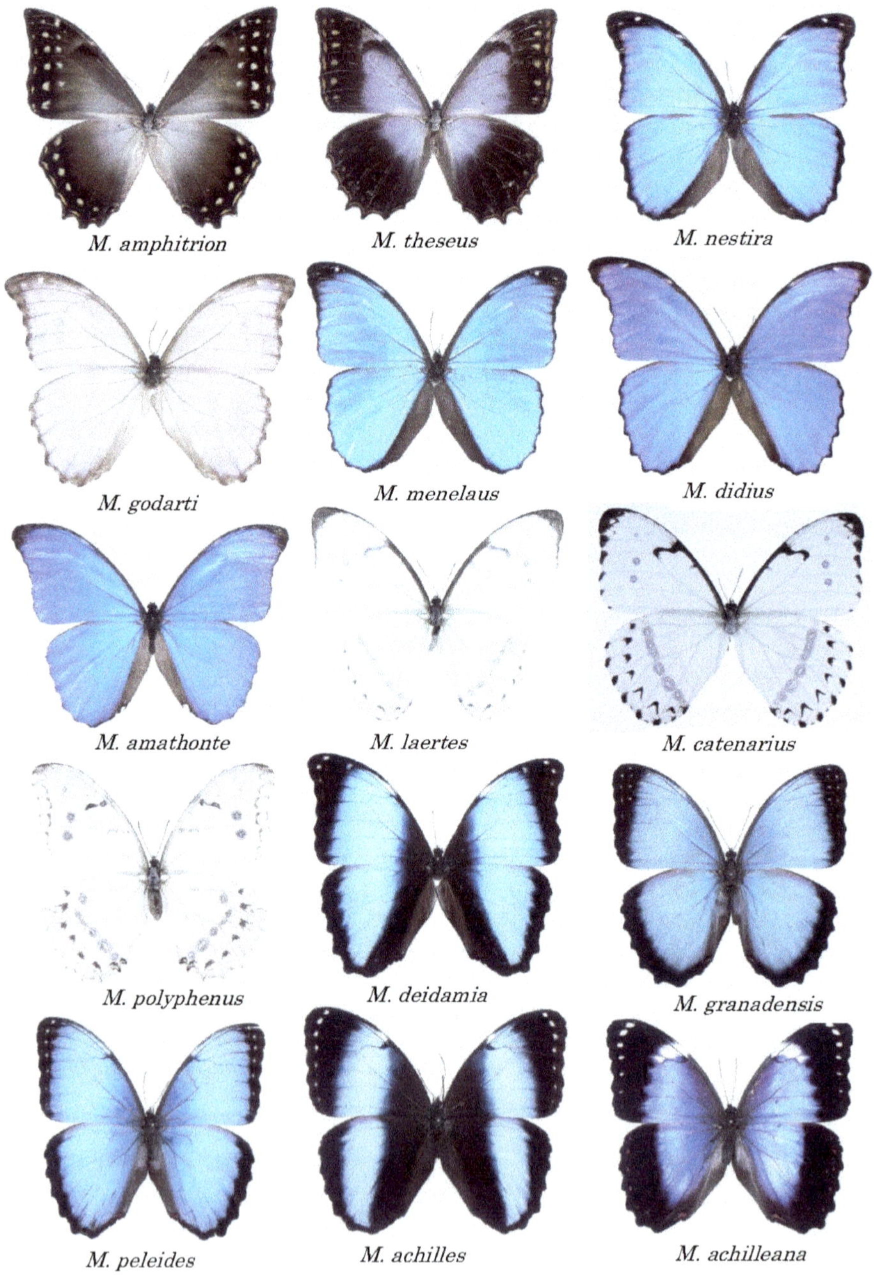

Plate 8 The male *Morpho* butterflies, photographed by N. Okamoto, in the possession of The Nature and Human Activities, Hyogo, Japan.

B.4 Structurally Colored Butterflies

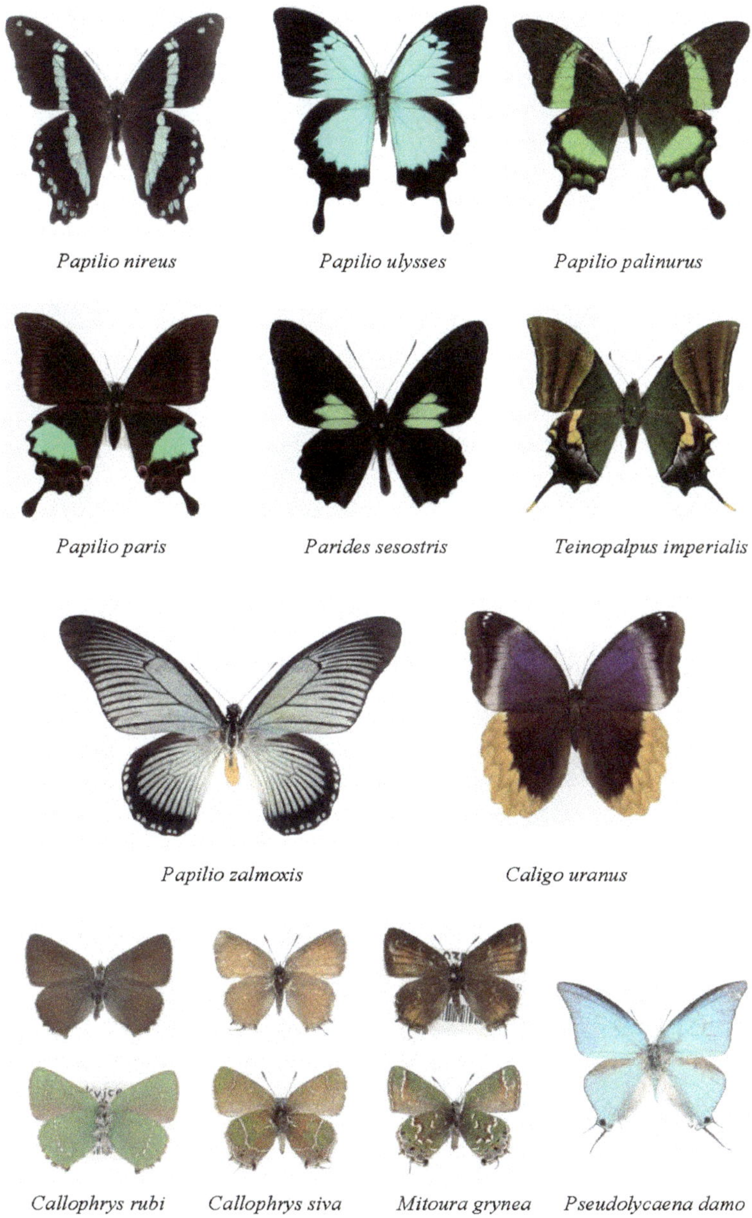

Plate 9 Various structurally colored butterflies, photographed by Dr. M. Kambe and N. Arakawa, in the possession of The Nature and Human Activities, Hyogo, Japan.

B.5 Scarabeid Beetles Under Circular Polarizers

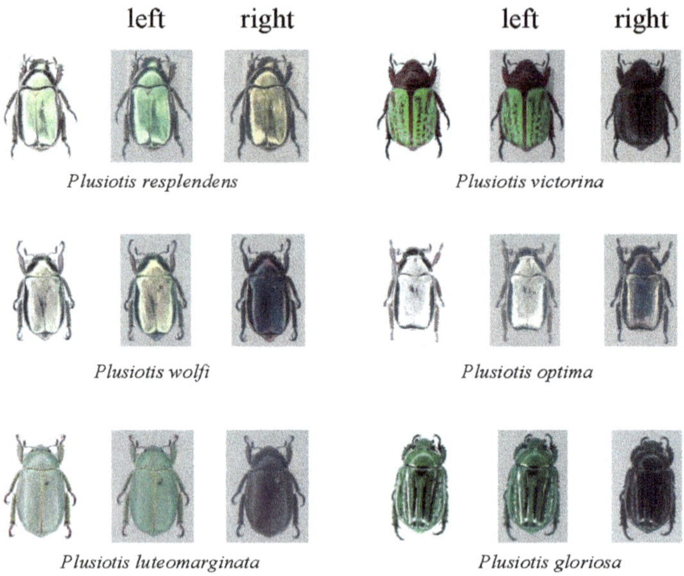

Plate 10 Various iridescent beetles viewed without (left) and with left- (center) and right-circular polarizer (right), photographed by E. Nakamura and N. Arakawa, in the possession of The Nature and Human Activities, Hyogo, Japan.

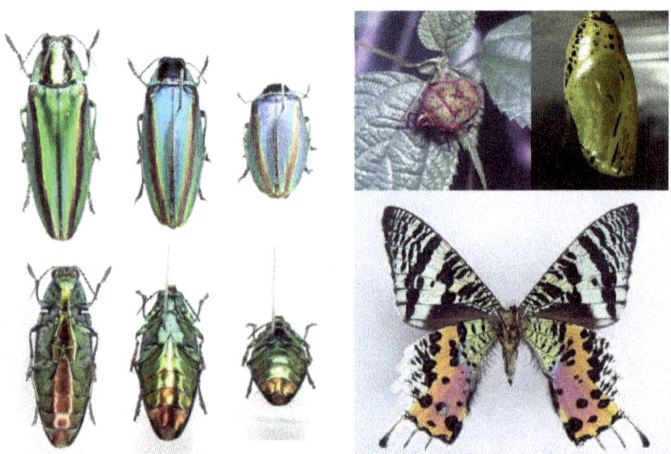

Plate 11 Left: viewing-angle dependence of dorsal and ventral sides of jewel beetle (S. Kinoshita, *Cell Technol.* **22**, 1113 (2003)). Right: shield bug (upper left), golden pupa of Danainae butterfly (upper right) (S. Kinoshita *et al.*, *J. Jpn. Soc. Colour Mater.* **75**, 493 (2002)) and the dorsal side of Madagascan sunset moth (lower) (Yoshioka and Kinoshita, 2007).

B.6 Structurally Colored Birds and Fish

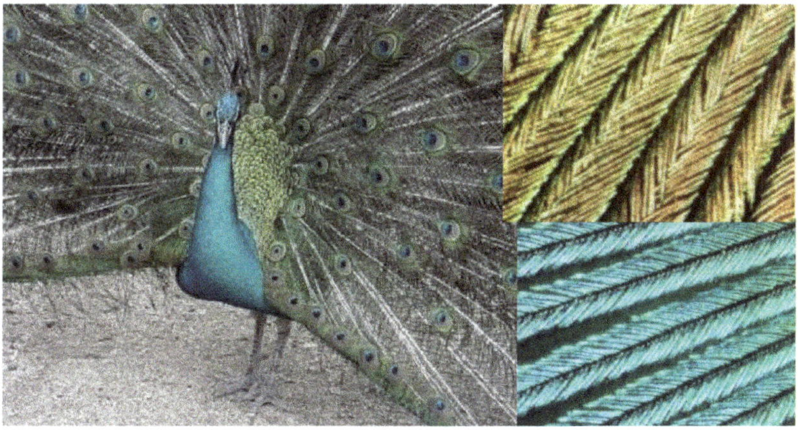

Plate 12 Peacock and feathers (Yoshioka and Kinoshita, 2000).

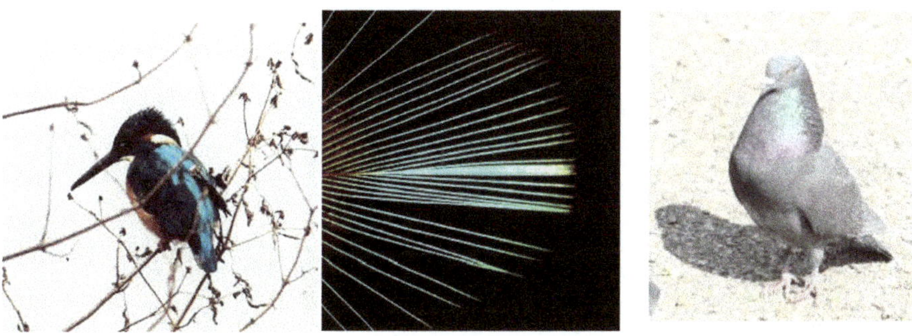

Plate 13 Kingfisher and its feather (left) (Kinoshita and Yoshioka, 2005b), and rock dove (right) (Yoshioka *et al.*, 2007).

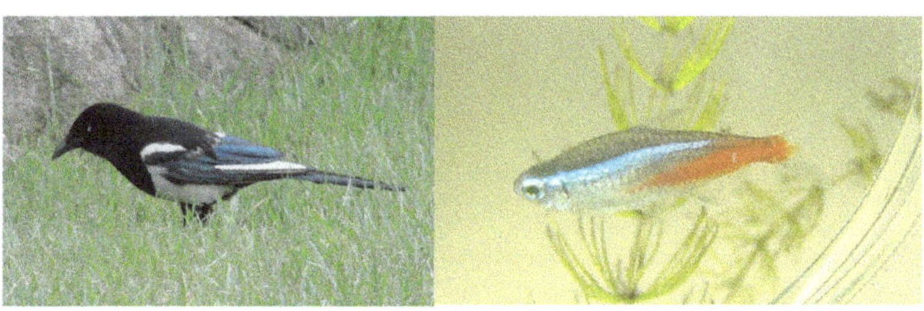

Plate 14 Magpie (left) and neon tetra (right). Photographed by Dr. S. Yoshioka.

B.7 Morpho Mimicry

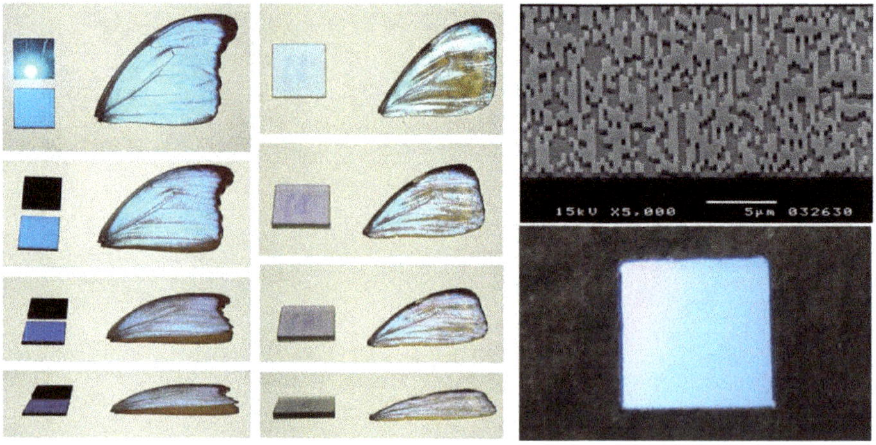

Plate 15 *Morpho* substrates fabricated by roughening the surface mechanically (left) and using electron lithography (right). (Reproduced from Yoshioka *et al.* (2004) with permission.)

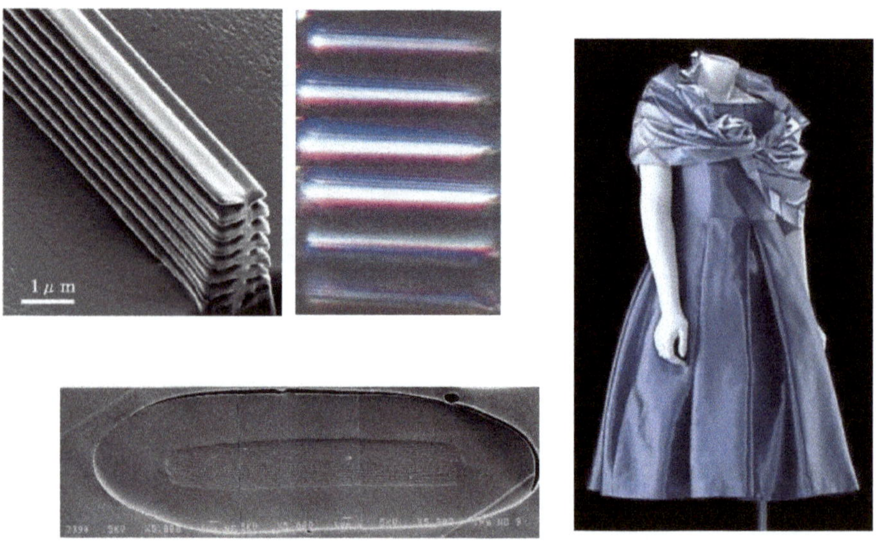

Plate 16 Upper left: *Morpho* "shelf" structure fabricated by focused-ion beam chemical vapor deposition method. (Reproduced from Watanabe *et al.* (2005a) with permission.) Lower right: A cross-section of a fiber containing a multilayer of 61 layers and a dress woven by the fibers. (Courtesy of Teijin Fibers Limited.)

Index of Scientific Names

Subject Index

www.ingramcontent.com/pod-product-compliance
Ingram Content Group UK Ltd.
Pitfield, Milton Keynes, MK11 3LW, UK
UKHW020350150926
8624IPUK00004B/93

9 789812 707833